HEALTHCARE TECHNOLOGIES SERIES 26

Wireless Medical Sensor Networks for IoT-based eHealth

IET Book Series on e-Health Technologies – Call for Authors

Book Series Editor: Professor Joel P.C. Rodrigues, the National Institute of Telecommunications (Inatel), Brazil and Instituto de Telecomunicações, Portugal

While the demographic shifts in populations display significant socio-economic challenges, they trigger opportunities for innovations in e-Health, m-Health, precision and personalized medicine, robotics, sensing, the Internet of things, cloud computing, Big Data, Software Defined Networks, and network function virtualization. Their integration is however associated with many technological, ethical, legal, social, and security issues. This new Book Series aims to disseminate recent advances for e-Health Technologies to improve healthcare and people's wellbeing.

Topics considered include Intelligent e-Health systems, electronic health records, ICT-enabled personal health systems, mobile and cloud computing for eHealth, health monitoring, precision and personalized health, robotics for e-Health, security and privacy in e-Health, ambient assisted living, telemedicine, Big Data and IoT for e-Health, and more.

Proposals for coherently integrated International multiauthored edited or coauthored handbooks and research monographs will be considered for this Book Series. Each proposal will be reviewed by the Book Series Editor with additional external reviews from independent reviewers. Please email your book proposal for the IET Book Series on e-Health Technologies to: Professor Joel Rodrigues at joeljr@ieee.org or joeljr@inatel.br

Other volumes in this series:

Volume 1	**Nanobiosensors for Personalized and Onsite Biomedical Diagnosis** P. Chandra (Editor)
Volume 2	**Machine Learning for Healthcare Technologies** D.A. Clifton (Editor)
Volume 3	**Portable Biosensors and Point-of-Care Systems** S.E. Kintzios (Editor)
Volume 4	**Biomedical Nanomaterials: From design to implementation** T.J Webster and H. Yazici (Editors)
Volume 6	**Active and Assisted Living: Technologies and a**pplications F. Florez-Revuelta and A.A Chaaraoui (Editors)
Volume 7	**Semiconductor Lasers and Diode-based Light Sources for Biophotonics** P.E Andersen and P.M Petersen (Editors)
Volume 9	**Human Monitoring, Smart Health and Assisted Living: Techniques and t**echnologies S. Longhi, A. Monteriù and A. Freddi (Editors)
Volume 13	**Handbook of Speckle Filtering and Tracking in Cardiovascular Ultrasound Imaging and Video** C.P. Loizou, C.S. Pattichis and J. D'hooge (Editors)
Volume 14	**Soft Robots for Healthcare Applications: Design, modelling, and control** S. Xie, M. Zhang and W. Meng
Volume 16	**EEG Signal Processing: Feature extraction, selection and classification methods** W. Leong
Volume 19	**Neurotechnology: Methods, advances and applications** V. de Albuquerque, A. Athanasiou and S. Ribeiro (Editors)
Volume 20	**Security and Privacy of Electronic Healthcare Records: Concepts, paradigms and solutions** S. Tanwar, S. Tyagi and N. Kumar (Editors)
Volume 23	**Advances in Telemedicine for Health Monitoring: Technologies, design and applications** Tarik A. Rashid, Chinmay Chakraborty and Kym Fraser
Volume 24	**Mobile Technologies for Delivering Healthcare in Remote, Rural or Developing Regions** P. Ray, N. Nakashima, A. Ahmed, S. Ro and Y. Soshino (Editors)

Wireless Medical Sensor Networks for IoT-based eHealth

Edited by
Fadi Al-Turjman

The Institution of Engineering and Technology

Published by The Institution of Engineering and Technology, London, United Kingdom

The Institution of Engineering and Technology is registered as a Charity in England & Wales (no. 211014) and Scotland (no. SC038698).

First published 2020

The Institution of Engineering and Technology
Michael Faraday House
Six Hills Way
Stevenage Herts
SG1 2AY, United Kingdom

www.theiet.org

British Library Cataloguing in Publication Data
A catalogue record for this product is available from the British Library

ISBN 978-1-83953-056-2 (hardback)
ISBN 978-1-83953-057-9 (PDF)

Typeset in India by MPS Limited

Contents

About the editor

Prof. Dr. Fadi Al-Turjman received his Ph.D. in computer science from Queen's University, Kingston, Ontario, Canada, in 2011. He is a full professor and a research center director at Near East University, Nicosia, Cyprus. Prof. Al-Turjman is a leading authority in the areas of smart/intelligent, wireless, and mobile networks' architectures, protocols, deployments, and performance evaluation. His publication history spans over 250 publications in journals, conferences, patents, books, and book chapters, in addition to numerous keynotes and plenary talks at flagship venues. He has authored and edited more than 25 books about cognition, security, and wireless sensor networks' deployments in smart environments, published by Taylor and Francis, Elsevier, and Springer. He has received several recognitions and best papers' awards at top international conferences. He also received the prestigious *Best Research Paper Award* from Elsevier Computer Communications Journal for the period 2015–2018, in addition to the *Top Researcher Award* for 2018 at Antalya Bilim University, Turkey. Prof. Al-Turjman has led a number of international symposia and workshops in flagship communication society conferences. Currently, he serves as an associate editor and the lead guest/associate editor for several well reputed journals, including the *IEEE Communications Surveys and Tutorials* **(IF 22.9)** and the Elsevier Sustainable Cities and Society **(IF 4.7)**.

Foreword

The Internet of Things (IoT) is enabling nowadays several applications/services in academic/industrial disciplines, and especially in healthcare and medical fields. Remarkably, due to the rapid proliferation of wearable devices and smartphones, the IoT-enabled technology is evolving healthcare from conventional hub-based system to more personalized eHealth systems. The successful utilization of IoT-enabled technology in eHealth will enable faster and safer preventive care, lower overall cost, improved patient-centric practice, and enhanced sustainability. Efficient IoT-enabled eHealth systems can be realized by providing highly customized access to rich medical information and efficient clinical decisions to each individual with unobtrusive monitoring. Wireless medical sensor networks (WMSNs) are at the heart of this concept, and their development is a key issue if such concept is to achieve its potential. Hence, the book is planned to be dedicated for addressing major challenges in realizing WMSNs in the forthcoming IoT-based eHealth systems. Challenges vary from cost and energy efficiency to security and service quality. To tackle such challenges, WMSNs must meet certain expectations and requirements, such as size constraints, manufacturing costs, resistance to environmental factors existing at deployment locations, etc. Based on that, the aim of our book is to focus on both design and implementation aspects in IoT-based eHealth applications that are enabled/supported by WMSNs in realizing patient-centric systems.

Chapter 1

Sensor-enabled smart suit electronic IoT design platform with emergency services application

Migdat Hodzic[1], James M. Brennan[2] and Enis Dzanic[3]

An integrated smart suit sensor and positioning system electronic Internet of Things (IoT) prototype has been developed to address the growing need for personal welfare monitoring of first-line responders, defenders, and workers exposed to industrial or other hazards, as well as other commercial and defense, and new applications in cloud-based IoT. The system provides a global positioning system (GPS) position map with coordinate data, current Greenwich mean time (GMT) readout, subject's heart rate, body temperature, and a long-wave thermal video camera that provides a forward-looking thermal image. Physiologic data and thermal imaging of the subject may be viewed by monitoring personnel using Internet browser connected to the system's static Internet protocol (IP) address. The system is Wi-Fi connected to a local network, which can be extended to enable secure connection to the Internet with incorporation of additional firmware. Details regarding hardware and software configuration are presented along with an appendix containing additional data. Source code for the software modules currently running on the prototype system is also available for interested parties or potential users and customers.

1.1 Introduction

Recent years witnessed very considerable development in the areas of various sensors for many related applications. In this context new IoT (Internet of Things) technologies [1], in particular Cloud-based IoT [2,3], emerged as a response to a growing need to connect a variety of devices in our homes, in the streets, cities, or sensor-enabled devices which attach to our body or uniforms. New low power wireless and wired sensor technologies have been developed and are used more and more in many old and lots of new applications [4]. Sensors range from

[1]Engineering Department, American University in Bosnia and Herzegovina, Sarajevo, Bosnia and Herzegovina
[2]BH Analytics, Santa Clara, California, USA and Sarajevo, Bosnia and Herzegovina
[3]Economics Department, University of Bihac, Bihac, Bosnia and Herzegovina

environmental, physiologic, range measurements, proximity sensors, and all the way to very sophisticated special-purpose sensors for industrial and defense use [5–7]. A variety of embedded and inexpensive platforms exist now for fast design and prototyping such as Raspberry Pi [8]. Besides new sensors and IoT technologies, new "smart" materials are also becoming available more and more, with a variety of functionality built-in, even some basic electronics built into the material fabric [9]. One specific and important area of sensor development is infrared (IR) and thermal sensors and cameras based on them [10]. These sensors now allow for very detailed IR or thermal image sensing and digital framing with a usable number of video frames [10]. In any useful IoT sensor-enabled smart suite, there is typically a need for positioning data and hence GPS sensors are also required. In the smart suite case, it would be required to have physiological data of the person wearing the suit, and this means a need for temperature, heart rate, blood pressure, outside environment pressure, humidity and temperature, and sensor for some dangerous gas presence (such as in mines). It is in this spirit that we developed simple IoT sensor-based smart suit design and testing electronic platform described in this chapter. We opted to incorporate only basic sensor components such as GPS with the antenna, temperature, and heart rate sensors, as well as thermal sensors. The platform is based on popular Raspberry Pi HW and SW computer board, with Wi-Fi and 4G built-in for Internet communications. In order to be able to demonstrate a smart suit design platform, we also incorporated ability to view the suit online using its static IP address. Wi-Fi is used for Internet connectivity. A variety of customizations are possible depending on the interest of final users and customers and their needs for specific sensors.

1.2 System components

A general system block diagram of the smart suit system electronics is shown in Figure 1.1. More specific component choices for our prototype and demonstration design are indicated in Figure 1.2. The thermal camera (FLIR Lepton brand) is connected to the host processor via SPI0 bus. The subject's heartbeat and surface body temperature sensors are connected to the host processor through the analog to digital converter (ADC) via SPI1 bus and processed by an algorithm to extract heart rate and average temperature. Additional inputs to the ADC are available for future expansion capabilities to provide respiration rate and activity monitoring. The host processor for the smart suit system is a Raspberry Pi 3 running a Debian Stretch Linux distribution. When the system powers up, it automatically starts a C++-based thermal imaging application as well as a Python-based Flask web server, which also inputs and formats all incoming data for presentation as a served webpage.

1.2.1 FLIR Lepton IR camera

The FLIR Lepton is an infrared camera system that integrates a fixed-focus lens assembly, an 80 × 60 long-wave infrared (LWIR) microbolometer sensor array, and incorporates signal processing electronics [10]. Easy to integrate and operate,

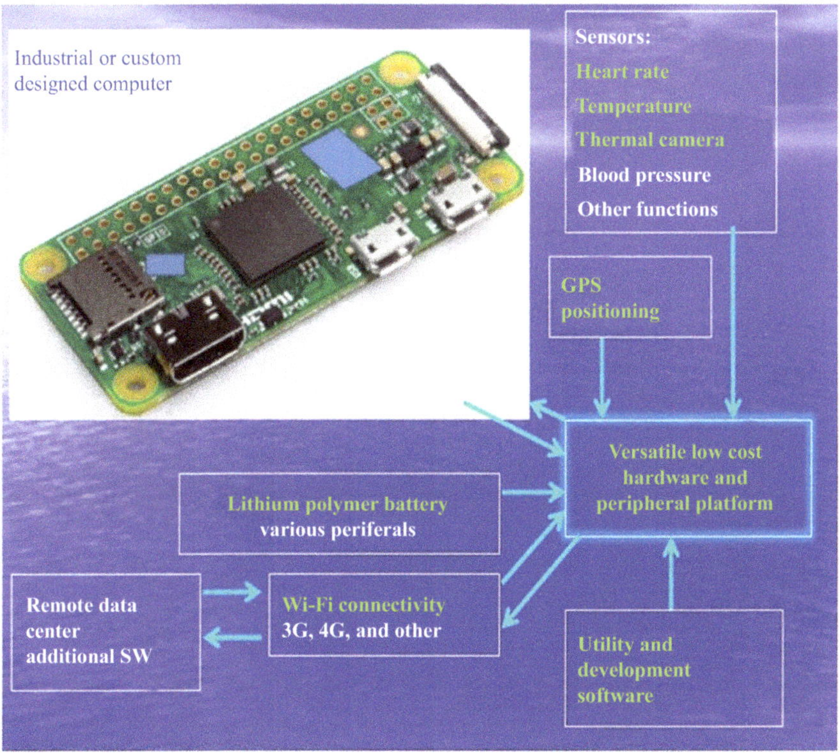

Figure 1.1 Smart suit system general electronic block diagram

Lepton is intended for mobile devices as well as any other application requiring a very small footprint, very low power, and instant-on operation. Lepton can be operated in its default mode or configured into other modes through a command and control interface (CCI). Using its default conditions, the FLIR Lepton camera outputs 60 video packets per frame, each 1,312 bits long, at approximately 25.9 frames per second. The minimum output data clocking rate to the camera is on the order of 2 MHz to allow it to keep up with its real-time image generation. The camera data output provides three repeated identical frames in a row, followed by a new frame making up another series of three frames. This activity will repeat indefinitely if not interrupted or an error has happened. It should be noted that this frame series format means that real-time new frame output is approximately at a 9 Hz rate so that actual frame processing only operates using only one out of three frames. The smart suit system processes thermal frames at about nine per second but, at present, only grabs one still frame per web page update. Software processing of the thermal frames includes false-color map encoding that normalizes the brightest pixels to the most intense coloring. This ensures that the image does not saturate thus masking less bright features of the scene.

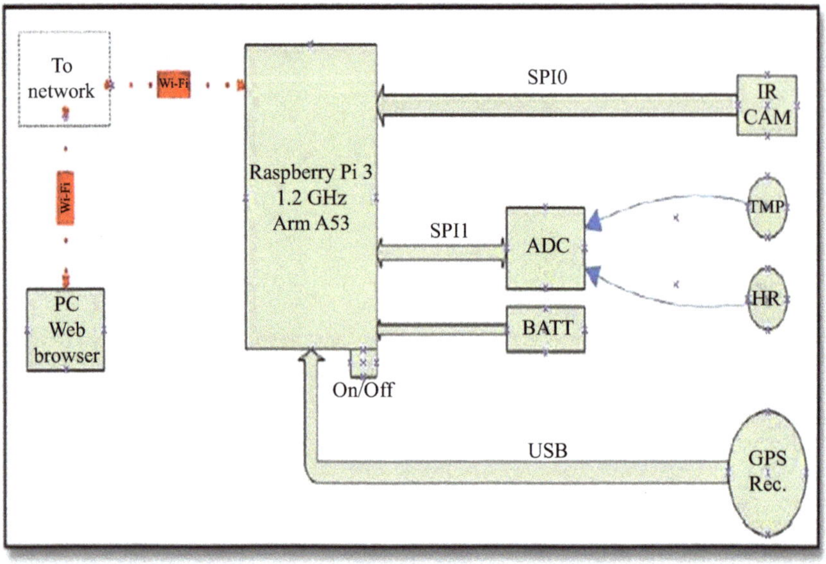

Figure 1.2 Smart suit system-specific block diagram

1.2.2 IR camera software

Image processing is done in the C++ language and includes algorithms to identify and synchronize processing starting with frame packet number one, formatting and processing each FLIR Lepton data line input into a 480 × 800 pixel image frame, color mapping the brightness of pixels, and then converting the frame into a JPG format for storage in memory. The frame stored in memory is accessed by the Python-based web server application approximately once every 2–3 s for output to the client display for the demonstration purposes.

1.2.3 Python-based flask web server

A Flask micro web server has been implemented to serve a web page containing human subject physiologic data, current GPS coordinate data, an up-to-date GPS location map, and the IR thermal image showing a color mapped scene ahead of the subject. This server is implemented in Python, which generates a new web page when a client application asks for an update (currently once every 2–3 s). Processing includes capturing GPS time and coordinates, acquiring subject's heart rate and external body temperature, computing and formatting all of these values into a python dictionary, and then sending the dictionary (data strings) to the client webpage view. The client webpage presently uses a timed refresh period to ask for an update from the server every 2–3 s. The data are collected in Java variables within the page and processed to form result text, a google API map of the location, and to render the thermal JPG image seen in Figure 1.3 (US Silicon Valley

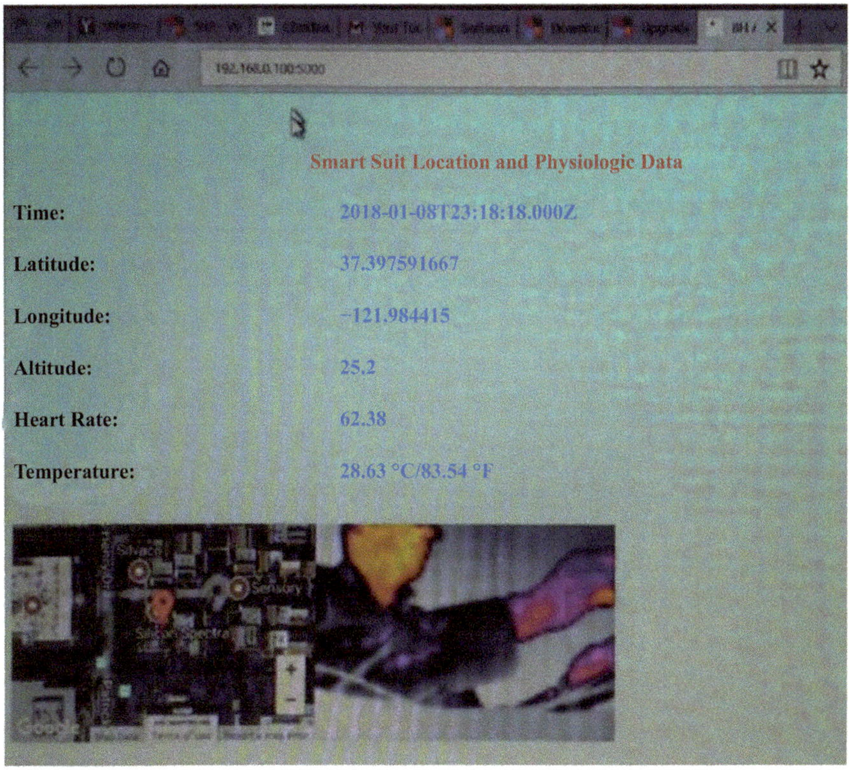

Figure 1.3 Client web page displaying collected data, map, and image located in Silicon Valley, USA

location) and Figure 1.4 (Sarajevo, Bosnia and Herzegovina location). At this point, we did not spend lots of time in making graphical user interface (GUI) more sophisticated. As Figures 1.3–1.5 indicate the GUI is just a basic one for the suit system demonstration and prototyping purposes. Figure 1.6 shows an additional thermal image of a person in front of the Lepton thermal image sensor built into the suit.

1.2.4 Raspberry Pi 3 Debian stretch operating system start

The smart suit system connection diagram shown in Figure 1.7 relies upon the underlying Linux operating system to host the thermal and Flask web server applications. At power-on, the system automatically starts both the Flask server and the Lepton Thermal Imager applications. The Flask web server startup process uses the CHRON daemon to read a script at startup. This is initially setup during development using the CRONTAB command and is done from within the directory **/var/spool/cron/crontabs/pi**. See Appendix A for the general procedure. During startup, the operating system will look for shell scripts in the user's home directory,

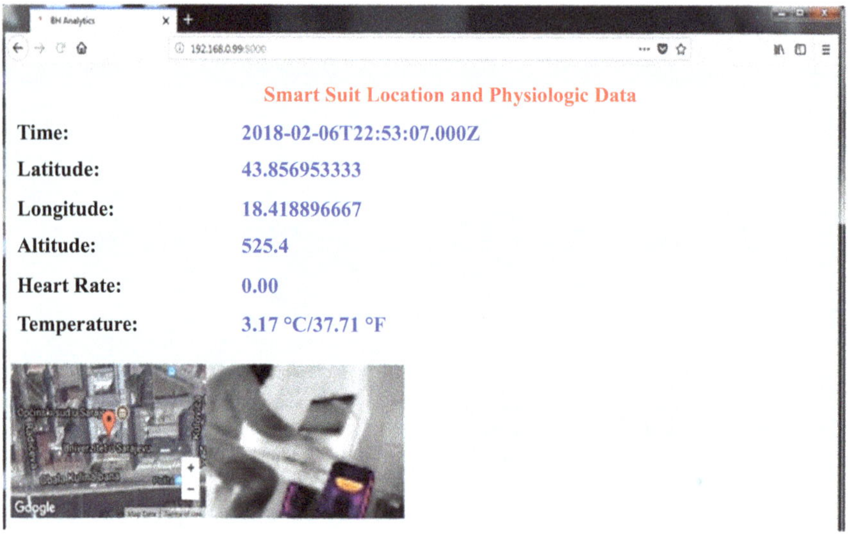

Figure 1.4 *Client web page displaying collected data, map, and image located in Silicon Valley, USA*

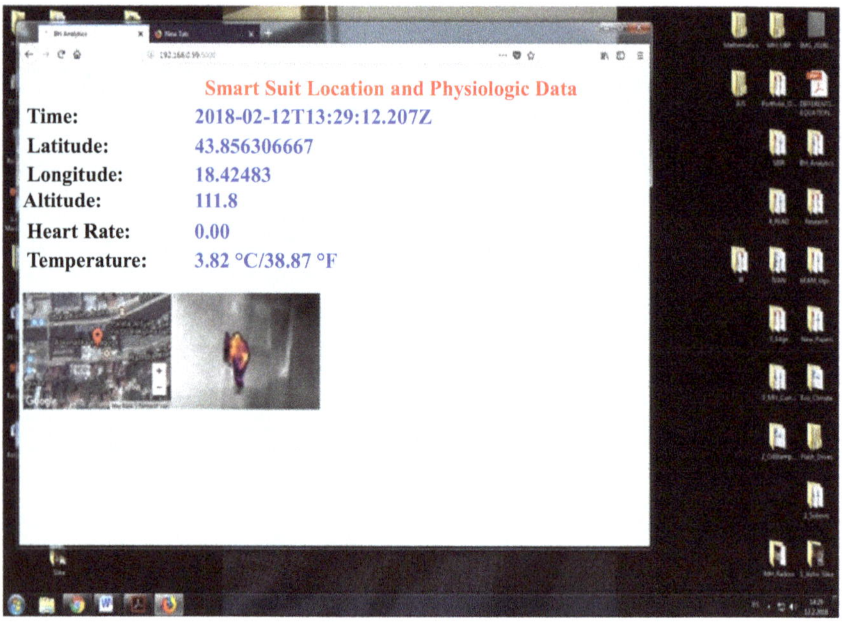

Figure 1.5 *Client web page displaying collected data, map, and image located in Ilidza TRZ factory, B&H*

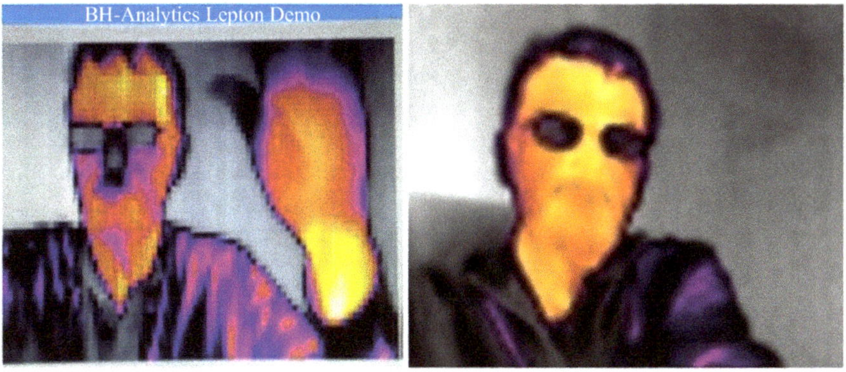

Figure 1.6 Thermal images of a person in front of Lepton thermal camera

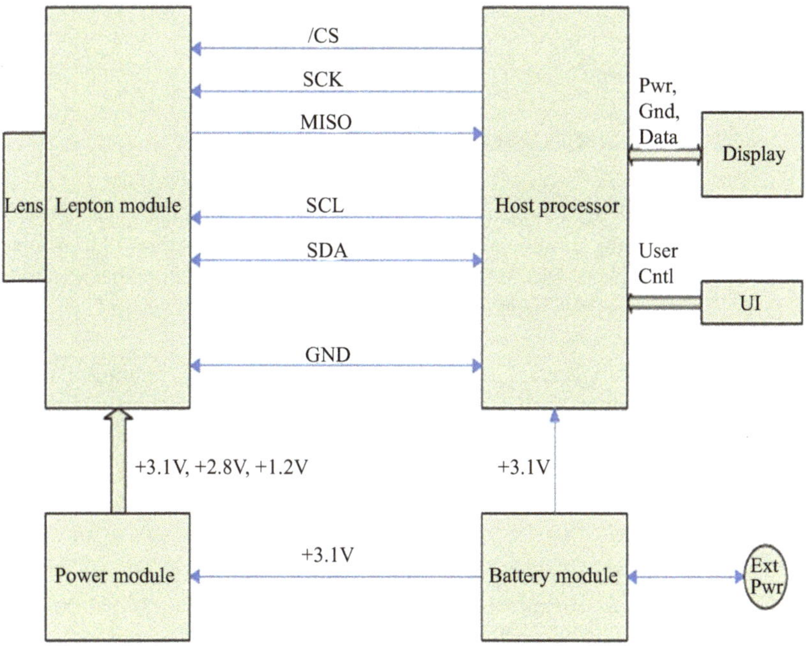

Figure 1.7 Smart suit system connection block diagram

in this case **/home/pi/startup.sh**. The script found at this location is used to start up the Lepton thermal imaging application. Content of startup.sh can be obtained if a customer requires this. It should be noted that the Lepton thermal imaging application named "Demo," as found in directory **/home/pi/Qt/Demo/build-Demo-kit2-Debug/**, is a stand-alone executable and has been built using the Qt4 development application. The setup of the Qt4 environment is referenced in smart suit system documentation. Another script that is automatically executed during the startup process allows system shutdown when a specific hardware pin is grounded. This script is found at **/etc/init.d/listen-for-shutdown.sh** and runs a Python script found at **/usr/local/bin/listenfor-shutdown.py**, which when invoked runs continuously in the background waiting for a hardware pin instituted shutdown interrupt signal. Once the system starts up and the required applications are each separately running, the Flask web server is ready to respond to a client application request if the Wi-Fi feature is operational.

1.3 System hardware

The connection diagram in Figure 1.6 shows the Raspberry Pi 3 I/O ports and general pins used to interface system hardware components. Numerous connectors are used to route sensor data to the required I/O pins. Depending on the number of required sensors, the I/O pins functionality can be customized for a specific application. The system has enough capacity to accommodate a number of additional sensors if required.

1.3.1 Hardware data collection and transfer

The Raspberry Pi 3 (RP3) uses 2 separate SPI busses to transfer data: (1) SPI0 to clock in data from the MCP3008 ADC and (2) SPI1 to clock in data from the FLIR Lepton thermal imaging module. This is done to provide the different clock speeds required by each device. The MPC3008 ADC uses a 200 KHz transfer clock to provide the lower demand heart rate and temperature measurements, which are collected at ten samples per second.

The FLIR Lepton thermal module, however, requires a minimum transfer clock of 2 MHz to provide its 27 fps image rate. The thermal module operates in an autonomous manner in its present configuration after it is powered on by providing serial data in an open-loop fashion when given a clock. No other programming is required to access its data via SPI serial bus. At present, a momentary power interrupt switch is placed in the thermal module's Vdd input to allow for operator reset of the device due to over temperature or other uncontrolled noisy conditions. A later version of the smart suit system will use a dedicated I/O control bit to periodically reset and resynchronize the module to prevent unexpected loss of sync. The MPC3008 requires a lower clock speed due to its internal analog-to-digital converter electronics conversion of the input signal into a 12-bit digital value. Two analog channels are currently used with the heart rate monitor requiring ten acquisitions per second, with temperature acquisition being converted at the same time for convenience as these parameters do not change rapidly. GPS information is input to the RP3 via a USB connection and provides NMEA standard 0183 output using a simple serial protocol. The GPS module used is an Adafruit Version 3 Ultimate Breakout, with a -165 dBm input sensitivity and capable of receiving 66 channels with 10 Hz updates. It is possible to connect an external active antenna to its input to increase its sensitivity to in excess of −185 dBm when the environment provides weak signal conditions. GPS coordinate information is provided to a Python program running in the background and feeding the Flask web server each time it packages data to send to the client web page. A hardware shutdown button is provided to ensure the orderly shutdown of the Linux operating system. The operation of this switch causes an interrupt that executes the listen-for-shutdown.sh shell script. The shutdown script is available in the smart suit detailed documentation.

1.4 Smart suit system

As indicated earlier two separate software modules and applications are automatically started during the system power-on operation:

- The Lepton Thermal Imager and
- The Flask server

1.4.1 Thermal imager module

When the Lepton Thermal Imager application is started, it creates a "LeptonThread" object to set up the operating parameters, such as SPI clock speed and frame size, then enters a continuous working loop. The loop inputs a line of data from the thermal camera consisting of row number plus 80 pixels while scanning for and then synchronizing to the first line count number of a frame. Once the first row is identified, the thread continues inputting lines of pixel data, storing them in an array, until the line count reaches the maximum (currently set to 60 lines), after which it scans the data array to find its max and min pixel values. Finally, it invokes an update image operation with arguments consisting of the data array, with maximum and minimum values for further processing in the main thread. When the main thread receives a completed frame, it pseudo colors the pixels (using a selected color map) based upon the maximum and minimum values present to form a completed image frame. The completed image frame is written as a high-quality JPG to working memory based upon a periodic timer event, which at present is limited to 1.8 s per frame.

1.4.2 Flask server module

When the Flask server is started, it first initializes the GPS system and senses that it is actually attached to the RP3. If the hardware is not present, it keeps looking for the attached hardware and will not continue until this important component part of the system is connected. When present, the GPS subsystem runs continuously in the background to populate time and coordinate data arrays for use by the web server. When the web server receives a request from the client application, it utilizes the populated GMT time and coordinates data arrays to construct a data structure (python dictionary). Additionally, the collected physiologic data (heart rate and temperature) are appended to the data structure. When all data have been incorporated, the web page template is rendered to contain the newly updated data, along with other java script operations to finalize the page. See Appendix A.1 for more details on system startup detailed scripts and Appendix A.2 on smart suit system application modules.

1.5 Wi-Fi setup and operation

Since the smart suit system uses a local area network to transport and to present its data, it must first be connected to the network. The smart suit system uses a fixed dedicated network IP address (e.g., 192.168.0.99), which will be different for each smart suit system, and must fall within the range of addresses used by the network configuration. Each smart suit system acts as its own web server and is accessed on a fixed IP address at port 5000 (such as 192.168.0.99:5000). Wi-Fi setup of the system requires that a specific file contains the Service Set Identifier "SSID" and passphrases "PSK" for the network. For a "headless" system (one without monitor, keyboard, and mouse), these two items must be known ahead of time and added to the configuration file. This file may be set up prior to connecting to the Wi-Fi network for the first time, which should allow for automatic connection thereafter when the smart suit system boots. In a "headless" system, this configuration file must be positioned in the root directory on the RP3, which then gets automatically transferred to the **/etc/wpa_supplicant** directory upon the first bootup. The easiest way to setup this file is to edit it directly on the micro-SD card using a PC. This may be accomplished by using a micro-SD adapter plugged into the PC. Note that the boot sector of the micro-SD card is readable by the PC (under Windows) because it has the correct FAT32 structure. A text editor (Notepad) may be used to edit this file (**wpa_supplicant.conf**) in the micro-SD cards root directory. One has to make sure that no nontext or other characters are inadvertently entered in the file. The detailed contents of this file are in smart suit system documentation and it is used when setting up the system at some particular location with a specific Wi-Fi Access Point. After booting, Wi-Fi configuration file will be relocated to RP3 directory: **/etc/wpa_supplicant/wpa_supplicant.conf**. If the Wi-Fi does not appear to start up during operation, it could be due to an error in the original wpa_supplicant. conf file (such as tabs, or other wrong characters, or formatting) that had been placed in the micro-SD card's/boot directory. If HDMI display output with keyboard and mouse is available for the RP3, then editing the **/etc/wpa_supplicant.conf** file directly with the correct parameters should restore Wi-Fi hardware operation. This file may contain multiple network entries to allow the RP3 Wi-Fi to automatically connect at a number of locations. Since the Smart Server System web server delivers its web page to a fixed (programmed in) IP address, for example, **http://192.168.0.99:5000**, the router (Wi-Fi Access Point) used must not automatically assign some other device to this fixed address value. It might be necessary to set a reserved static IP address in the router for this purpose, but one needs to check the local router set up first. When finally connected to the network, one should use a newer Edge Microsoft web browser, or an updated Firefox browser to view the web page at the above IP address.

It is not recommended to use Chrome on PC, as it caches the thermal image then only uses the first one received with none of the new updates being viewable. It is possible to use a smart phone or tablet web browser to see the web page at the above IP address. This may later be useful in using a tablet directly as one of the smart suit components, for example, for some control purposes [11]. Figure 1.8 shows a partial list of configuration files.

1.6 Implementation

1.6.1 Components

The smart suit system platform components are shown in Figure 1.9, and their typical cost is summarized in Table 1.1. The total component cost is less than $150 in small quantities. These components were chosen because they were readily available and more choices exist today at smaller prices, so we can assume the total cost to be less than $100 in large quantities. More sensors can be added as well per the customer's requirement. In any case, we did not optimize either the cost or the

Name	Date modified	Type
overlays	23.4.2019 10:48	File folder
bcm2708-rpi-0-w.dtb	10.11.2017 5:57	DTB File
bcm2708-rpi-b.dtb	10.11.2017 5:57	DTB File
bcm2708-rpi-b-plus.dtb	10.11.2017 5:57	DTB File
bcm2708-rpi-cm.dtb	10.11.2017 5:57	DTB File
bcm2709-rpi-2-b.dtb	10.11.2017 5:57	DTB File
bcm2710-rpi-3-b.dtb	10.11.2017 5:57	DTB File
bcm2710-rpi-cm3.dtb	10.11.2017 5:57	DTB File
bootcode	10.11.2017 5:56	VLC media file (.b
cmdline	8.1.2018 5:08	Text Document
config	11.1.2018 6:57	Text Document
config1	28.12.2017 17:51	Text Document
COPYING.linux	10.11.2017 5:57	LINUX File
fixup.dat	10.11.2017 5:56	DAT File
fixup_cd.dat	10.11.2017 5:56	DAT File
fixup_db.dat	10.11.2017 5:56	DAT File
fixup_x.dat	10.11.2017 5:56	DAT File
issue	7.9.2017 17:11	Text Document
kernel	10.11.2017 5:57	Disc Image File
kernel7	10.11.2017 5:57	Disc Image File
LICENCE.broadcom	10.11.2017 5:56	BROADCOM File
LICENSE.oracle	7.9.2017 17:11	ORACLE File
occidentalis	11.11.2017 6:27	Text Document
start.elf	10.11.2017 5:56	ELF File

Figure 1.8 An excerpt from smart suit set of configuration files

Table 1.1 Component cost in small quantities

Description	Unit cost
Raspberry Pi 3—Model B—ARMv8 with 1G RAM	$35.00
Aluminum heat sink for Raspberry Pi 3 15 × 15 × 15 mm	$1.95
Pi Model B+ Pi 3 case base—smoke gray	$5.00
Adafruit ultimate GPS breakout-66 channel W/10 Hz updates—Version 3	$39.99
GPS antenna—external active antenna 3–5 V 28 dB 5 m	$12.95
SMA to uFL u.FL/IPEX RF adapter cable	$3.95
USB to TTL serial cable—debug/console cable for Raspberry Pi	$9.95
Pulse sensor amped	$25.00
TMP36—Analog temperature sensor	$1.50

design configuration. The aim was to have an electronic demonstration and marketing platform as well as a general electronic design platform. The biggest challenge in the project was to design this platform in a very short period of time which was only 3 months. Additional 2 weeks were needed to implement the components into the suit itself and test it. All the components were embedded inside the suit with addition of a couple of pockets to hold the components and make them available for set up and system activation/reset. The GPS sensor was placed up in the shoulder of the suit, and its antenna on the other shoulder. The thermal camera was situated in the upper pocket. The computer module with the battery is in an inner pocket secured with Velcro; heart rate and temperature sensors were located in one of the suit sleeves. Also, cabling was embedded throughout the suit to connect all the components. The computer module has both 4G as well as Wi-Fi modules, and Wi-Fi was used to connect the suit with Internet via local Wi-Fi access point. A Micro SD card had all the drivers and software required to run the suit. The suit was tested in a number of locations around Sarajevo, Bosnia and Herzegovina and Silicon Valley, USA. For example, Figures 1.4 and 1.5 show two streets in Sarajevo area and Figure 1.3 shows a street in Silicon Valley, with Google Map in lower-left corner indicating the street, as well as showing a person inside the building with his temperature and heart rate data and his thermal image obtained from the camera in the suit upper pocket. These data were obtained by logging into the suit "web site" which showed what the suit condition at that point in time. The platform is suitable for a specific sensor(s) extension as it may be required by a specific application at hand.

1.6.2 Application

The smart suit platform as described in this chapter is suitable for many different applications, both for defense and for a variety of commercial applications [11]. The suit in the right upper corner of Figure 1.9 was supplied for demonstrating purposes, and a smaller and lighter suit (jacket) can be also used. The components from Table 1.1 were physically implemented in that demo suit using proper wiring. One could have also used various wireless versions of these components. The suit

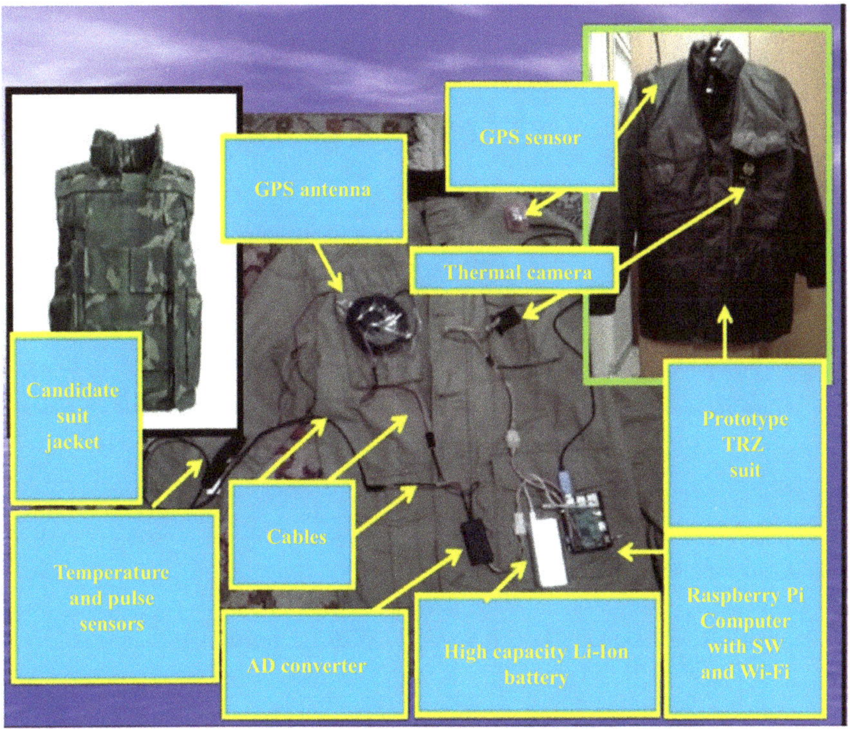

Figure 1.9 Smart suit prototype and components layout

was demonstrated to a number of potential users such as local police, mountain rescue groups, as well as civil protection service and several defense application users. Each of the potential customers indicated their own specific requirements, in particular sensor-related details. Our smart suit design platform can accommodate adding additional sensors and integrating them via embedded computer and software developed for the Platform. Sensors can be either wired or wireless, and each customer may supply their own specific suit or jacket. Figure 1.10 shows some possible applications and a variety of different suits and also a possible remote connectivity arrangements whereas the suit would be connected to some customer control center from where the suit parameters and the movement of the person wearing it could be observed. One of the applications which emerged from potential customers following demonstration of our design was a need to have personnel positions and their health as well as kinetic (moving, not moving, running, and walking) information available at all times within a company or other campus, or a large building. For this, the Wi-Fi would be good enough. For field applications one

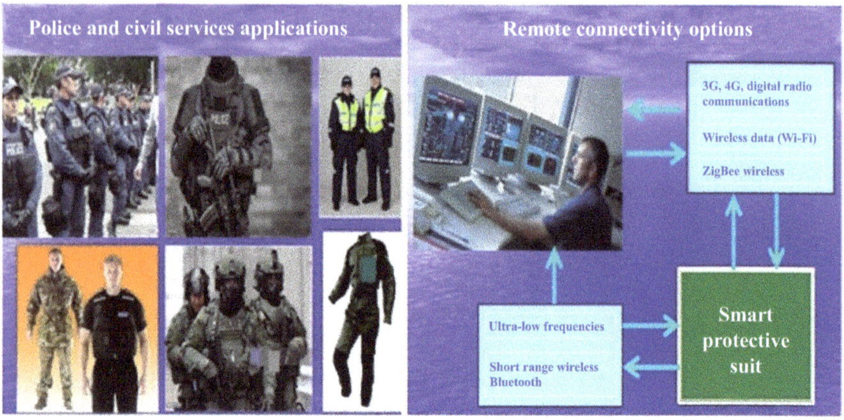

Figure 1.10 Applications and remote connectivity

would need 3G or 4G for wide-area wireless networking. Our design has both of these options built-in.

1.6.3 Mountain rescue services emergency response application

As described in Section 6.2, our smart suit design has a wide applicability across various commercial, security, and emergency services applications. One specific area of interest has emerged from our application surveys in Sarajevo area in Bosnia and Herzegovina came from a dedicated mountain rescue unit, which is of interest due to a number of surrounding mountains and Olympic skiing tradition, with a sizeable amount of snow and skiing activities during the winter season. Per our discussion with a local mountain rescue group, our basic suit design is an excellent technological base which would require some sensor additions and fine tuning. In particular, due to the remoteness of the applicable terrain where mountain rescue might transpire, Wi-Fi may not be the best choice for audio, physiological, and visual or thermal data communications, hence mobile GSM 3G, 4G, or 5G would be a preferable choice. In addition to GSM communication, a digital VHF radio infrastructure already used by the mountain rescue unit is under consideration. This is important as there are potential blind spots for GSM coverage in some deep ravines and canyons in the area. Our design accommodates that option as well, whereas a suitable data modem would be connected instead of (or on top of) Wi-Fi device, see Figures 1.1 and 1.10. Mountain rescue teams would be able to communicate real time data from the terrain, as they move around in their rescue effort. As this activity transpires a control center nearby (Figure 1.10) would be able to access the team physical condition, their heart rate, and temperature, indicating the condition of the team, whereas additional sensors would be used to evaluate vital functions of the persons being rescued, once they are located. It would be pretty trivial to add additional sensors, such as a standard video camera to have a real time video transmission as the search is progressing. In addition to the above, an additional integrated or a separate voice communication device could be added to the rescue team suits, which can further facilitate the effectiveness of their efforts. As the prototype suit is designed, the rescue team will be used to test its various features in real rescue or simulated conditions on the real terrain, which can assist in fine-tuning the suit design. Figure 1.10 shows a few possible rescue uniforms and suits which can be used to implement our design for mountain rescue application.

1.7 Conclusion

In this chapter, we present a prototype design of a smart suit (jacket) which uses GPS, temperature, heart rate, as well as thermal information sensor embedded into a specific suit for demonstration purposes. The suit has its own Internet address and as such can be accessed remotely and its condition can be observed in real time every second. Real data transfer which the platform can accommodate is much larger. Other sensors can be added as well. Wi-Fi is also viable for the data communications, as well as 3G or 4G mobile communications if required by a specific application. These applications range from a variety of commercial to specific defense areas. The smart suit as implemented can be a part of a larger system such as of a Smart City with a number of city services interconnected and smart suit as a part of these services.

Appendix A

Software startup scripts and modules

A.1 System startup detailed scripts

Crontab

At system powerup, the Flask web server is started using the cron daemon, which is setup within directory **/var/spool/cron/crontabs/pi**. This daemon has already been setup for operation in released code. Development programming uses the terminal crontab command:

```
crontab -u pi -e
```

This will bring up an editor allowing changes to be made to the crontab (cron table). The following is then entered on the last line of the file that opens (Note: press i for insert):

```
@reboot sleep 20; /usr/bin/python3
/home/pi/webpy/servflask.py
```

When finished editing type “esc” to exit the edit mode, then save and exit the editor by typing “wq”. The file will run at startup and delay bootup for 20 s to allow Python to execute the script servflask.py found in directory /home/pi/webpy. After the 20 s delay is over, the python script should have run and the bootup process can continue.

startup.sh When the system starts up, it executes the startup.sh script found in the user's home directory /home/pi/startup.sh. This shell script is used to start up the Lepton thermal imaging application and contains the following:

```
#! /bin/sh
#Start Lepton
/home/pi/Qt/Demo/build-Demo-kit2-Debug/./ Demo
```

listen-for-shutdown.sh

During system startup, a script allowing shutdown when I/O Port pin 3 is grounded is executed. Note that shutting down the Operating System in this manner causes a processor halt. When halted, the system may be re-started again by grounding I/O Port pin 3. The shutdown script contains the following:

```
#! /bin/sh
# BEGIN INIT INFO
# Provides: listen-for-shutdown.py
# Required-Start: $remote_fs $syslog
# Required-Stop: $remote_fs $syslog
# Default-Start: 2 3 4 5
# Default-Stop: 0 1 6
# END INIT INFO
case "$1" in
start)
echo "Starting listen-for-shutdown.py"
/usr/local/bin/listen-for-shutdown.py
;;
stop)
echo "Stopping listen-for-shutdown.py"
pkill -f /usr/local/bin/listen-for-shutdown.py
;;
*)
echo "Usage: /etc/init.d/listen-for-shutdown.sh
start|stop"
exit 1
;;
esac
exit 0
```

listen-for-shutdown.py

Python script invoked from startup shell script listen-for-shutdown.sh. This script sets GPIO3 as an interrupt source that triggers when Raspberry Pi port pin 5 is grounded, causing system shutdown. Contents of Python script listen-for-shutdown.py is: Contents of listen-for-shutdown.py is:

```
#!/usr/bin/env python
import RPi.GPIO as GPIO
import subprocess
GPIO.setmode(GPIO.BCM)
GPIO.setup(3, GPIO.IN, pullq_up_down=GPIO.PUD_UP)
```

```
GPIO.wait_for_edge(3, GPIO.FALLING)
subprocess.call([“killall”, “python3”])
subprocess.call([“killall”, “python3”])
subprocess.call([“sudo”, “shutdown”, “now”])
subprocess.call([“echo”, “POWER OFF
!!!!!!!!!!!!!!!!!!!!!!!!!!!!!!!!!”], shell=True)
```

wpa_supplicant.conf

A first-time boot-up of the system transfers the wpa_supplicant.conf file (if found) from the root directory on microSD card to the Linux system /etc/wpa_-supplicant directory. After the system is started, look for this file in directory etc/wpa_supplicant/wpa_supplicant.conf. This file is used to configure the Wi-Fi with its operational credentials and has the following format:

```
ctrl_interface=DIR=/var/run/wpa_supplicant GROUP=netdev
update_config=1
country=BA
network=
ssid=“Your SSID name”
psk=“Your password”
key_mgmt=WPA-PSK
```

A.2 Smart suit system application modules

Thermal imager module

The thermal imager executable is invoked during the startup procedure when startup.sh, located at /home/pi/startup.sh, is run. This shell script issues the run operation (./Demo) on this executable file in directory /home/pi/Qt/Demo/build-Demo-kit2-Debug/. Contents of the startup.sh script are:

```
# /bin/sh
# Start Lepton
/home/pi/Qt/Demo/build-Demo-kit2-Debug/./Demo
```

The Demo project was developed using Qt4 and generates the executable file named “Demo.” Qt is a cross-platform development environment but is used natively on the Raspberry Pi 3 platform under Debian Linux distribution. This means that the entire development environment operates on the Pi and its build output is saved directly to a local development directory. Program execution by the startup shell file invokes the executable out of this target directory. Once started, the application initializes the Qt-based MainWindow parameters to setup the image for display, starts a periodic timer, and then starts the Lepton thread reading the Lepton thermal camera image. The Lepton thread runs in the background, taking in and processing pixel data at the thermal camera’s normal data rate of 27 frames per second. New frames are generated at only nine frames per second, however, with frames being saved to an array in memory. The periodic timer is, at present, setup to save the presently available image to a JPG format in memory, which is accessible by the web server when an image is needed. Thermal Imager Module operation is outlined in the following diagram in Figure A.1.

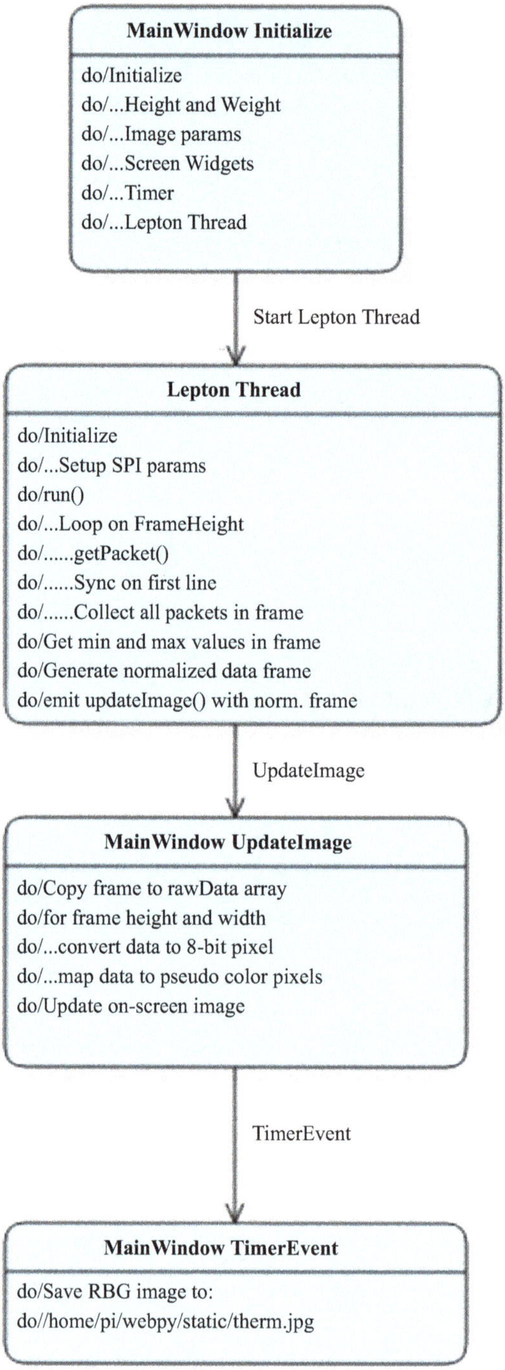

Figure A.1 Diagram of thermal imager application demo

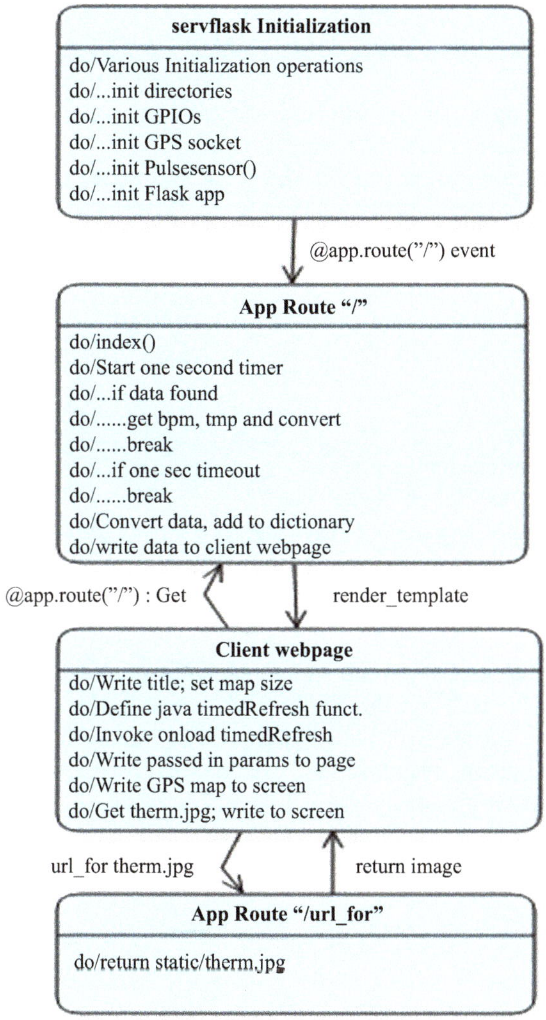

Figure A.2 Diagram of flask server module application

Qt4 creator

The software development framework from open source Qt4 has been used to edit and build the Thermal Imager module to produce executable "Demo." Installation of Qt4 on the Raspberry Pi is done using the following steps from terminal (use LXTerminal to build, compile, and run Qt):

1. Verify that current time and date are set on the Raspberry Pi
2. sudo apt-get update

3. sudo apt-get upgrade
4. sudo apt-get install qtcreator
5. Open Qtcreator and go to: Options ¿ build & run ¿ tab tool chain ¿ button add ¿Choose GCC
6. Set compiler path: /usr/bin/arm-linux-gnueabihf-gcc-4.6
   Debugger: /usr/bin/gdb Mkspec : default
7. Go to menu help ¿ about plugins and Uncheck device support ¿ remote linux

Restart Qt creator

Go to tools ¿ options TAB ¿ build & run ¿ Qt versions ¿ add "/usr/bin/qmake-qt4"

After code has been edited, open LXterminal and make the current directory containing the project files the current directory. To build project use the following:

1. qmake -project
2. qmake <yourprojectname>.pro
3. make (//For compilation of the code)
4. sudo ./< yourprojectname> (//For running the program)

Command qmake only needs to be done the first time to generate the make file, then after code changes are made only command make is necessary to build the project.

Flask server module smart suit web server software development is undertaken using Python version 3 and the Flask server environment. The Raspberry Pi 3 operating system comes with both Python v2.7.13 and v3.5.4 installed. To run programs under Python 3, however, the user must explicitly enter "python3" on the command line (entering just python will invoke version 2.7). Note that the startup script that invokes the Flask server application first calls Python 3 (/usr/bin/python3), before Flask. Flask operation is outlined in the following diagram in Figure A.2.

References

[1] Al-Turjman F., Altrjman C., Din S., and Paul A. "Energy monitoring in IoT-based ad hoc networks: An overview." Elsevier Computers & Electrical Engineering Journal. 2019, vol. 76, pp. 133–142.
[2] Al-Turjman F. "Cognitive routing protocol for disaster-inspired Internet of Things." Elsevier Future Generation Computer Systems. 2019, vol. 92, 1103–1115.
[3] Al-Turjman F. "Cognitive-node architecture and a deployment strategy for the future sensor networks." Springer Mobile Networks and Applications. 2019, vol. 24, no. 5, pp. 1663–1681.
[4] Hodzic M. and Muhic I. "Internet of Things: Current technological review." Periodicals of Engineering and Natural Sciences. 2014, vol. 2, no. 2.

[5] Accenture Corporation. *Delivering public service for the future: Navigating the shifts. Corporate handbook.* Austin, TX, USA; 2012.
[6] European Commission. *Smart wearables: Reflection and orientation paper.* Directorate-General for Communications Networks, Content and Technology. Brussels, Belgium: EU; 2016.
[7] Research and Markets. *Wearable electronics—Market analysis, trends, and forecasts.* Global Industry Analysts, Inc. San Jose, CA, USA; 2016.
[8] Raspberry Pi Foundation. Raspberry Pi 3 Model B (Reduced Schematics). Cambridge, UK; 2015.
[9] Coyle, S., Wu, Y., Lau, K.T., De Rossi, D., Wallace, G. and Diamond, D. "Smart nanotextiles: A review of materials and applications." Mrs Bulletin. 2007, 32(5), pp. 434–442.
[10] FLIR Corporation. Lepton Engineering Datasheet. Document Number: 500-0659-00-09 Rev: 203. Wilsonville, OR, USA; 2017.
[11] Hodzic M. Smart suit presentation. Author's private archive. 2019.

Chapter 2
Medical sensor networks impact in smart cities

Bhawana Rudra[1]

Management and prediction of the chronic disorders play a major role in health. Aging-related diseases represent one of the most relevant challenges for developed countries. The emerging technologies such as Internet of Things (IoT) support healthcare system infrastructure development. IoT is the collection of sensors, actuators, and processors that are connected to Internet to communicate with each other. Smart healthcare applications developed using IoT can deliver comprehensive care to its acute and long-term and sometimes community-based patients. The wireless body sensor network (WBSN) is a popular wearable device technology of IoT used for healthcare system, where the patients are monitored using a tiny-powered and lightweight wireless sensor nodes. The usage of IoT brings convenience to patients and doctors and can be applied in various medical areas. The patients attached with sensors can be used to measure vital signs and other biometric information and problems associated with that patient and can provide a better quality of care and resources. We discuss about various healthcare monitoring systems by using low-cost wireless sensors and already existent IoT technology as communication platform. We focus even on open challenges, and possible new research trends are also discussed.

2.1 Introduction

The development and usage of mobile technology bring in the existence of smart healthcare solution to the people who are residing in the urban area. The Internet of Things (IoT) is expanded in energy, transport, security, and infrastructure. A unified healthcare system gathers all information together, shares and analyzes the data, and gives a solution to the healthcare problems. In 2015, a world report has mentioned that the older age people's population is increasing and most of the people will live for more than 60 years. They have many diseases such as heart problems, strokes, respiratory disorders, dementia, etc. The survey conducted in the United States and Europe states that the older aged people prefer homely

[1]National Institute of Technology-Surathkal, Karnataka, India

environment rather than being in a hospital. As a result, developing a smart city is the main focus for research. According to the National Healthcare Expenditure projections 2012–2022, the total expenditure of the smart healthcare increases from US$ 99.5 billion to US$ 144.9 billion [1–3].

The current lifestyle and food habits are increasing many diseases such as cardiovascular disease (CVD), diabetes, and heart stroke. World Health Organization said that due to CVD, yearly 31% of people are getting affected worldwide and day by day the number of diabetic people and the cost of healthcare services are increasing. To serve the patients, we need a smart healthcare system. IoT technology interconnects humans and devices anytime and anyplace. With the help of this technology, devices gather the information from the users and transfers when required. IoT provides good healthcare services by combining information technology and telecommunications. With IoT users, information can be transferred between various locations to diagnose the diseases and to provide proper medication mainly in rural areas. This technology reduces the cost, travel time, and hospital stays. Wireless body area network (WBAN) is used to connect wearable devices with sensors. These devices placed inside or outside of the body and connected with the Internet. With this device, patients can be monitored at remote locations also. MQTT protocol connects various devices to share the information. Compared to other protocols, MQTT is a light weighted protocol. A person's health data will be compared with existing data for further analysis with the help of a pattern matching algorithm. Identified disease information will be sent to the caretaker, and the same information will be sent to the doctor in emergency cases for further treatment. IoT healthcare systems are facing challenges in security, privacy, mobility, and scalability [4].

Many applications such as network automation, efficiency enhancement, error reduction, etc. use AutoID. AutoID center released the electronic product code (EPC) network. The EPC permits tracking the things moving from one location to another location. This gives an idea for the implementation of IoT. The radio frequency identification (RFID) further gives the idea for developing IoT. In 2005, IoT proposed as the combination of computing and sensor-based technologies [5–9]. This combination allows the object to be tagged, sensed, and controlled over the networks. This provides interaction and communication among linked devices.

2.2 Smart city

A smart city is simply an urban area which uses various sensors to collect the data from citizens, assets, and various devices, and these data will be used to monitor transportation system, air pollution, hospitals, crime detection, schools, libraries, etc. To run a smart city, data are required which are gathered from individuals and things. We need systems with networking, monitoring equipment, sensors, data analytics, computational techniques, and also humans to make data useful. Collaboration of people, government authorities, technology providers, local enterprises, and service delivery will lead to a successful smart city. Technologies play a very important role in smart city to reduce the cost of medical treatment. But here the challenges with data collection are security and privacy. There are many possibilities to overcome these challenges and to create smart cities [1,2,10].

According to the United Nations, by 2050, the population of urban area will be increased by 68%. So, taking care of patients with this increased number of people will be difficult, and it can be simplified by providing remote monitoring which can be done with smart healthcare. Juniper published a report on smart city rankings based on four categories: safety, health, mobility, and productivity, in that Singapore ranked as no 1 smart city and Seoul is in next place as shown in Figure 2.1. These two cities are adopted digital services to monitor their citizen's healthcare and health-related information. Seoul developed U-Health Strategy to monitor aged citizens by sending monitoring devices to their houses to access the patient's information. This helped to reduce the number of people unnecessarily going to hospitals [11].

	Mobility	Health	Safety	Productivity
1	Singapore	Singapore	Singapore	Singapore
2	San Francisco	Seoul	New York	London
3	London	London	Chicago	Chicago
4	New York	Tokyo	Seoul	San Francisco
5	Barcelona	Berlin	Dubai	Berlin
6	Berlin	New York	Tokyo	New York
7	Chicago	San Francisco	London	Barcelona
8	Portland	Melbourne	San Francisco	Melbourne
9	Tokyo	Barcelona	Rio de Janeiro	Seoul
10	Melbourne	Chicago	Nice	Dubai
11	San Diego	Portland	San Diego	San Diego
12	Seoul	Dubai	Melbourne	Nice
13	Nice	Nice	Bhubaneswar	Portland
14	Dubai	San Diego	Barcelona	Tokyo
15	Mexico City	Wuxi	Berlin	Wuxi
16	Wuxi	Mexico City	Portland	Mexico City
17	Rio de Janeiro	Yinchuan	Mexico City	Rio de Janeiro
18	Yinchuan	Hangzhou	Wuxi	Yinchuan
19	Hangzhou	Rio de Janeiro	Yinchuan	Hangzhou
20	Bhubaneswar	Bhubaneswar	Hangzhou	Bhubaneswar

Figure 2.1 Smart city ranking in 2015

IoT can do location sensing in which RFID tags are used for tracking the location. IoT can control the traffic in the city using devices and networks. IoT can also give information about environmental changes such as pollution and disaster forecast. It can give alarm in an emergency situation. Real-time data about the patient can be determined using IoT devices. Remote monitoring can be done for appliances control using IoT for energy conservation. The architecture will give the necessary security and privacy. Ad hoc network helps in widening the network.

2.3 Smart healthcare in smart cities

Many advancements have taken place in the medical field that is from traditional equipment to wireless reprogrammable devices. It includes even the emergence of IoT in medical field where the devices are connected to cell phones. This is helpful to monitor the devices and patients remotely. The parameters are recorded by the backend system for the analysis of data to provide appropriate medicine or the required staff to attend the patient. This record helps to know the patient's condition and react to the critical issues. The research work is in progress to develop medical devices that are smart and cost effective. An IoT Stack Protocol architecture was developed by Bui and Zorzi for healthcare applications [1,2,11–13].

The challenges such as sensing and visualization were addressed to improve the healthcare along with cost reduction [10]. IoT-based healthcare technique was used by Muhtadi and Albuquerque for analyzing problems such as heart rate, ECG, etc. They even discussed about the available technologies for medical IoT service providers, researchers, and for developers. Two criteria for the development of medical IoT devices are identified by Alansari *et al.* [14], that is, economic property and quality of life. An IoT Fog-based healthcare solution was introduced by Mahmud *et al.* The framework will analyze the use of iFog simulator with respect to power consumption and latent reduction. Sharma *et al.* have developed a k-health framework that can monitor diseases such as diabetes and BP along with considering the privacy of the patients. Eco health platform was developed to share real-time data between doctor and patient [15].

Many smart cities around the globe are using IoT to improve the health of their public. Some of the initiatives around the globe are Barcelona deployed sensors in the street light to monitor air pollution. Chicago used sensors to capture data on noise levels, humidity along with air quality. Washington, D.C. used IoT to handle wastewater and leakage problems by using underground sensors. Smart pumping system of wastewater uses sensors to detect the problem and also to send the feedback to the operators who are at pumping station for better outcomes. Spanish city used 12,500 sensors in 2010 in and around the Santander city to check the air pollution. This city is even helping the programmers to develop apps that will help the citizens, etc. Not only these, many cities such as Norway, Oslo, Dublin, and Ireland are using sensors embedded on bikes and buses to check pollution levels in the city [5–7,16–19].

Smart healthcare system cures the disease with the help of digital and mobile devices. IoT in healthcare collects the patient's data remotely with the help of sensors. These data can be used by the doctors for better treatment. With this, the time and cost of both patients and hospitals will be reduced. IoT with AI will give the best results and many companies are investing to develop wearable devices with IoT and AI technologies. Patients are getting problem when doctors are not available in the hospital, and to solve this problem, robots can be developed to treat the patients [20–24].

Around the globe, the aging population is increasing and many diseases such as asthma, diabetes, and heart-related issues are also increasing in every age group. But the doctors and nurses are less compared to that of patients. According to one of the reports, India is running out of medical professionals of at least 6 lakh doctors and 10 lakh nurses. This report clearly indicates the requirement of IoT in the country. Back in 2009, the former president of China explained in his inaugural speech of IoT research center that the need of advanced IoT models and its application of healthcare in the country to maintain the health of the people [3,25–28]. ISO considers Healthcare as one of the Key factor of the smart city organization. And they give smart city scores with good weightage to Smart healthcare. Many countries including the United States, Saudi Arabia, and Chile focusing on building good Smart healthcare to get a good score and in order to become a smart city [15–17,29,30].

2.4 Technologies used in smart healthcare

Sensors play a major role in eHealth systems. Sensors collect the information such as glucose levels, BP, etc. of a patient and send it to the doctor for diagnosis purposes. With 3D pills, technology user can take a pill which is developed with a sensor to know his health information. Connectivity is a major challenge with sensors [8,20–23]. Smart cities and communication between users and doctors will depend on sensors. While transmitting private data, it is also very important to provide privacy. Data anonymization technique is used to provide security for the patient's data.

2.4.1 *Artificial intelligence*

Using IoT devices, the data are sent to the healthcare. Artificial intelligence (AI) can be used to analyze the data to give more appropriate treatment for the concerned patient. AI apps can also be used to access the current medical condition of the patient.

2.4.2 *Blockchain*

Electronic healthcare data and records can be protected using the Blockchain technology where the data cannot be modified. It can also help in linking the services such as payments and insurance.

2.4.3 Internet of Everything

Internet of Everything (IoE) in healthcare will increase the productivity. Using IoE, patient can print the pills at home and without traveling to the hospital. This will be a huge achievement in healthcare treatment.

2.5 IoT services in healthcare

System architecture of IoMT consists of device layer, Cloud layer, and Internet-connected layer. This architecture connects patients with their doctors. With this, the doctor can continuously monitor patient's BP, heart rate, temperature, brain hemorrhage, etc. This architecture uses sensors to collect data from humans. Sensor output will be in an electrical format and it will be processed and sent to the user terminal with the help of microprocessors. The user terminal is connected to Fog layer. With the help of communication protocols, such as Bluetooth, ZigBee, etc., the Fog layer allows the user device to carry the stored data [25–27,31,32]. Cloud layer is connected with the Fog layer. This cloud layer stores and processes the data and also contains patient data backup such as reasons to visit the hospital and the number of time hospital visits.

System architecture of IoT healthcare consists of network transmission, information perception, and application service. Information perception layer contains sensors to monitor health. The data which are collected by sensors will be transmitted through the networks with the help of wireless technology and stored in the cloud. Remote healthcare service is provided by the application service layer [33–36].

2.5.1 Remote patient monitoring

In remote patient monitoring, patients are monitored at any time and any place with the help of IoMT. These systems are very useful for old age homes, hospitals, etc. With this, problems can be monitored easily and it reduces waiting time and cost, and improves the treatment quality by monitoring patients continuously. Remote healthcare systems can also be operated with cellphones, smart watches, tablets, etc. In this, sensors directly send the data such as glucose levels, pulse rate, and heart rate to users in emergency cases. Remote monitoring system has three tiers: wearable sensors, user terminal, and cloud layer. Wearable sensors collect the user information and send to cloud layer to store and process with the help of Bluetooth, 2.4 GHz wireless, and Wi-Fi [34–39]. In the user terminal, data will be accessed by the persons who are working at the hospital. Remote monitoring system will share the data with doctors and patients with the help of sensors, and with these data, the doctor can treat the patient.

2.5.2 Telehealth

Telehealth provides communication between doctors and patients in the form of audio and video. With this, doctors can see the signs and treat the patient without going to their house. Telehealthcare system contains tele meeting mode, save and forward mode, and video meeting mode. In save and forward mode, electronic medical record (EMR) is used to store medical and diagnosis information of a patient [8,9,32,40–43]. EMR sends the patients' reports to the doctor for treatment. Telemedicine reduces the waiting and travel time. But this service is not good for the problems which need treatment immediately. In telemeeting, patient and doctor can discuss about health problem through Internet. Video meeting uses audio and video for discussion between doctor and patient.

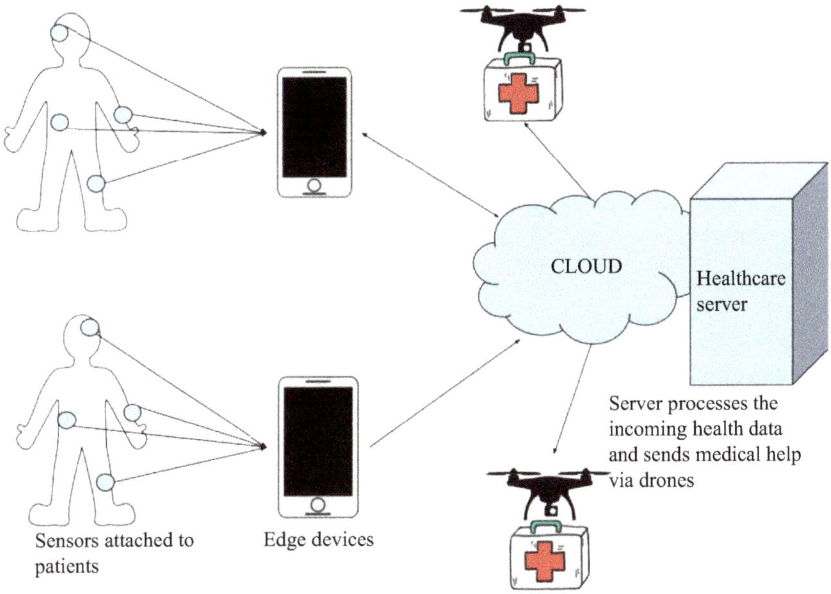

2.5.3 *Wearable devices for IoMT solutions*

Many people are using Smart wearable devices to know their health or fitness status without taking help from professionals. One of the popular wearable devices is fitness band, and many people are using this to track their fitness and new technologies are added to these devices by companies such as Under Armour and Fitbit. Not only these many, other devices also available in the market, and those are as follows: [4,6,10,12,14,16,17]

- **Smart pulse oximeter:** With this device, a person can check his pulse rate, respiration rate, oxygen saturation (SO_2), and perfusion index. This device also checks the percentage of hemoglobin and oxygen in blood. This device can easily operate by placing it on earlobe or fingertip. Persons who need to check their oxygen levels continuously such as mountain climbers, pilots, or athletes will use this product more.

- **Smart watches**: People can check their fitness with these watches as it also allows to call and sends text messages. These device batteries will work long time. It also monitors sleep pattern, calories burned, heartbeat, etc.
- **3D printed pills:** US FDA approved these pills in 2015. Spritam is a first smart pill used by children and adults for seizures and epilepsy. These pills are like other pills which dissolve in the body with sip of water. These pills are developed using ZipDose technology. With this technology, pills are developed with the required dosage of medicine for the treatment.
- **Smart contact lenses:** Google developed this device to help poor people who are having diabetics and poor eyesight. Smart lenses use tears to calculate glucose levels of a person.
- **Smart bras:** Women's breast health can be monitored with smart bras. These devices are connected with sensors. These sensors check breast tissue and if any cancer symptoms are seen, then it alerts the user. This device also stores the information and suggests the users about their breast health with the help of app. It also monitors about heart rate, breathing count, etc.

These devices monitor the patients and collect the data such as temperature, pulse rate, and heart rate and send it to the doctor through the cloud. Wearable devices use sensors to collect the data from the user. Size of these devices are very small but more efficient. These devices are directly connected to cellphones so user can check his data continuously. Wearable devices work with small size batteries, so the power consumption is less compared to other wireless networks. Devices such as smart shoe, smart bracelet, smart watch, smart shirt, etc. are very useful for aged people to check their clinical signs while doing any physical activity. Smart devices also help users to check the details of calories burned and heart rate while doing exercise. The main challenge with wearable devices is communication between device and network. As per the experts of cybersecurity, the data which are gathered from smart devices is a treasure for hackers. Providing security is very important to overcome this problem.

2.5.4 E-textiles in healthcare

The technical textile field has developed quickly. The recognized textiles are applied to medical uses. In the United States and the European Union, the newly invented smart shirts monitor the health of the person. The monitored data are processed using a monitoring unit which is inserted into the lower band of the bra. The data are read using a watch-shaped monitor worn on the wrist. This product is washable and easy to use. Newly invented fabric can monitor vital signs such as heart rate and temperature. The objective of the study of E-textiles is to develop a smart glove that can cure hypertension and estimate the performance of the same device [4,12,15,18].

2.5.5 *Cancer treatment*

In a clinical trial, the American Society of Clinical Oncology (ASCO) got to know that 300+ patients are taking the treatment for cancer on neck and head. In this trial, they used Bluetooth devices to calculate the blood pressure, and these data will be sent to the symptom tracking app. This app will send the data to the patient's physician for treatment. This smart system using patients' data is called as CYCORE. When compared to the people who are visiting doctors regularly, CYCORE got less symptoms of cancer. According to the ASCO President, Bruce E. Johnson, this smart system helping the patients and their caretakers to find the side effects quickly to give best treatment [35–37].

2.5.6 *Smart continuous glucose monitoring*

Continuous glucose monitoring (CGM) device is used to monitor the blood glucose levels of diabetic patients by considering readings regularly. US Food and Drug Administration approved the first CGM System in 1999, and now many CGMs are available in the market. Smart CGMs send the blood glucose level data to an app on smart phone or Apple Watch so that patients can easily check their information. The wearable devices can sense the level of the glucose and it can track the patient's health based on other parameters and the collected data can be transferred via IPV6 to the appropriate healthcare providers. The device consists of a glucose collector, a detector to monitor the glucose level, and a mobile phone. Using the device patient can decide about their food system and physical activities. Smart

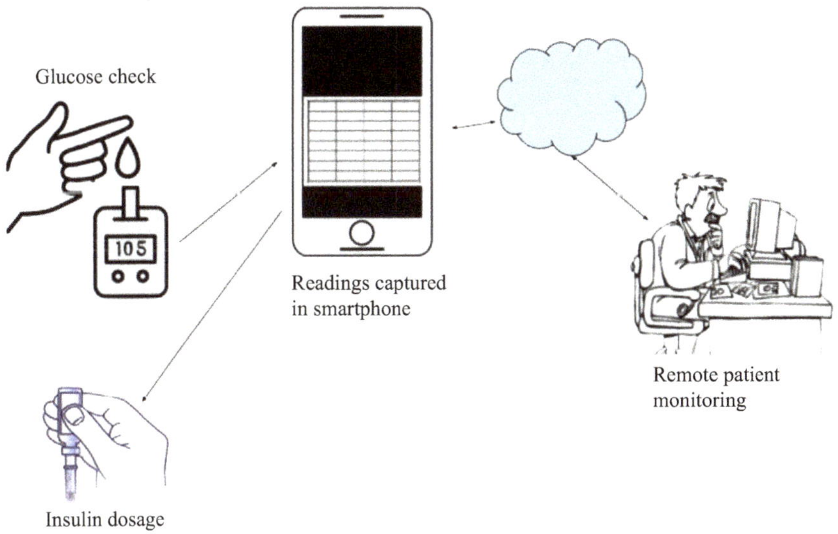

insulin pens recommend the type of insulin injection and the amount of insulin needed for diabetic patients. Insulin pens can also be connected with a smart phone app to help patients to calculate their glucose levels [38–44].

CGM system monitors the glucose levels of diabetic patients but OpenAPS (Open Artificial Pancreas System) monitors the patients' glucose levels and delivers the required amount of insulin to the patients. Dana Lewis and Scott Leibrand started Open APS in the year 2015, and they developed their own software to complete the loop by taking the data from CGM and a system that consists of Raspberry Pi.

2.5.7 Connected inhalers

Connected inhalers are one of the smart technologies to help asthma patients to control their problem by providing proper medication. Propeller Health is the biggest producer of smart inhaler technology. Propeller inhaler attached a sensor to the inhaler and it connected to an app and it helps the people who are suffering from asthma. Propeller Sensors detect inhaler use and this information will be shared with patients' doctor and it also reminds the patients when to take their medication [34,35,38–41].

2.5.8 Ingestible sensors

As per World Health Organization, most of the persons are not taking medicines as prescribed by the doctors. To overcome this problem, Proteus Digital Health Company has created pills with sensors, when a patient takes these pills it will dissolve and generates a signal. Sensor will receive that signal and sends the data to app, with this patient can find that he has not taken medicine as per doctor's suggestion. The first drug tracking system is developed by Proteus and Otsuka Pharmaceutical Company in late 2017 with the drug name ABILITY MYCITE [8,9,24,26,27,30].

2.5.9 Connected contact lenses

Smart lenses are desired application in IoT in the context of medical system. Google Life Sciences declared that it would build a new model called smart contact lens that could estimate the glucose level in tear. If the blood glucose level is not within the range of the limit of a healthy person, then it gives the notification to the appropriate person and using this required decision can be made. Many people have thought that measuring the glucose level using tear may not be accurate but Verily, a subsidiary of Google Company has proven that the project is scientifically correct. The treatment for the Presbyopia and cataract surgery recovery are the current research study of the Verily using the contact lenses. Swiss Company has invented nonharming smart contact lenses named Triggerfish, which will notice the changes in the eyeball that can lead to glaucoma. The Triggerfish is now CE marked and FDA approved for sale in various countries [8,35–37,45–47].

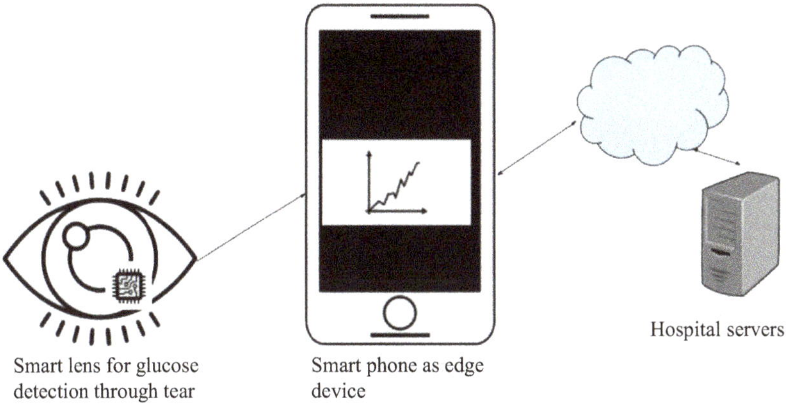

2.5.10 Apple Watch app

The Apple Watch can monitor the mood and cognition of the patient who is wearing the device. It sends the gathered information to the healthcare via app. The professionals can insight into more depth. It is found to be more robust and reliable. Takeda Pharmaceuticals USA and cognition Kit Limited were combined to explore an Apple Watch app to monitor the depressive disorder of the patient. The result is presented in the CNS summit in November 2017 [8,9,46,48,49].

2.5.11 Coagulation testing

Roche developed a coagulation system that works with Bluetooth and detects how quickly blood clots in the human body. The patients can do self-testing which minimizes the risk of strokes and bleeding. It has the feature of reminding the patient when the person crossed the threshold level of clot [8,9,45–47].

2.5.12 Apple's research kit

A new feature is added to the API Research kit by Apple Company. The newly added feature is called as Movement disorder API. The target is to make the process continuous and automatic. The data collected using this can deliver the records either daily or at any specific regular intervals. It is used in a variety of studies such as Arthritis study and epilepsy study. In 2017, Apple launched CareKit, and it helps to develop apps for the medical healthcare industry [8,48,49].

2.5.13 ADAMM asthma monitor

ADAMM is a monitoring device that senses the symptoms of asthma in a very early stage and helps the user to manage the disease well before the attack. When the device detects the problem, it vibrates to notify the patient who is wearing it. It has the feature of sending the information to the diagnostic center to notify the physician. Using the device, the patient can also set reminders according to the treatment plan [8,9,45–47].

2.5.14 Wheelchair management

The smart wheelchairs with complete automation will help disabled people. Using this application, the patient's monitoring can also be done. This device can also access the location of the patient to give more security [35,37,48,49].

2.5.15 Electrocardiogram monitoring

Electrical activity of the heart is recorded using electrocardiography. It includes heart beat rate, recognizing basic rhythms, diagnosing myocardial ischemia, etc. The ECG monitor consists of a wireless transmitter and a receiver. This can detect drastic changes and variations in heart activity. In real time, the data are transferred to mobile phones and to healthcare providers via a network. The IoT uses algorithms for continuous ECG monitoring [8,9,35,43,45].

2.5.16 Hand hygiene compliance

Increase in the number of deaths can be prevented using hand hygiene. It can also decrease the extra days of stay in the hospital and thereby decrease the cost. The absence of infection in the patient must detect early to make healthcare more economic and more safety for the patients. This can be achieved using hand hygiene [8,9,49].

2.5.17 Blood pressure monitoring

The BP monitoring device consists of few apparatuses with network-based communication capabilities. Blipcare is such a device that uses a home Wi-Fi network to record and upload the data. This machine also has LCD display to show BP level. This can be monitored continuously using the proposed wearable device [6,8,12,13,25,50,51].

2.5.18 Body temperature monitoring

The device includes an embedded sensor to record body temperature. Jian and colleagues proposed a device that uses a home gateway to monitor body temperature. This gateway uses infrared detection to send the recorded data [9,22,23,45,46].

2.6 IoT advantages in healthcare

The data collected using IoT devices helps in determining the need of an individual more accurately. The treatment of advice can be sent to the patient using mobile technology so that visiting time to the healthcare center and waiting times at the hospital can be eliminated. The device monitors the patient on a regular basis and alerts healthcare providers in case of emergency. IoT can do automatic checks and screenings to help the healthcare providers in reducing the number of errors that they could do. Number of human interventions and interactions become less as more machine to machine interaction and result in less healthcare cost. IoT helps in removing the possibility of human errors. Real-time availability of the patient's data to the physician helps in remote monitoring and assists the patients. IoT assists the needy patient in giving the suggestion, and hence the problem of availability of doctors could be solved [3,8–10,12,13,25,27,50].

2.7 Challenges

- **Security and privacy:** The IoT device is sending the data from the user place to destination through the network. While the data transfers in the network, it is possible for a hacker to observe or attack the data by altering the original data which may lead to a serious issue in healthcare management. The person may fail to get enough support when needed from the smart healthcare due to the false data created by the third party which can threaten the life of the person. Hence there is a strong data transfer protocol is required to transfer the data.
- **Mobility and scalability:** The person who wears IoT devices can move from one place to another place. So, connectivity is always required for the continuous monitor of the person. The larger the number of the device, the amount of sharing the data is huge in the network. So, the scalability should be maintained.

2.7.1 Security solutions

For masquerade attack, Bruce proposed an efficient and cost-effective device which may be wired or wireless. Li. C. *et al.* proposed two possible defenses for the attacks that happen on wearable devices. Ren has proposed an approach to detect the clone attack. Yu has presented the solution to the accountability and recoverability of attack, and Liang X has proposed a solution to data injection attack using a distributed prediction-based secure and reliable framework. In 2013, Lu presented two schemes for privacy attack namely an attribute oriented authentication scheme and an attribute-oriented transmission scheme. Garkoti in 2014 proposed a new model that combines the functionalities of digital watermarking and auditing support to enable the detection of intra cloud and extra cloud attack whereas Shen has proposed the solution to traffic analysis attack [5,6,8,23,30,43,49].

2.8 Conclusion

The advanced development in the electronic devices, mobile technology, and its applications made a revolution in providing the healthcare facility. It has changed the traditional healthcare approach to a smart healthcare approach. This approach of smart healthcare will help users to monitor themselves as well as informing to the responsible authorities which results in the total improvement in the treatment of the patient. Although solutions exist for the security attacks, still the vulnerability exists and appropriate solutions are required.

References

[1] Al-Turjman F., Ulusar U., and Nawaz M. "Intelligence in the Internet of Medical Things era: A systematic review of current and future trends." Computer Communications. 2020; 150: 644–660.

[2] Al-Turjman F., and Alturjman S. "Context-sensitive access in Industrial Internet of Things (IIoT) healthcare applications." IEEE Transactions on Industrial Informatics. 2018; 14(6): 2736–2744.

[3] Baker S. B., Xiang W., and Atkinson I. "Internet of Things for smart healthcare: Technologies, challenges, and opportunities." IEEE Access Journal. 2017; 5: 26521–26544.

[4] Bačić Ž., Jogun T., and Majić I. "Integrated sensor systems for smart cities." Tehnički Vjesnik. 2018; 25: 277–284.

[5] Jawhar I., Mohamed N., and Al-Jaroodi J. "Networking architectures and protocols for smart city systems." Journal of Internet Services and Applications. 2018; 9: 26. https://doi.org/10.1186/s13174-018-0097-0

[6] Ganesh Prabhu S. Continuous monitoring of human health using internet of things. International Journal of Advanced Research. 2017. Available from: https://www.nxp.com/docs/en/white-paper/IOTREVHEALCARWP.pdf

[7] Chatterjee P., and Armentano R. L. "Internet of Things for a smart and ubiquitous eHealth system." *2015 International Conference on Computational Intelligence and Communication Networks (CICN)*, Jabalpur, 2015; 903–907, DOI: 10.1109/CICN.2015.178.

[8] Gautam K., Puri V., Tromp J. G., Van Le C., and Nguyuen N. G. "Internet of Things and healthcare technologies: A valuable synergy from design to implementation." International Journal of Machine Learning and Networked Collaborative Engineering. 2018; 02(3): 128–142.

[9] Ghani A. "Healthcare electronics: A step closer to future smart cities." ICT Express, 15 February 2018. https://doi.org/10.1016/j.icte.2018.01.009

[10] Kalaiselvi G. "Internet of Things – The current trend in connected world: A comprehensive study on healthcare applications using IoT." International Journal of Engineering Science Invention. 2319–6734.

[11] Pandharkame H. P., and Mudholkar P. "Smart hospitals using Internet of Things (IoT)." International Journal of Scientific and Engineering Research. 2017; 8(5): 2229–5518.

[12] Kim R.-H. "Cure performance and effectiveness of portable smart healthcare wear system using electro-conductive textiles." *6th International Conference on Applied Human Factors and Ergonomics (AHFE 2015) and the Affiliated*

Conferences, AHFE. Procedia Manufacturing. 2015; 3: 542–549. DOI: 10.1016/j.promfg.2015.07.264
[13] Taştan M. "IoT based wearable smart health monitoring system." Celal Bayar Üniversitesi Fen Bilimleri Dergisi. 2018; 343–350. DOI: 10.18466/cbayarfbe.451076
[14] Novotný R. Kuchta R., and Kadlec J. "Smart city concept, applications and services." Journal of Telecommunications System and Management. 2014;3(2). DOI: 10.4172/2167-0919.100011
[15] Sakib S. M. S. "*Smart hospital management system using Internet of Things*."
[16] Hancke G. P., de Carvalho e Silva B., and Hancke Jr. G. P. "The role of advanced sensing in smart cities." Sensors (Basel). 2012; 13(1). DOI: 10.3390/s130100393
[17] Cook D. J., Duncan G., Sprint G., and Fritz R. "Using smart city technology to make healthcare smarter." *Proceedings of the IEEE*. Institute of Electrical and Electronics Engineers. 2018; 106(4): 708–722. DOI: 10.1109/JPROC.2017.2787688
[18] Obinikpo A. A., and Kantarci B. "Big sensed data meets deep learning for smarter health care in smart cities." Journal of Sensors and Actuator Networks. 2017; 6(4). DOI 10.3390/jsan6040026
[19] Vermesan O., Friess P., Guillemin P., *et al. IoT beyond the HYPE: Research, innovation and deployment*. River Publishers Series in Communications. http://www.riverpublishers.com/pdf/ebook/RP_E978-87-93237-98-8.pdf
[20] Islam S. M. R., Kwak D., Kabir M. H., Hossain M., and Kwak K.-S. The Internet of Things for health care: A comprehensive survey. IEEE Access. 2015; 3: 678–708. DOI: 10.1109/ACCESS.2015.2437951
[21] Hammoudeh M., and Arioua M. "Sensors and actuators in smart cities." Journal of Sensors and Actuator Networks. 2018; 7. DOI: 10.3390/jsan7010008
[22] Du R. "Wireless sensor networks in smart cities: The monitoring of water distribution networks case." *Thesis*. Licentiate Thesis Stockholm, Sweden, 2016.
[23] Laya A. "The Internet of Things in health, social care, and wellbeing." *PhD Thesis*. Information and Communication Technology School of Information and Communication Technology, KTH Royal Institute of Technology, Stockholm, Sweden. 2017.
[24] Talari S., Shafie-khah M., Siano P., Loia V., Tommasetti A., and Catalão J. P. S. "A review of smart cities based on the Internet of Things concept. Energies. 2017; 10: 421. DOI:10.3390/en10040421.
[25] Yuehong Y., Zeng Y., Chen X., and Fan Y. "The Internet of Things in healthcare: An overview." Journal of Industrial Information Integration. 2016; 1. http://dx.doi.org/10.1016/j.jii.2016.03.004
[26] Arulananthan C., and Hanifa S. M. "Smart health potential and pathways: A survey." IOP Conference Series: Materials Science and Engineering. 2017; 225: 012065.

[27] Katake R., Kute B., Ranjane S., and Jaiswal S. C. "Survey of health monitoring management using Internet of Things (IOT)." International Journal of Science and Research (IJSR). 2016; 5(11): 1144–1147.
[28] Sriram N. "Survey on security of Internet of Things in eHealth and clouds." International Journal of Innovative Computer Science and Engineering. 2017; 4(4): 25–30.
[29] Pawar S. P. "Smart city with Internet of Things (sensor networks) and big data." 2013.
[30] Sundaravadivel P., Kougianos E., Mohanty S. P., and Ganapathiraju M. "Everything you wanted to know about smart healthcare." IEEE Consumer Electronics Magazine. 2018: 7(1). DOI: 10.1109/MCE.2017.2755378
[31] Csanád Csáji B., Kemény Z., Pedone G., Kuti A., and Váncza J. "Wireless multi-sensor networks for smart cities: A prototype system with statistical data analysis." IEEE Sensors Journal. 2017; 17(23): 7667–7676.
[32] Al-Turjman F. "Intelligence and security in big 5G-oriented IoNT: An overview." Future Generation Computer Systems. 2020; 102(1): 357–368.
[33] Alías F., and Ma R. "Alsina-Pagès review of wireless acoustic sensor networks for environmental noise monitoring in smart cities." Hindawi Journal of Sensors. 2019, Article ID 7634860. https://doi.org/10.1155/2019/7634860
[34] Islam S. M. R., Kwak D., Kabir M. H., Hossain M., and Kwak K.-S. "The Internet of Things for health care: A comprehensive survey." IEEE Access. 2015;3. DOI: 10.1109/ACCESS.2015.2437951
[35] Jeong J.-S., Han O., and You Y.-Y. "A design characteristics of smart healthcare system as the IoT application." Indian Journal of Science and Technology. 2016; 9(37). DOI: 10.17485/ijst/2016/v9i37/102547
[36] Yu L., Lu Y., and Zhu X. J. "Smart hospital based on Internet of Things." Journal of Networks. 2012; 7(10): 1654–1661.
[37] Kulkarni A., and Sathe S. "Healthcare applications of the Internet of Things: A review." (IJCSIT) International Journal of Computer Science and Information Technologies. 2014; 5(5): 6229–6232.
[38] Csáji B. C., Kemény Z., Pedone G., Kuti A., and Váncza J. "Wireless multi-sensor networks for smart cities: A prototype system with statistical data analysis." IEEE Sensor Journal. 2018;17(23). DOI: 10.1109/JSEN.2017.2736785
[39] Azzawi M. A., Hassan R., and Bakar K. A. A. "A review on Internet of Things (IoT) in healthcare." International Journal of Applied Engineering Research. 2016; 11(20): 10216–10221.
[40] Sharma C., and Sunanda. "Survey on smart healthcare: An application of IoT." *International Journal on Emerging Technologies* (Special Issue NCETST-2017). 8(1): 2249–3255.
[41] Dewangan K., and Mishra M. "Internet of Things for healthcare: A review." *International Journal of Advanced in Management, Technology and Engineering Sciences*. 2018.
[42] Saminathan S., and Geetha K. "A survey on health care monitoring system using IoT." International Journal of Pure and Applied Mathematics. 2017; 117(17): 249–254.

[43] Shah J., Soni S., Darji F., and Chandak S. "Smart hospital using IoT." IOSR Journal of Engineering. 9: 26–32. ISSN (e): 2250-3021, ISSN (p): 2278-8719.

[44] Alqahtani F. H. "The application of the Internet of Things in healthcare." International Journal of Computer Applications. 2018; 180(18): 19–23.

[45] Kim R.-H. "Cure performance and effectiveness of portable smart healthcare wear system using electro-conductive textiles." *6th International Conference on Applied Human Factors and Ergonomics (AHFE 2015) and the Affiliated Conferences, AHFE 2015*, Vol. 3. https://doi.org/10.1016/j.promfg.2015.07.264

[46] Gkouskosa D., and Burgosb J. "I'm in! Towards participatory healthcare of elderly through IOT." *International Workshop in Universal Design for IOT Smart Health (UDISH2017), Procedia Computer Science 2017*. Vol. 113. https://doi.org/10.1016/j.procs.2017.08.325

[47] Gómeza J., Oviedob B., and Zhumab E. "Patient monitoring system based on Internet of Things." *The 7th International Conference on Ambient Systems, Networks and Technologies (ANT 2016), Procedia Computer Science, 2016.* Vol. 83. https://doi.org/10.1016/j.procs.2016.04.103

[48] Santosa A., Macedoa J., Costaa A., and Nicolau M. J. "Internet of Things and smart objects for M-Health monitoring and control." *CENTERIS 2014 – Conference on ENTERprise Information Systems/ProjMAN 2014 – International Conference on Project MANagement/HCIST 2014 – International Conference on Health and Social Care Information Systems and Technologies*. Vol. 16. https://doi.org/10.1016/j.protcy.2014.10.152

[49] Yuehong Y., Zeng Y., Chen X., and Fan Y. "The Internet of Things in healthcare: An overview." Journal of Industrial Information Integration. 2016; 1. https://doi.org/10.1016/j.jii.2016.03.004

[50] Bhat M. I., Ahmad S., Amin A., and Ashraf S. "E-health with Internet of Things." International Journal of Computer Science and Mobile Computing. 2017; 6(6): 357–362.

[51] Idwan S. "Smart solutions for smart cities: Using wireless sensor network for smart dumpster management." *2016 International Conference on Collaboration Technologies and Systems (CTS)*, Orlando, FL. 2016; 493–497. DOI: 10.1109/CTS.2016.0092

Chapter 3

The use of CRISPR as a diagnostic tool for healthcare in the IoT era

Abdullahi Umar Ibrahim[1,2], *Zubaida Sa'id Ameen*[1,2] *and Mehmet Ozsoz*[1,2]

This chapter is an overview of clustered regularly interspace palindromic repeat (CRISPR) technology and the application of CRISPR sensors with Internet of Things (IoT) in healthcare system. It explains the molecular mechanism of CRISPR technology which is a natural adaptive immune system used by bacteria and archaea against viruses. In genetic engineering, scientist imitates this natural approach to design a novel system which utilizes a single guide RNA (SgRNA) that hybridizes (i.e., bind) with a specific target for cleavage. The binding approach of this system serves as the main mechanism for various CRISPR sensors. Further emphasis is made on the application of IoT in healthcare system using CRISPR sensors. In the future, IoT will change healthcare system using all the data collected from different CRISPR sensors to increase efficiency, lower cost, and provide better life outcomes to patients.

3.1 Introduction

Internet of Things (IoT) has emerged as one of the trending technologies in the last decade as its application and principles are adopted in different fields of technology and sciences such as healthcare and medicine, industries, and environmental sciences and engineering [1]. IoT and Industrial Internet of Things (IIoT) have evolved as the leading alternative approaches that can revolutionize healthcare system to e-healthcare system. IoT in healthcare system circles around devices that can connect all different participants (patients, medical doctors, nurses, clinicians, surgeons, etc.) and allow information to be exchanged in order to obtain clinical decisions. Integration of IoT in the healthcare system will contribute to an effective and faster preventive healthcare system, in addition to providing sustainability and reduction in costs [1].

[1]Biomedical Engineering, Innovation and Information Technologies Center, Near East University, Nicosia, Mersin 10, Turkey
[2]Research Center for AI and IoT, Near East University, Nicosia, Mersin 10, Turkey

Detection of nucleic acid is a significant molecular diagnostic approach that has been continuously growing for the past years [2]. Detection of diseases has now become easier since the discovery of clustered regularly interspace palindromic repeat (CRISPR) Cas system which is an adaptive immune mechanism used by bacteria against viruses. Scientist harnessed this mechanism as a gene-editing tool which utilizes a single guide RNA (SgRNA) to form a complex with Cas system to target complementary DNA or RNA sequence. This technology is used to edit (insert or delete) human genes for diagnosis and therapies. Molecular diagnostic approach is regarded as a sensitive and precise method for the detection of pathogens associated with diseases and characterization. The application of IoT systems in CRISPR Cas devices will assist scientists to monitor various reaction mechanisms to help make better decisions and analysis on the CRISPR Cas system [3].

Scientists have already utilized CRISPR Cas9 with electronics to form a CRISPR–Chip, and the biosensor uses the gene-targeting capacity of catalytically deactivated CRISPR-associated protein 9 (dCas9) complexed with a specific single-guide RNA and immobilized on the transistor to yield a label-free nucleic-acid-testing device whose output signal can be measured with a simple handheld reader [4]. The discovery of Cas12 and Cas13a cleavage activity has offered a broad range of diagnostic applications to detect viruses, bacteria, and cancer mutations related to p53 protein [2]. Scientists are currently utilizing the same mechanism behind the powerful gene-editing tool CRISPR to develop and modify cheap devices that can quickly diagnose infections such as Hepatitis, Dengue, Ebola, and Zika viruses [5]. The integration of CRISPR Cas system sensors with IoT can be employed to assist in tracking cancer progression and identifying the expression patterns of genes which in turn could help to specify potential targets to enhance the global therapeutic possibilities.

3.1.1 Internet of Things in healthcare

Internet of Things in Healthcare (IoTH) is all about connecting data or information from sensors embedded in electronic and other devices such as sensors, monitors, mobile phones, etc. to the Internet or exchange of information from one device or sensor to another [1]. The application of wireless technology in different fields has shown a positive transformation as this technology has been utilized for monitoring devices. Merging healthcare with IoT system will link all the participants (people and devices) to a network capable of performing diverse healthcare activities such as monitoring, diagnosis, carrying out remote surgeries, and operations over the Internet [6]. The application of IoT in hospitals is to capture a steady stream of patient's data from monitors and medical devices and feed it automatically to electronic medical records. Wireless technology has been applied to healthcare system to connect all resources which include hospitals, clinics, rehabilitation centers, doctors, surgeons, nurses, assistive devices, ambulances, etc. along with patients [7].

A network manager is responsible for running the server which is fully equipped with a centralized database. A processing proxy is liable for carrying out data analysis, evaluation, detection of critical events, and generating or coming up with rehabilitation strategies. The system is connected to the Internet which is supported by radio frequency identification (RFID) technology-based program [7,8]. One of the first applications of IoT in healthcare includes smart rehabilitation to reduce the challenges of scarce or lack of resources facing the increasing population of elderly people globally [9]. Another application of IoT in healthcare includes detection of facial disorders using cameras which take images of infected faces and process through the Internet. This system works as an artificial intelligent dermatologist that can detect faces of infected people [10].

The integration of deep learning approaches with IoT improved application performance. The use of convolutional neural network and deep generative models (DGMs) can assist clinicians and pathologists in signal and image analysis and classification for making better medical decisions [11]. In order to reduce user fatigue, miss operation, facilitate rehabilitation, and improve user capabilities, Jacob [12] designed artificial muscle intelligence system with deep learning (AMIDL) to monitor real-time human intentions using EEG (electroencephalogram) sensors.

To control paralyze patient's intentions, a muscle-inspired algorithm is combined with transcutaneous electrical nerve stimulation (TENS) machine to perform six different movements on the affected limb and gesture recognition by utilizing communication aids. One of the challenges related to the diagnosis of diseases caused by pathogens and genetic diseases is detecting the best genes responsible for the disease. A classification of DNA microarray to reduced genes is achieved using an artificial bee colony (ABC) and these genes are employed as input data to train artificial neural networks (ANNs) to classify the DNA micro-arrays [13]. An improved ABC with a forward neural network utilized magnetic resonance images of the brain to classify them into normal and abnormal categories [14].

3.1.2 CRISPR and CRISPR in nature

CRISPR is an acronym that stands for "clustered regularly interspaced short palindromic repeats" which is first discovered in bacteria as an adaptive immune mechanism against bacteriophages (i.e., viruses). CRISPR along with Cas systems are used by scientists to precisely target and edit genes of interest in both prokaryotic and eukaryotic genomes [3,15,16].

Just like the way our system employed T-cells to fight and neutralize foreign invaders, different microbial species (such as bacteria and archaea) utilize CRISPR to fight viruses. When a bacteriophage attacked and injected its DNA into a bacterial cell, the viral DNA replicates inside which leads to the death of the bacterial cell [37]. Overtime bacteria adopted an immune mechanism known as CRISPR/Cas system which recognizes viral DNA and destroys it via cleavage. This process occurs in three distinct steps, namely adaptation, recognition, and interference [17,18,19].

3.1.2.1 Stages

Adaptation is the first stage, and here bacterial cell first comes in contact with viral DNA through injection. Inside the bacterial cell, there is an array or locus called CRISPR locus which is made up of repeats and different DNA sequences known as spacers and Cas genes that are located adjacent to the array. The system utilizes Cas1, Cas2, and Cas9 to locate the protospacer adjacent motif (PAM) sequence in the viral DNA, cleaves it and places it as a spacer in the leader end of bacterial CRISPR array, and stores it as memory or arsenic in case of the second attack or passes it to next generation [3,17,20]. Recognition stage occurs when virus injects its DNA into a bacterial cell for the second time. Bacterial CRISPR array transcribed its stored spacers to form pre-CRISPR RNAs which are noncoding RNAs and combine with transactivating RNA (TracRNA) through base pairing to form a matured CRISPR RNA [20,21].

In the interference stage, the Cas genes in the array are translated into Cas9 protein which possesses the endonuclease domain. Cas9 along with matured CRISPR RNA unwind the viral DNA and when it finds a matching sequence (which is complimentary to matured CRISPR RNA) it will cleave the DNA and there by destroying the Viral DNA and preventing it from undergoing replication. Bacterial CRISPR/Cas9 system is able to recognize viral DNA due to the presence of PAM sequence known as “NGG” in some bacterial species. PAM sequences are different from the Guanine-Thymine-Thymine (GTT) in the CRISPR loci [17,20,21].

3.1.3 CRISPR in genetic engineering

The discovery of CRISPR/Cas system in prokaryotic has led scientists to emulate or mimic the same mechanism and approach to deliberately insert or delete a DNA sequence from the eukaryotic genome. As the name implies, Gene Editing or Genetic Engineering is a technique that involves deliberate modification of genes such as deletion or insertion of DNA sequence. In the past, scientists utilized Zinc Finger Nuclease (ZFN), transcription activator-like effector nucleases (TALENS), and Adeno-associated virus (AAV) to edit genome, but these techniques possess a wide range of limitations such as complexity, high cost, and difficulties for multiple editions [16].

CRISPR/Cas system technique has become a trending gene-editing tool due to its precision, accuracy, simplicity, and ability to edit gene at multiple sites over the past approaches. CRISPR Cas9 system is the most utilized class which contains Cas9 endonuclease and a guide RNA which is a fusion of both transactivating RNA (TracRNA) and CRISPR RNA (CR-RNA). Cas9 enzyme is mostly isolated from different bacterial species such as *Staphylococcus aureus*, *Streptococcus pyogenes*, *Streptococcus thermophilus*, and *Brevibacillus laterosporus*. A single guide RNA is a synthetic form of RNA designed by scientist which contains 100 nucleotides and the first 20 nucleotides in the sequence guide the Cas9 to the target sequence [22].

The cleavage of the target DNA sequence occurs via two domains known as RuvC domain and HNH domain. RuvC domain cleaved the opposite strand of the target DNA sequence while the HNH domain cleaved the complementary strand. After cleavage, cells mediated repair mechanism known as non-homologous end joining (NHEJ) which is an easy mechanism where cells add or delete indels (nucleotide to repair the damage). This repair mechanism has limitations as it can lead to a frameshift mutation and it is mostly utilized in order to disrupt gene function. Homology directed repair (HDR) is an efficient repair mechanism that utilizes homologous donor that undergoes modifications to repair the damage [23] (As shown in Figure 3.1).

NHEJ is an imprecise repair mechanism that generates indels through the addition of nucleotide or sticking of DSB end together. HDR is a precise repair mechanism which uses a donor template to copy the information across the break during the repair process.

3.2 The use of CRISPR-based biosensor as a diagnostic tool for point of care

Some of the challenges of assays used by scientists to detect diseases include lack of sensitivity and specificity. Some of the assays require sophisticated systems and tedious sample/reagent treatments, which rely on well-established laboratories with

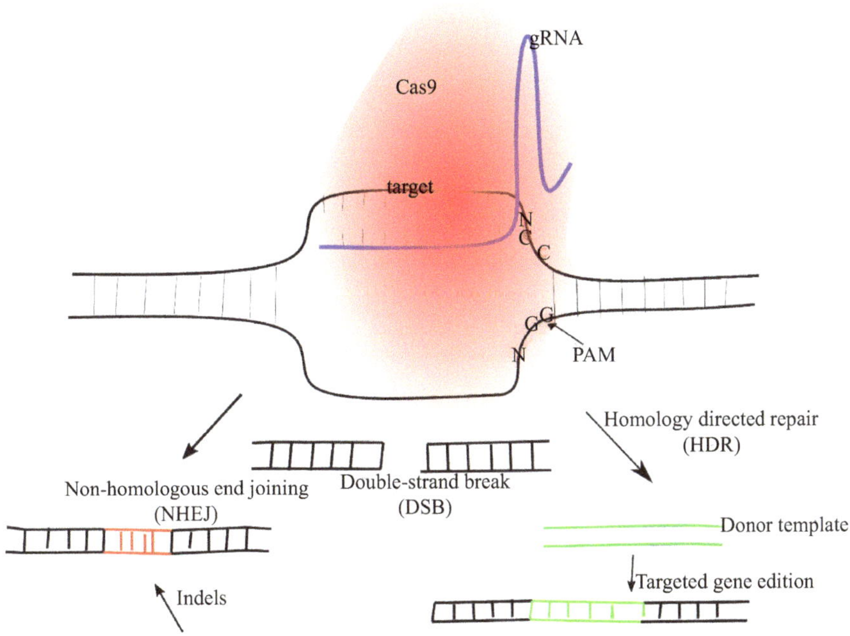

Figure 3.1. NHEJ and HDR

dedicated instruments or well-trained operators. Clinicians employed antigen-based method to detect some pathogenic diseases but this approach requires minimal equipment with lower sensitivity and assay development can take time. The advantage of CRISPR/Cas system for mediated diagnosis and therapies over other approaches is due to its high accuracy and specificity. These technologies can be harnessed and integrated into a chip device as a biosensor and can be used in a point of care to generate readout results from cells and molecules [4]. When these CRISPR sensors are connected to IoT system, the data generated from cells and molecules can be easily assessed and interpreted.

In the past, scientists rely on polymerase chain reaction (PCR) for amplification of DNAs which is somehow tedious and requires different steps. To avoid or bypass the amplification of DNA using PCR, scientists employ DNA strand displacement amplification under the isothermal condition to obtain more copies of DNA for precise and specific identification of target (such as cells, bacteria, and viruses) for point of care diagnosis. Coupling of isothermal amplification and specificity of Cas system labeled with a fluorescent protein to recognize the target and produce detectable signal makes it an ideal, robust, sensitive, and fast approach for diagnosis [24]. Connecting the whole isothermal and Cas system to IoT will enable fast and accurate means of detecting target DNA, and subsequently, a real-time remedy to certain issues will be provided globally (as shown in Figure 3.2).

Samples are obtained from the patient who undergoes isothermal amplification. Guided DNAs/RNAs are tagged or labeled with fluorescent protein or reporter which will result in color change or detection of the signal after binding and cleavage of guide RNA with the target sequence.

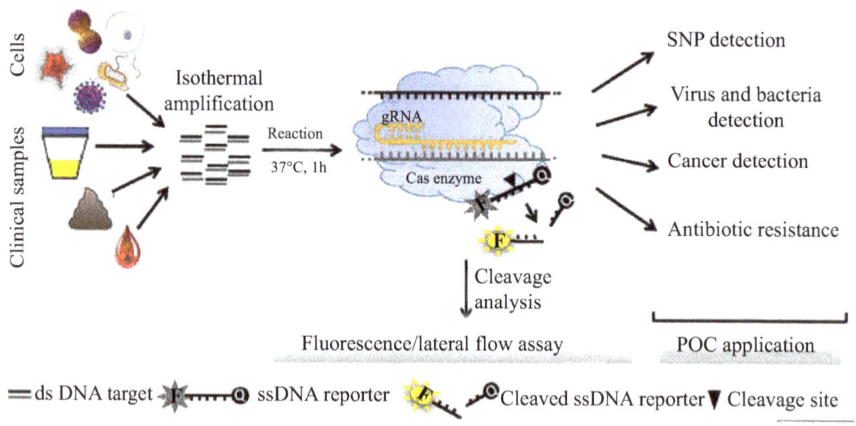

Figure 3.2. Mechanism of detection of nucleic acid using CRISPR/Cas system. Adapted from [38]

3.2.1 CRISPR Cas9 and dCas9

CRISPR/Cas9 has many biological applications in diagnostics which include nucleic acid detection with high sensitivity and selectivity. Also, the integration of nucleic acid amplification techniques with gRNA Cas9 complex in biosensors is highly important in diagnosis. Scientists from the University of Toronto, Harvard, Massachusetts Institute of Technology (MIT), and Cornell University collaborated together and developed a paper-based chip sensor for rapid detection of Zika virus. If this chip is integrated with IoT system, many global issues on Zika viral infections will be solved. This chip employed isothermal RNA amplification and reverse transcription method with toehold switch sensors [25]. The system utilizes CRISPR/Cas9 system to bind single-strand DNA through PAM and form double-strand DNA which is cleaved by the Cas9 endonuclease and thus activates toehold switch. This research has shown to discriminate between African and American Zika virus strains. Because gRNA can detect and bind specifically to bacterial, viral, or any DNA of interest and Cas9 can cleave those sequences. This technique was highly efficient for the detection of Zika virus. Data generated from such devices if available on the IoT system will further be used to discriminate Zika virus and other viral infections globally at any time [25].

Most of the CRISPR biosensors are highly selective because they can detect single base substitution at femto molar (fM) or atto molar (aM) concentrations. Cas9 cleavage when merged with nucleic acid amplification to locate precise nucleic acid sequences can be used for genotyping pathogens and discriminating SNPs. One example is the combination of isothermal exponential amplification reaction with CRISPR/Cas9 known as CAS-EXPAR. In this CAS-EXPAR strategy, CRISPR/Cas9 cleaved the target DNA generating DNA fragment which was amplified to produce large signals which were detected with real-time fluoresces monitoring. This amplification technique does not need primers like other amplification methods. The primers were generated after Cas9/gRNA cleavage [2,26].

dCas9 is a Cas9 enzyme having mutations in the cleavage domains (D10A in the RuvC domain and H840A in the HNH domain) thereby losing the capability to cleave target dsDNA. However, dCas9 still maintains the effective ability to bind specifically to target DNA under the guide of sgRNA. dCas9 has been applied in gene regulation, epigenetic engineering, genome imaging, genetic screening, and different areas to study the relationship between gene and its function [27]. Epigenome involves all the chemical modifications of DNA and histone proteins which control the expression of genes within the genome. It has been difficult to apprehend the correlation among epigenetic modifications and gene expression at particular genomic loci because of inefficient tools. dCas9 fused to epigenetic proteins has enabled epigenetic modifications at a unique locus. DNA methylation as well as histone modification play a very crucial role in the regulation of gene expression. The fusion of dCas9 with methyltransferases has allowed the repression of certain genes; therefore, it can be used for gene silencing and gene upregulation [28]. Also, the fusion of dCas9 with different histone-modifying enzymes can lead to gene activation or repression [29]. The study of gene function has been a big issue for scientists; imagine a platform where all this information from dCas9 sensors will be readily available to the world through IoT systems and the relationship between gene structure and function can be better explained.

Another CRISPR/Cas9 biosensor was developed for the detection of *Mycobacterium tuberculosis* DNA. Here in this system, a pair of dCas9 were linked to luciferase enzymes which were directed to target DNA by gRNA and upon binding to the target DNA, luminescence is released due to activity of the luciferase enzyme. In the presence of *M. tuberculosis* DNA, there was an increase in luminescence when compared to the control *Escherichia coli* [30]. In point of care where fast and quick decisions are needed, combining such biosensors to IoT systems will help scientists make better decisions for the management of diseases.

Other applications of CRISPR/Cas9 as biosensor involves genome imaging of live cells. Genome imaging has been a challenge to scientists because the tools required to visualize and understand the relationship between DNA structure in the nucleus with gene function and cell behavior are complicated and not applicable for living cells; however, dCas9 has made it possible to visualize any specific genomic sequence in nucleus. dCas9 imaging technique has the advantage of nucleotide base-pairing interaction between gRNA and any target gene sequence; therefore, it is now possible to visualize any genomic sequence in living cells. The first application of dCas9 for live-cell imaging was carried out by [31]. In this work, dCas9 was fused to a green fluorescent protein EGFP and gRNA against coding MUC4 gene on chromosome 3 and noncoding genes (introns of MUC4 gene) in living cells of humans. Thereafter, various approaches were established to improve signals during live-cell imaging with dCas9 [32].

Similarly, another application of Cas9 biosensor is the development of a CRISPR-mediated analog multievent recording apparatus (CAMERA) platform for recording cellular events in bacteria and mammalian cells using base editors and Cas9 nucleases [31]. Integrating genome imaging with IoT platforms will enable monitoring the interaction between drugs and living cells in patients, easy medical consultation, medical images, and videos for diagnosis hence better quality of life outcome.

Recently, a CRISPR–Chip was fabricated using several conventional microelectromechanical systems based on graphene field-effect transistor and catalytically deactivated Cas9 (dCas9) for detection of Duchenne Muscular Dystrophy (DMD). The CRISPR–Chip was used to analyze DNA samples from HEK293T cell lines expressing Blue Fluorescent Protein (BFP), as well as clinical samples of Duchenne muscular dystrophy DNA. gRNA was designed against BFP and DMD and after incubation with the sample, the sensor's response was monitored and measured using a commercial reader (Agile R100; Nanomedical Diagnostics). CRISPR–Chip was able to scan the entire genome within 15 min and generated a sensitivity of 1.7 fM without the need for any amplification technique [33]. If such devices are readily available at different geographic locations and connected to IoT system, communication will be easier and fast clinical response against any diseases will be made.

CRISPR/Cas9 technology has taken over the next generation biosensing technologies since CRISPR/Cas9 diagnostics has been successfully utilized in genome editing, detection of pathogen and genotyping, epigenetic engineering, genome imaging, discrimination of single nucleotide polymorphisms (SNPs), and detection of cancer mutation [2]. Therefore, CRISPR dCas9 sensors have applications beyond gene editing, and future applications will include fabricating point of care sensors connected to IoT platform that can be used in healthcare systems [33].

3.2.2 Cas12 (Cpf1)

Cas12 is a single guide RNA guided endonuclease of class 2 CRISPR/Cas system. It is also known as Cpf1 which means CRISPR/Cas from Prevotella and Francisella 1. It is a class 2 type V CRISPR/Cas system used by some bacteria as an adaptive immune system against viruses. The mechanism of immunity is somehow similar with Cas9 with some variations. The Cpf1 CRISPR array is transcribed to form mature CrRNA without the need for additional transactivating RNA (TracrRNA). These mature CrRNA binds to target via PAM rich in thymine (TTTN) unlike NGG (rich in guanine) in Cas9. After binding, Cpf1 exerts a staggering cut on both 5′ end of the targets as a result of the RuvC domain present in the Cas endonuclease without the HNH domain [34].

3.2.2.1 Mechanism of Cas12a as a sensor

The mechanism of Cas12 and Cas13a is somehow similar as both exhibit collateral cleavage activity of targeted nucleic acid and nontargeted strand. Cas12a exhibits RNA guided DNase activity as it can collaterally destroy (cut) both targeted double-strand DNA (DSDNA) and also nontargeted single-strand DNA (SSDNA). It is very important to note that Cas12a cannot transcleave single-strand RNA (SSRNA) reporter or targeted SSRNA sequence which is specific for Cas13a. Thus, Cas12a only possesses DNA-activated DNase activity but not RNA-activated RNase activity [34]. Since Cas12 can detect the presence of DNA through its collateral activity, such sensors can be used to detect pathogenic DNA especially for point of care applications, and when connected to IoT system, better decisions can be made by physicians and other governmental and nongovernmental agencies to prevent diseases.

3.2.3 Cas13a (C2C2)

It is the most recent discovered type II Cas system and it shows remarkable and unique characteristics that distinguish it from other type II Cas systems, thus it offers significant roles in the molecular diagnostic toolbox. It was discovered through bio informative experiment designed for the computational pipeline by both MIT and Harvard researchers. Unlike Cas9 and Cas12, Cas13a possesses no detectable RuvC and HNH domains like Cas9 and RuvC like Cas12 rather, it possesses two higher eukaryotic and prokaryotic nucleotide (HEPN) binding domains. These unique domains have shown to possess RNA cleavage activity and cleave single-stranded RNA and use single guide RNA (guided CrRNA which contains spacer with 28 nucleotides in length) as a guide to navigate and recognize the target, and the spacer binds to the target through PAM rich in uracil residues in multiple regions within the RNA and cuts it off (unlike Cas9 and Cas12 which possess DNA cleavage activity and cleave only DNA target) [35].

Like dCas9, Cas13a can be mutated to become dCas13a and thus loses its RNA cleavage activity. dCas13a can still bind to specific target RNA and thus can be employed to serve as an RNA-guided RNA binding protein. The mismatch tolerance in Cas13a is different from the one in Cas9 which can tolerate 3–5 mismatches while Cas13 can only tolerate single base mismatch [36].

3.2.3.1 Mechanism of Cas13a

The collateral activity was harnessed to cut labeled RNA reporters for the detection of off-target NA which may come from bacteria, viruses, or eukaryotic cells. It cleaves SSRNA and uses single CRISPR RNA as a guide. Cas13a from *Leptotrichia wadei* (LwaCas13a) where the crRNA–Cas13a complex has a bi-lobed architecture, consists of a nuclease (NUC) lobe and a crRNA recognition (REC) lobe. This process begins with the co-expression of Cas13a and pre-CrRNA. The pre-CrRNA is able to bind to the CrRNA recognition (REC) lobe of Cas13 where it is cleaved to become a matured CrRNA. Once a matured RNA bound to the Cas13 CrRNA complex, it is able to bind specifically to the homologous target RNA sequence. Binding to the target SSRNA, Cas13a allows nonspecific RNA activity to be exhibited. When target bind Cas13 undergoes conformational changes that activate it nonspecific RNA abilities. The conformational change results in the formation of the catalytic active site as the HEPN 1 domain moves toward the HEPN domain (placing catalytic histidine and arginine residues into contact) leading the catalytic site exposed and not internal like the catalytic site of Cas9, and thus Cas13 is able to cleave free SSRNA in solution [16]

3.2.3.2 Cas13a as biosensor

The RNA cleavage activity of Cas13a makes it ideal for the detection and sensing of the presence of transcripts. Scientists exploited the collateral cleavage activity of this Cas protein to develop a robust, precise, and sensitive diagnostic tool for in vitro detection of RNA associated diseases with the specificity of single-base mismatch. CRISPR-Cas13 biosensing system has been applied for detecting Zika virus, Dengue virus, bacterial isolate, antibiotic-resistant gene, human DNA genotype, and cancer mutation [5]. Hence, diseases associated with RNA can be detected easily and IoT platforms will provide adequate mechanisms for the prevention of such diseases.

3.2.3.3 SHERLOCK

Due to collateral mechanism of Cas13 is also termed as collateral cleavage, a process where Cas13 cleaved target RNA using RNA guide and in some circumstances, it cuts any RNA it encounters [39], scientist created SHERLOCK which stands for specific high sensitivity enzymatic reporter LOCK-unlocking. Scientists took samples from patients with a possible viral infection. They amplified the levels of RNA in it. They added reporters that are sensitive to Cas13. An engineered CRISPR-Cas13 is added to the sample which is programmed with a guide RNA that is designed to find only viral RNA and bind to it. Cas13 activates its cleaving mechanism and starts slicing nearby RNA including the reporters. Since each end of the reporter carries different labels, Cas13 separates these two signatures creating a unique signal within the sample [5]. This sensor is highly efficient for detecting viruses and can be used for disease prevention and environmental monitoring when connected to IoT system.

3.3 Conclusion

Since molecular diagnostics is crucial in life sciences, food engineering, and environmental monitoring as well as biosecurity, there is a significant need for developing devices with accurate, easy, and fast nucleic acid detection. In the future, a CRISPR-based biosensor integrated with IoT can be used to detect and classify any type of pathogen including its morphology and drugs it can be susceptible to. It will be able to detect cancer and other genetic diseases such as sickle cell anemia by designing the guide RNA that will match the target of interest. IoT can be used to monitor the progression and retraction of disease and monitor the epidemiology of the disease. Integration of IoTH in disease diagnosis can be used to generate further useful information and transfer it to physicians and pharmacists for therapy. Also, integration of CRISPR sensors and IoT-based platform might also provide fast medical response where emergency cases are available, especially in rural areas or where medical personnel are not available.

References

[1] Ulusar, U. D., Al-Turjman, F., and Celik, G. "An overview of Internet of Things and wireless communications." *In 2017 IEEE International Conference on Computer Science and Engineering (UBMK).* Samsun, Turkey: IEEE, October 2017, pp. 506–509.

[2] Li, Y., Li, S., Wang, J., and Liu, G. CRISPR/Cas systems towards next-generation biosensing. Trends in biotechnology. 2019.

[3] Barrangou, R., Fremaux, C., Deveau, H., *et al.* "CRISPR provides acquired resistance against viruses in prokaryotes." Science. 2007; 315(5819): 1709–1712.

[4] Uppada, V., Gokara, M., and Rasineni, G. K. "Diagnosis and therapy with CRISPR advanced CRISPR based tools for point of care diagnostics and early therapies." Gene. 2018; 656: 22–29.

[5] Gootenberg, J. S., Abudayyeh, O. O., Lee, J. W., *et al.* "Nucleic acid detection with CRISPR-Cas13a/C2c2." Science. 2017; 356(6336): 438–442.

[6] Tarouco, L. M. R., Bertholdo, L. M., Granville, L. Z., *et al.* "Internet of Things in Healthcare: Interoperability and security issues." In *2012 IEEE*

international conference on communications (ICC). IEEE, June 2012, pp. 6121–6125.

[7] Fan, Y. J., Yin, Y. H., Da Xu, L., Zeng, Y., and Wu, F. "IoT-based smart rehabilitation system." IEEE Transactions on Industrial Informatics. 2014; 10(2): 1568–1577.

[8] Floerkemeier, C., Roduner, C., and Lampe, M. "RFID application development with the Accada middleware platform." IEEE Systems Journal. 2007; 1(2): 82–94.

[9] Yuehong, Y. I. N., Zeng, Y., Chen, X., and Fan, Y. "The Internet of Things in Healthcare: An overview." Journal of Industrial Information Integration. 2016; 1: 3–13.

[10] Al-Turjman, F. M. "Towards smart eHealth in the ultra large-scale Internet of Things era." In *2016 23rd Iranian Conference on Biomedical Engineering and 2016 1st International Iranian Conference on Biomedical Engineering (ICBME)*. November 2016. Tehran, Iran: IEEE, pp. 102–105.

[11] Al-Turjman, F., Zahmatkesh, H., and Mostarda, L. "Quantifying uncertainty in internet of Medical Things and big-data services using intelligence and deep learning." IEEE Access. 2019; 7: 115749–115759.

[12] Jacob, S., Menon, V. G., Al-Turjman, F., Vinoj, P. G., and Mostarda, L. "Artificial muscle intelligence system with deep learning for post-stroke assistance and rehabilitation." IEEE Access. 2019; 7: 133463–133473.

[13] Zhang Y. D, Wu L., and Wang S. "Magnetic resonance brain image classification by an improved artificial bee colony algorithm." Progress in Electromagnetics Research. 2011; 116: 65–79.

[14] Mavrovouniotis, M., and Yang, S. Evolving neural networks using ant colony optimization with pheromone trail limits. In 2013 13th UK Workshop on Computational Intelligence (UKCI). September 2013; Guildford, United Kingdom: IEEE, pp. 16–23.

[15] Cui, L., Vigouroux, A., Rousset, F., Varet, H., Khanna, V., and Bikard, D. "A CRISPRi screen in *E. coli* reveals sequence-specific toxicity of dCas9." Nature Communications. 2018; 9(1): 1912.

[16] Liu, X., Wu, S., Xu, J., Sui, C., and Wei, J. "Application of CRISPR/Cas9 in plant biology." Acta Pharmaceutica Sinica B. 2017; 7(3): 292–302.

[17] Bhaya, D., Davison, M., and Barrangou, R. "CRISPR-Cas systems in bacteria and archaea: versatile small RNAs for adaptive defense and regulation." Annual review of genetics. 2011; 45: 273–297.

[18] CRISPR 101. https://www.synthego.com/resources/crispr-101-ebook

[19] Song, G., Jia, M., Chen, K., *et al.* "CRISPR/Cas9: A powerful tool for crop genome editing." Crop Journal. 2016; 4(2): 75–82.

[20] Garneau, J. E., Dupuis, M. È., Villion, M., *et al.* "The CRISPR/Cas bacterial immune system cleaves bacteriophage and plasmid DNA." Nature. 2010; 468(7320): 67.

[21] Jinek, M., Chylinski, K., Fonfara, I., Hauer, M., Doudna, J. A., and Charpentier, E. "A programmable dual-RNA–guided DNA endonuclease in adaptive bacterial immunity." Science. 2012; 337(6096): 816–821.

[22] Ma, X., Zhang, Q., Zhu, Q., *et al.* "A robust CRISPR/Cas9 system for convenient, high-efficiency multiplex genome editing in monocot and dicot plants." Molecular Plant. 2015; 8(8): 1274–1284.

[23] Hsu, P. D., Lander, E. S., and Zhang, F. "Development and applications of CRISPR-Cas9 for genome engineering." Cell. 2014; 157(6): 1262–1278.

[24] Ye, L., Wang, C., Hong, L., *et al.* "Programmable DNA repair with CRISPRa/i enhanced homology-directed repair efficiency with a single Cas9." Cell Discovery. 2018; 4(1): 46.

[25] Pardee, K., Green, A. A., Takahashi, M. K., *et al.* "Rapid, low-cost detection of Zika virus using programmable biomolecular components." Cell. 2016; 165(5): 1255–1266.

[26] Huang, M., Zhou, X., Wang, H., and Xing, D. "Clustered regularly interspaced short palindromic repeats/Cas9 triggered isothermal amplification for site-specific nucleic acid detection." Analytical Chemistry. 2018; 90(3): 2193–2200.

[27] Xu, X. and Qi, L. S. "A CRISPR–dCas toolbox for genetic engineering and synthetic biology." Journal of Molecular Biology. 2019; 431(1): 34–47.

[28] Morita, S., Noguchi, H., Horii, T., *et al.* "Targeted DNA demethylation in vivo using dCas9–peptide repeat and scFv–TET1 catalytic domain fusions." Nature Biotechnology. 2016; 34(10): 1060.

[29] Hilton, I. B., D'ippolito, A. M., Vockley, C. M., *et al.* "Epigenome editing by a CRISPR-Cas9-based acetyltransferase activates genes from promoters and enhancers." Nature Biotechnology. 2015; 33(5): 510.

[30] Zhang, Y., Qian, L., Wei, W., *et al.* "Paired design of dCas9 as a systematic platform for the detection of featured nucleic acid sequences in pathogenic strains." ACS Synthetic Biology. 2017; 6(2): 211–216. https://doi.org/10.1021/acssynbio.6b00215

[31] Chen, B., Gilbert, L. A., Cimini, B. A., *et al.* "Dynamic imaging of genomic loci in living human cells by an optimized CRISPR/Cas system." Cell. 2013; 155: 1479–1491. https://doi.org/10.1016/j.cell.2013.12.001

[32] Wang, H., La Russa, M., and Qi, L. S. "CRISPR/Cas9 in genome editing and beyond." Annual Review of Biochemistry. 2016; 85: 227–264. https://doi.org/10.1146/annurev-biochem-060815-014607

[33] Hajian, R., Balderston, S., Tran, T., *et al.* "Field-effect transistor." Nature Biomedical Engineering. n.d. https://doi.org/10.1038/s41551-019-0371-x

[34] Zetsche, B., Gootenberg, J. S., Abudayyeh, O. O., *et al.* "Cpf1 is a single RNA-guided endonuclease of a class 2 CRISPR-Cas system." Cell. 2015; 163(3): 759–771.

[35] Aman, R., Ali, Z., Butt, H., *et al.* "RNA virus interference via CRISPR/Cas13a system in plants." Genome Biology. 2018; 19(1): 1.

[36] Cox, D. B., Gootenberg, J. S., Abudayyeh, O. O., *et al.* "RNA editing with CRISPR-Cas13." Science. 2017; 358(6366): 1019–1027.

[37] Ishino, Y., Krupovic, M., and Forterre, P. "History of CRISPR-Cas from encounter with a mysterious repeated sequence to genome editing technology." Journal of Bacteriology. 2018; 200(7): e00580-17.

[38] Zeng, L., Mukama, O., Lu, X., Cao, S., and Lin, D. "Strand displacement amplification for multiplex detection of nucleic acids." In *Modulating Gene Expression-Abridging the RNAi and CRISPR-Cas9 Technologies*. IntechOpen.

[39] Ibrahim, A. U., Al-Turjman, F., Sa'id, Z., and Ozsoz, M. Futuristic CRISPR-based biosensing in the cloud and internet of things era: an overview. Multimedia Tools and Applications. 2020; 1.

Chapter 4

Evaluation of mobile patient monitoring: a study in practice

Saad Eddin Abdulaal[1], Ali Sawtari[1], Sinem Akman[1], Hamza Alhajiibrahim[1], Sabareela Victory Moro[1], Fadi Al-Turjman[2,3,4], Ilker Ozsahin[1,4] and Dilber Uzun Ozsahin[1,4]

Over the years, providing good medical care has been an issue for patients with heart and lungs diseases, hypertension, and diabetes because of lack of medical personnel, lack of access to hospital, cost, etc., which has led to increase in mortality rate and has affected the socioeconomic structure of any society. This concern has led to the development of strategies and approaches to improve the healthcare software. One of the approaches is the development of mobile health apps or devices for treatment or management of these diseases or conditions. These mobile apps or devices are portable and can be used for the function of the treatment of the disease or sickness.

Mobile patient monitoring allows patients or individuals to monitor their health status using a device or health app wherever they are. In this work, we developed a device that can be used for monitoring the patients' health condition especially for babies and elderly persons and assistance will be needed to get accurate results which could be sent to the doctor or medical personnel for interpretation. A patient monitoring device is designed to provide a portable device that monitors the patient's condition which is less expensive and saves time and energy to improve the quality of the patients' lives.

4.1 Introduction

Mobile healthcare monitoring allows patients to monitor themselves anywhere in spite of the location or distance rather than going to an expensive hospital or hospital

[1]Department of Biomedical Engineering, Near East University, Nicosia / TRNC, Mersin-10, Turkey
[2]Department of Artificial Intelligence, Near East University, Nicosia / TRNC, Mersin-10, Turkey
[3]Research Center for AI and IoT, Near East University, Nicosia / TRNC, Mersin-10, Turkey
[4]DESAM Institute, Near East University, Nicosia / TRNC, Mersin-10, Turkey

to waste time and energy. The advantage of mobile healthcare monitoring is that it provides efficiency prior to the examination of a doctor, nurse, or healthcare provider. Mobile healthcare monitoring uses devices that do not involve the introduction of instruments into the body (noninvasive) using wearable sensors [1].

According to the World Health Organization (WHO), the increase in the elderly population could have a positive impact on the healthcare and social welfare of any society. Healthcare has become increasingly expensive because of the increase of drug price, medical instruments, and hospital care which have led to the development and adoption of techniques and strategies to improve individuals' health at a low price so that there are ease and comfort among children, elderly, and individuals/areas who have limited access to healthcare [1,2].

Mobile technology is very important because of its accessibility to the rural areas, elderly, and babies who are less educated. It is also cost effective which helps to reduce the number of patients who visit the hospital frequently. Mobile technology helps in the communication between patient and healthcare providers, monitor patients' health, and collect data that can be used to make an informed decision about the patients. It can also be used to check the progress or monitor various health conditions, diseases, and illnesses [3,4]. Lives have been improved through mobile healthcare because people are now been responsible for their health and environmental hygiene.

A study showed that the use of mobile technology has brought about the detection of diseases early, improve medication, and treatment of the disease. This mobile technology uses mobile phones and text messages to remind patients about their medication, doctor's appointment, and consultation. It helps the healthcare givers or providers to make good decisions based on the information provided by the patient [5]. Patients need to have knowledge of self-management techniques for these devices and applications to work effectively for example (blood glucose measurements in diabetes management). The use of the mobile health apps helps to reduce the workload on the medical personnel where physical examination is not needed and saves time [6,7].

Also, the production and quality of some of these mobile health apps which are not properly verified are later launched into the market and cause the patient or health consumer to misuse the app to treat or manage a condition [8].

Our work is created for patients who are babies and elderly to monitor their blood pressure and saturation percentages within their comfort zones without visiting the hospital or healthcare center.

4.2 Literature review

There have been recent studies done to monitor the patients' health using different apps; however, their approaches weren't accurate.

Different patients' health devices were examined and these were easy to use, convenient, and very effective but could lead to technical problems, lack of sufficient network for transmitting biosignals from patients to healthcare givers [9]. Another article discussed about how wearing sensors helps to transfer data from the device to the doctor or healthcare provider which is inexpensive and does early detection but these devices or systems can only be used or operated by someone who has a personal knowledge about the disease or health condition [10]. Another approach was using digital technologies which were effective on a small scale but were not effective for a major problem on large-scale implementation [11]. A portable active assistive living (AAL) device is used by an increasing number of elderly people but arterial blood pressure (ABP) and electrocardiogram (ECG) which are important components for the monitoring of the patient are not available in the clinical market [12].

Unlike the above-mentioned studies, our study enables the patient to carry out his daily activities while checking the progress of the patient's health and this can be done anywhere and anytime which saves time and cost and can be operated by anybody.

4.3 Mobile health monitoring device approach

This shows the components used in this device and the architecture of the device. This device is mainly used by babies and the elderly.

4.3.1 Components

1. Arduino uno circuit

   It is a microcontroller board chip that has a transmitter and a receiver end which are both connected to each other using the radiofrequency waves.
2. Radiofrequency modules (transmitter and receiver)

   The radiofrequency between the transmitter and the receiver varies between 30 kHz and 300 GHz. It receives signals from its own transmitter alongside a pair of encoder and decoder. The transmission is done through the encoder and the signals are received by the decoder. The transmitter circuit contains the temperature, pulse, and humidity sensor.
3. Temperature sensor

   It consists of two ends whereby one end consists of three wires which are connected to the Arduino and the other end is placed under the armpit of the patient to measure the temperature.
4. Moisture sensor

   It is also called the humidity sensor which measures moisture in the diaper if the diaper needs to be changed, the sensor sends an alert that the diaper needs to be changed. The moisture sensor has three pins which consist of a pin at zero voltage with respect to the power supply and ground plane of the circuit board (GND), voltage common collector (VCC) pin, and data pin which is used as a signal interface, connected to the Arduino.
5. Power supply

   Without a power supply, this device can never work. It uses 5V batteries which are used to power the Arduino circuit.

6. Display screen
   A liquid crystal display (LCD) 2 × 16 is used to display the information which is to be sent to the doctor or healthcare provider.
7. Light-emitting diode
   Light-emitting diodes (LEDs) are semiconductors that emit light. When the light is emitted, it sends an alert/notification that the diaper needs to be checked or changed.
8. Wires and transistors
   They are used to connect the parts inside the circuit and sensors. The resistors help to reduce high current which could lead to electrical shock.
9. Buzzer
   It has a Piezo speaker which sends an alert to check if the temperature of the patient is high, normal, and dry.

Figure 4.1 shows the final version of the mobile patient monitoring device.

4.3.2 Architecture

Figure 4.2 shows the component and build-up of this work.

The mobile health monitoring devices use a power supply which sends current to the Arduino circuit transmitter that measures the temperature, pulse, and amount of water in the patients and the signals are picked up by the radio frequency and sent to the Arduino receiver which checks the temperature of the patient again and if the temperature or humidity of the patient is high or low, these sensors pick a signals and initiate the buzzer. The role of the buzzer is to send a message through the LED if it is safe or unsafe and the result is displayed on the display screen.

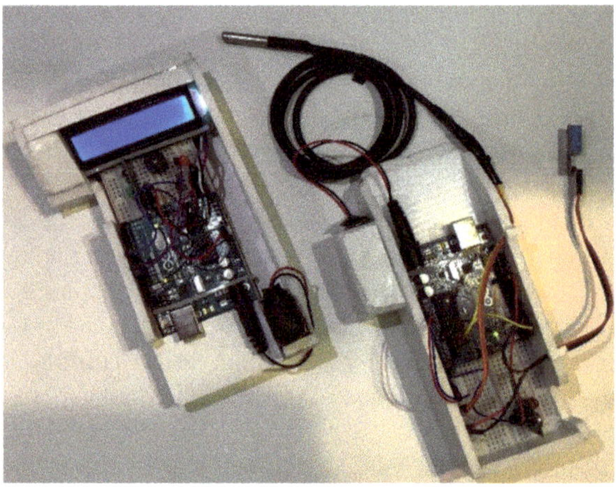

Figure 4.1 Mobile patient monitoring device

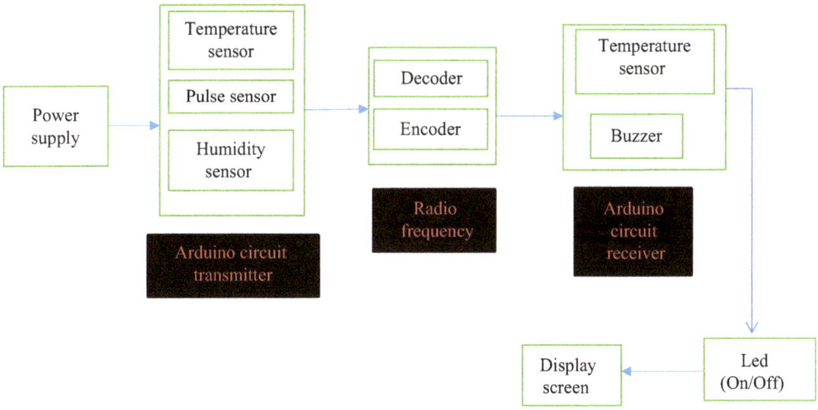

Figure 4.2 Architecture of the mobile patient monitoring

4.4 Discussions

Mobile patient monitoring in the next decade will be frequently used in the healthcare sector because it helps in the reduction of workload on the doctor, nurses, and healthcare givers. Mobile health devices and applications are made to help save costs and reduce the number of people who visit the hospital frequently. Patient monitoring also makes the lives of patients and physicians more convenient where the patients help the doctors or healthcare providers to get the necessary results and make the right decision quickly about the drugs and treatment to be given enhancing better services and accurate treatment for the patient. Also, the use of digital technology has made keeping patients, data and information very safe and reducing medical errors. With mHealth apps and devices, patients can know when to take their medications, chat with a doctor without leaving home or whichever location he or she is in which saves time on both the patient and doctor and helps the doctor to spend more time on emergencies.

Better communication is enhanced through an online communication whereby doctors can see their patients, have a chat with them, check their health status, and also check for symptoms. Through mobile health monitoring, patients who live far away from hospital and health centers can have access to remote medical services. However, mobile health monitoring can be inefficient when there is lack of accessibility due to poor bandwidth connectivity especially for those in remote or rural areas, confidentiality privacy could also pose as a threat when the information gotten from the patients which are to be entrusted to the doctors or physicians with their health information are not kept private. Inaccurate information and lack of regulation and approval by regulatory bodies could also affect mobile health monitoring.

Our work is to provide a device that saves time and cost, gives accuracy, and gives comfort to the patient without going to the hospital or healthcare centers.

Although it is very efficient, more sensors could be added to the device to help in the interpretation of diagnosis better, mobile phones could be attached to it to send emails and SMS messages to remind patients on their consultation, doctor's appointment, and medication. The communication provided can be improved to have better results.

4.5 Conclusions

This device was created to improve the lives of patients especially the elderly and babies because they need the most attention because of their physical state which makes them unable to move from home to hospital and hospital back home. This device which is mobile can be used to check or monitor blood pressure and saturation percentages wherever they are not going to the hospital. Assistance is needed by another individual to help them check or monitor their condition efficiently so that the results can be recorded and sent to the doctor or healthcare provider. The aim of this work is to provide a portable mobile app that saves time, energy, and cost, and a quality health lifestyle is been assured within the comfort of the patients.

References

[1] Majumder S., Mondal T., and Deen M. J. "Wearable sensors for remote health monitoring." Sensors (Basel). 2017; 17(1): 130.
[2] Ayatollahi H., Ghalandar Abadi M., and Hemmat M. "Web and mobile-based technologies for monitoring high-risk pregnancies." BMJ Health & Care Informatics. 2019; 26(1): e000025.
[3] Frontini R., Sousa P., Carvalho M., Alves R., Ferreira R., and Figueiredo M. A "Mobile-based monitoring sleep system integrated in a mHealth program." European Journal of Public Health. 2019; 29(1): ckz034.003.
[4] Khan A., and Madden J. "Mobile Devices as a Resource in Gathering Health Data: The Role of Mobile Devices in the Improvement of Global Health." International conference on computational science and computational intelligence. 2016.
[5] Stojmenovic M., Gedeon T., Qi H., and Buhari S. M. "Health informatics: Applications of mobile and wireless technologies." Wireless Communications and Mobile Computing. 2019; Article ID 6518784.
[6] Cannon C. "Telehealth, mobile application and wearable device are expanding cancer care beyond walls." *Seminars in Oncology Nursing*. 2018; 34(2): 118–125.
[7] Anderson K., Burford O., and Emmerton L. "Mobile health apps to facilitate self-care: A qualitative study of user experience." PLoS One. 2016; 11(5): e0156164.
[8] Rossi M., and Bigi S. "MHealth for diabetes support: A systematic review of apps available on the Italian market." Mhealth. 2017; 3: 16.

[9] Pawar P., Jones V., van Beijnum B.-J. F., and Hermens H. "Framework for the comparison of mobile patient monitoring systems." *Journal of Biomedical Informatics*. 2012, 544–556.
[10] Muznya M., Henriksenc A., Giordanengo A., *et al*. "Wearable sensors with possibilities for data exchange: Analyzing status and needs of different actors in mobile health monitoring systems." International Journal of Medical Informatics. 2020; 133: 104017.
[11] Sarmento A., Vignati C., Paolillo S., *et al.* "Qualitative and quantitative evaluation of a new wearable device for ECG and respiratory Holter monitoring." International Journal of Cardiology. 2018, 231–237.
[12] Ciani O., Cucciniello M., Petracca F., *et al.* "Lung cancer App (LuCApp) study protocol: A randomised controlled trial to evaluate a mobile supportive care app for patients with metastatic lung cancer." BMJ Open. 2019; 9: e025483.

Chapter 5

Image-based IoT measurement techniques in disease diagnosis

S. Vijayalakshmi[1]*, Savita*[1] *and Balamurugan Balusamy*[1]

A fast-growing technology with application in science and engineering is Internet of Things (IoT) with digital image processing. Digital image processing is a procedure in which image is taken as an input and provides an output which is also an image but enhanced by applying some technology as per requirement. The main focus of image processing is to take out meaningful information from an input. The use of image processing with IoT is not only in a single field but it spreads all over the area like remote sensing, medical, agriculture, defense, color processing, and industry also. In this chapter, we explain how IoT and image processing can help to find out the disease. In the medical field, it is a very important task to find out disease before diagnosing it. Various types of diseases are found in humans, plants, and animals, so it is important to find out these diseases and provide the best treatment for the disease. In medical, disease diagnosis is a difficult, costly, and time-consuming task, so the use of image processing is used in this field to make it simple and less time-consuming. Disease diagnosis procedure based on image processing with IoT involves some simple steps like image acquit ion, preprocessing, segmentation, and classification.

5.1 Introduction

In the present time, the use of Internet of Things (IoT) and image processing is increasing in every single application. In this chapter, the title "Image-based IoT measurement techniques in disease diagnosis" clearly states that how an image with IoT plays an important role in the measurement of something. The term "image" is something that can define the object in detail. Here we are using the image to know the object well. A large number of application fields where the image with IoT is used as a measurement tool like in the medical field (disease diagnosis), aquaculture field (to measure various parameters of animals), and in agriculture filed (to measure plants growth and disease). The disease detection task in the medical field

[1]School of Computer Science and Engineering, Galgotias University, Greater Noida, India

is a very critical and time-consuming process, but the use of IoT and image processing makes this task easy and less complicated. The various diseases like cancer, melanoma, HIV, psoriasis, infection, and stroke are manually treated by medical experts. Similarly, the diseases related to plants and animals are also handled by medical expertise. The manual process is very time consuming as well as costly, so the use of image processing makes this process easy. In an image-based IoT measurement system, the image of an object with IoT is used to examine the detailed information about the object. In infectious disease detection, the experts need the blood sample to know about the stage and types of a particular disease but this process leads to a lot of harm to humans. The use of an image-based IoT measurement system in skin disease detection makes the process easy and it is also not harmful. First, take the image of the skin by using any suitable imaging modality and then put this image into a computer system for analysis purposes. The same process can be applied to plants and animals. But the image-based IoT measurement system depends on an image analysis system. The image analysis system is something that observes the feature of an image and then generates the result. The next process is the measurement of image features' parameters. The result of the computer-assisted tool depends on the image parameters so the correct measurement is very important. In computer-assisted tools, the role of a smart imaging modality that is used to capture the image is also very important because the image is the first thing on which the whole process depends. In image processing, various IoT-enabled imaging modalities are available like computed tomography (CT), X-ray, ultrasound, single-photon emission computed tomography, and positron emission tomography. A lot of IoT-based smart mobile applications are also available which the person can use to find the disease at home without the help of medical experts. The various mobile applications that are available also work for animals and plants. Leaf Doctor, LeafScan, and Plantix Skye are IoT-powered mobile apps that work on plant leaf image to find the correct plant disease. iCare health and DERMA are the E-health monitoring mobile applications that are used for humans. By using iCare mobile health application, a person can check heart rate, blood pressure, and oxygen level at home without a medical expert. DERMA care is used to check the skin-related disease. In this chapter, we explain the application field where image-based IoT measurement is used. The rest part of the chapter includes the image analysis system process, IoT-enabled imaging modality, image-based mobile application, and image analysis system such as WinRHIZO, I-ROP, and IMAFISH, in which WinRHIZO is a smart image analysis system that is used to analyze the plant roots, and the analysis system "IMAFISH" is used in smart aquaculture to measure the different morphometric and meristic characteristics of fish.

5.2 Literature review

An image-based IoT measurement system is one of the important tools in the current time that works on images and Internet of objects to find out more and detailed information about the object. The use of image processing and IoT is not limited to a single field but it can be used everywhere like education, medical, defense, object tracking, and remote sensing. Like image processing and IoT, its purpose is also not limited but can be used for different purposes such as identification of single object which is not clearly visible, looking inside the human anatomical structure with the help of images and differentiate objects according to their pattern. In this chapter, the image-based IoT measurement techniques in disease diagnosis are being discussed where the image of an object is used to find out various diseases in human beings, plants, and animals. A similar process can be followed by an image-based IoT measurement system for the detection of disease in humans, plants, and animals. A lot of work is being done by researchers in this field and many measurement systems are also developed for image analysis purposes. The image-based IoT measurement system is used in skin disease detection for humans, plants, and animals. In [1–3], the authors have presented a model that helps in the detection of plant skin diseases. The disease can affect the plant leaf, root, and stem. To improve the production and quality of food, it is very important to grow the healthy plant and find out the correct disease at the early stage itself. This model first examines the plant leaf that is damaged by insects, bacteria, and other types of disease by capturing the color image of plant leaf with the digital camera. The next process is extracting the useful image features so that by analyzing these features the expert system can detect the type and stage of plant disease. Here the expert system comprises IoT and image processing techniques like *k*-means clustering, artificial neural network, and many more techniques that detect the correct disease by analyzing the symptoms in the form of image features. The authors in [4–6] have explained a process that helps in diagnosing the skin disease in humans. The skin disease is a harmful disease that affects a person's daily life such as problems in movement, losing confidence, and many other problems that affect the person's development. Lots of skin diseases like skin cancer, ringworm, psoriasis, sunburn, hives, and dermatitis are caused by infection. The detection of their harmful diseases at an early stage is very necessary so that the treatment can be provided at the right time. The treatment by medical experts is very expensive and time-consuming and hence the application of image processing in this field is very helpful. The image of the affected area of skin is captured by using the appropriate IoT-enabled image modality system and computes the feature of normal and affected areas so that the image analysis system can analyze these features and clearly identify whether the change in skin is normal or damaged by any disease. If the area of skin is damaged due to the disease, the analysis system can detect the type and stage of that particular disease. In [7,8], IoT-based mobile application has been developed to assist the user to examine the sudden changes in the skin such as identifying rashes on the skin due to infection and changes in size and color of a mole. The user can use this IoT and image-based mobile application to check the skin anywhere just by capturing the skin image by using a camera mobile without the expert's help. This application is tested on more than six different types of skin diseases such as skin cancer, psoriasis, acne, scabies, eczema, and seborrhea dermatitis and the accuracy of the application is found to be 90%.

The use of an image-based IoT system is not only used to detect the plant and human disease but also very helpful in animal disease detection. In [9–12], the authors explain about the advanced technology of IoT and image processing in aquaculture to monitor the water environment, production, and telediagnosis of animal diseases. In aquaculture, fish disease is a big problem because these diseases can spread to other animals through the water quickly and easily. Hence the rapid action is required to detect and control the disease. In [13,14], the microscopic images of diseased fish are taken for further analysis purpose.

5.3 Applications of IoT with image processing in disease identification

5.3.1 Role of IoT in skin disease identification

A disease related to a very complex active organ with three layers of the epidermis, dermis, and the hypodermis, that is, skin bumps. Humans, animals, and plants are frequently and commonly affected by skin diseases.

American Academy of Dermatology (AAD) has presented a report on the burden of skin disease in 2013 and found that one among four people is affected by skin disease [15,16]. In humans, skin disease not only affects the skin but also spoils an individual's personal life affecting loss of confidence and creating problems in movement. There is no single reason responsible for the skin disease. Various common conditions that can cause skin disease are virus, immunodeficiency, fungus, and contact with infected skin disease person [17]. The skin disorders can be permanent, temporary, age-related and internal, painless and painful, situational and genetic, and minor as well as life-threatening [18,19]. In Table 5.1 and Figure 5.1, some skin diseases are listed.

Before the disease appears, skin also gives some signs that we can call symptoms which may be helpful to find out the correct disease. But some problems with skin may not result in skin disease for example blisters from new shoes. Some common symptoms but not limited to on our body that we can feel or see are skin bumps due to infection which can be the same color skin or different in color, painful rashes, changes in size and color of moles, tinea fungal infection,

Table 5.1. Classification of skin disease

Permanent	Temporary	Internal	Age-related
Seborrheic eczema	Acne	Carbuncle	Hemangiomas
Moles	Hives	Cellulitis	Measles
Melanoma	Warts		Impetigo
Rosacea	Fungal nail infection		Dermatomyositis
Lupus	Cold sore		Seborrheic keratoses
Psoriasis	Candidiasis		Shingles

Figure 5.1. Different skin disease

acne, and dandruff. So, it is very important to find out the correct skin disease on the initial stage to provide the right treatment. Here the combination of two most significantly used technologies such as image processing and IoT plays a very important role in identifying correct disease for providing treatment thereby reducing the time consumption [20–22]. Image processing and IoT are not only used to find out the skin disease but also help in analyzing the color and texture of the skin [23–25].

A simple model of image-based IoT system is designed for skin disease identification involves the following steps:

1. Image acquisition
2. Preprocessing
3. Segmentation
4. Classification
5. Result

This model can be used for the detection of various skin diseases such as blister, actinic keratosis, latex allergy, psoriasis, and many more diseases. The process is shown in Figure 5.2.

1. Image acquisition: Image acquisition plays a vital role in identifying correct skin disease because if the skin image is not in correct or satisfactorily formed, then there may be a possibility for incorrect results which may affect the rest of the entire system. A skin image of the suspected area of the body is captured by an imaging instrument. The skin image can be in RGB form or grayscale which depends on the image analysis system or method. If the image is in satisfactory form, then only it can be provided to the system as in the input form.
2. Preprocessing: In this second step, the unnecessary portion of the image is removed. Skin images have hair and pigments that can create a problem in achieving a correct result. Hair and pigments can be taken as noise and hence noise removal is done. If the image quality is also not good, this may be low and medium, then improvement of quality of the image is needed. Appropriate filters and enhancement techniques are applied to achieve this.

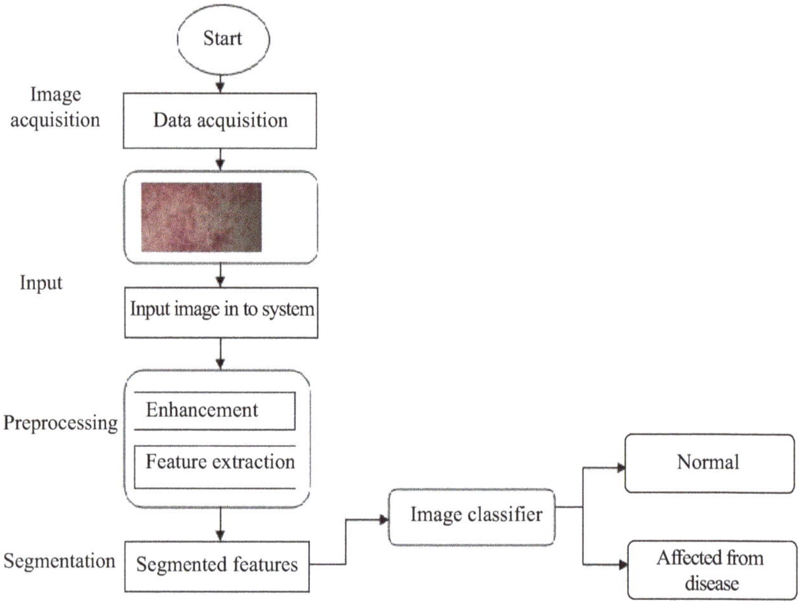

Figure 5.2. Simple model of image-based IoT system for skin disease detection

3. Segmentation: The extraction of the region of interest (ROI) from the whole image which is to be taken for further processing is known as segmentation. In this process, the specific part or lesion is segregated from the whole image to reduce the time complexity.
4. Classification: The complete ROI is carefully analyzed during classification. The statistical features of the image are studied and then categorized into the abnormal and normal area. The real disease is identified by observing the features of the abnormal area. The various classification algorithms using an image-based measurement system can be used.

The below-given process is to be followed to detect skin disease.

5.3.2 Cancer detection by using image processing and IoT

The second-largest cause of death and the uncontrolled growth of abnormal cells in the body is known as cancer, sometimes also called malignancy which can affect any part of the body. According to WebMD, there exist more than 100 types of cancer but symptoms for cancer may be varying depending on the type of cancer. The International Agency for Research on Cancer (IARC) works under the World Health Organization published a report GLOBOCAN 2018 (the global burden of cancer). According to the database, it is found that 18.1 million new cases of cancer and 36 new types of cancer are found. Cancer not only affects the skin but also affects the blood and circulatory system, the lymphatic system, the immune system, and the hormone system. There are lots of treatment processes to treat cancer. Some are chemotherapy, hormone therapy, precision medicine, and stem cell transplant. The traditional method for treatment of cancer involves the surgeons to examine the medical image of the affected body part. But due to the huge data set this process is laborious. Advanced technologies of image processing with IoT are applied to reduce time consumption and human error. Computer-aided diagnosis system aids the surgeons to identify the different types of cancer. In a computer-aided diagnostic system, the first step is to record the image of the suspected area of the body by using any suitable IoT-enabled imaging modality (image acquisition) and various algorithms are applied to the images to distinguish the cancerous and noncancerous region (image segmentation). The image features like texture and shape which are biomedical knowledge-based features are computed to characterize the regions. In the last, a trained classifier system is used to classify the cancer. But this segmentation process is a very critical and important step to identify the location of cancer and also helps in treatment of cancer. To identify cancers such as breast cancer, lung cancer, skin cancer, and other various types of cancers, different types of imaging modalities are used. In breast cancer detection, which is originating from the breast tissue, mammography screening is used to detect cancer in the breast. The other imaging modalities that are useful in the detection of breast cancer are X-ray mammograms, ultrasound imaging, and intensity-modulated radiation therapy. In the image-based IoT CAD system, X-ray mammogram captures the region of suspected area in the breast and at that time image processing segmentation technique focuses on two main goals: try to locate the exact suspicious area and classification of cancerous mass into benign or malignant. The most commonly used segmentation techniques are thresholding, region-based segmentation, morphological, and image texture feature-based techniques. Intensity-modulated radiation therapy is used for the breast cancer treatment at an early stage by giving a high dosage of radiation but less damage to the surrounding normal tissue. The second most dangerous cancer in male is prostate cancer. The early and accurate detection of prostate cancer boundaries can help in providing treatment on time. For detection of prostate cancer, a real time and less costly imaging technique like ultrasound is used. The detection of the prostate nodule from the ultrasound image becomes a difficult process due to the prostate size and presence of speckle noise. For the detection of prostate cancer edge, the classical techniques are used: nonlinear filtering and second-order derivative edge detection method. Here in the given example, the ABCDE classifier as shown in Figure 5.3 is used to classify cancer in normal and melanoma form.

The third most common cancer is lung cancer. The computed imaging modality is very sensitive to lung cancer. Two-step processes are involved in the automatic detection of lung cancer: nodule candidate detection and false-positive reduction. The combination of different techniques like supervised clustering, support vector machine and Bayesian network provides the better result in image based computer aided diagnostic tool. A new approach radiation therapy for lung cancer treatment is intensity-modulated radiation therapy which works well on the

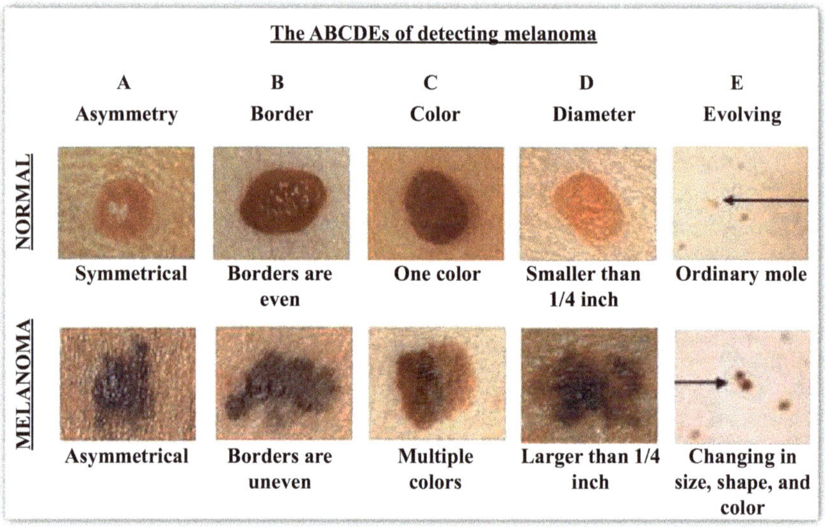

Figure 5.3 ABCDE classifier of cancer

accurate detection of target volume otherwise it can damage the surrounding normal tissues if not detected easily.

5.3.3 IoT-powered plant disease and cassava identification

Rice, corn, and cassava are the biggest source of carbohydrates. Cassava is a staple food that fulfills the basic diet of more than a billion people. In the world, Nigeria and Thailand are the two countries, which are the largest producer and exporter of cassava. The two varieties of cassava, bitter and sweet, are available, which contains antinutritional factors and toxins. Farmers grow the bitter variety to detect animals, pests, and thieves. The biggest source of carbohydrates is taken into a boiled form and the extracted starch is used for animal feed and industrial purpose. The other name of cassava in the United State is Yucca. Cassava presented by slave merchants in the sixteenth century at the western coast of Africa. Some risk factors of agriculture are virus and bacterial disease, climate change, and destructive insects that damage the crop. The different parts of cassava plant are shown in Figure 5.4.

1. This plant is vulnerable to many diseases like bacterial, fungal, oomycete, vein mosaic virus, and many more pests. The cassava mealy bug and green mite are two most threatening pests that damaged the cassava plant in 1970. The identification of these diseases at an early stage and to overcome the disease spread is a very complicated task. In most of the countries, there is no much support from the agriculture extension organization to find out the agriculture disease and even if it is available it is too costly. The IoT-powered smartphone application is the new technology in the field of agriculture to

Figure 5.4 The different parts of cassava plant

find out the disease at an early stage but this application is based on image recognition. In image-based IoT disease detection measurement system, the result depends on the image features. The combined result of image recognition and feature extractions gives promising result. The feature extraction task is time consuming and needs the expert's knowledge for the qualitative result. So there is a need to make the process more reliable in which computation is low providing the promising result. The researchers Amanda Ramcharan, Kelsee Baranowski, Peter McCloskey, Babuali Ahmed, James Legg, and David P. Hughes introduced the applicability of transfer learning in order to create a model for deep learning convolutional neural network to minimize the computation process [26]. The following steps are included in the method given by the researchers:

(i) Image of a cassava leaf is captured by Sony cyber shot 20.2 megapixel digital camera as shown in Figure 5.5. To train a deep learning model, many cassava disease images were recorded using cassava genotype.

In this dataset, images of cassava affected with three different types of diseases are recorded, which is categorized into six class:

Class 1: One class of 398 images of cassava brown streak image
Class 2: Second class of 388 images of mosaic disease
Class 3: Third class of 386 images of brown leaf spot
Class 4: Fourth class of 309 images of green mite damage
Class 5: Fifth class of 415 images of red mite damage
Class 6: Six class of healthy leaf images.

(ii) This dataset was analyzed by transfer learning from convolution neural network Inception v3 which was implemented in TensorFlow and trained from ImageNet. This model easily identifies the disease by

Figure 5.5 (a) Cassava brown streak image, (b) mosaic disease, (c) brown leaf spot, (d) green mite damage, (e) red mite damage, and (f) healthy leaf

analyzing the symptoms. It also identifies the type of disease and pest damaged leaf separately without including the complex task of feature extraction.

(iii) But this method gives accurate results on the cropped leaf rather than the whole leaf.

5.3.4 Malaria detection by using blood sample images with IoT

Malaria is a life-threatening infectious disease caused by plasmodium parasite and transmitted from one human to another by the bite of infected mosquitoes that can transfer from mother to baby by blood transfusion. World Health Organization published an annual report on malaria and according to the report published in November 2018, 219 million new cases of malaria have been found inside the 87 countries and around 443,000 people die from malaria disease. Some common symptoms of malaria are fever, cough, sweating, headache, nausea, and chest pain. Figure 5.6 shows how malaria transmits from one person to another person:

The manual method of malaria detection by using the microscope and the Giemsa process is time consuming as well as not accurate also. There is a need of an automated diagnostic system that can diagnose the disease correctly at an early stage so that the treatment can be done on time. Medical image processing is used here to detect malaria from blood cells. A method was given by Varsha Waghmare

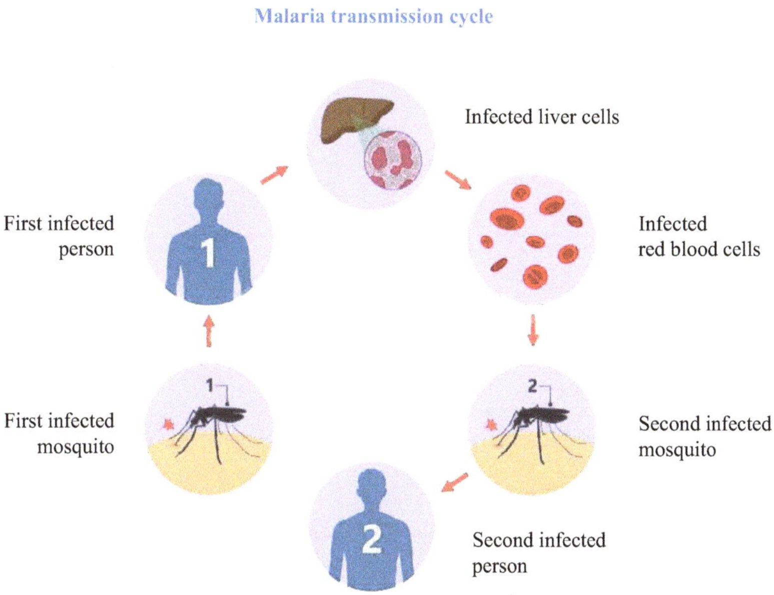

Figure 5.6 Malaria transformation from one person to another

and Syed Akhter that uses images of blood cells of patient to identify that patient is affected by malaria or not [27–30]. The framework of the proposed method of image-based IoT measurement system for malaria detection is given in Figure 5.7.

1. The first step is data collection. In preparation of data set for this method, JSB stain method (Jaswant Singh–Bhattacharji stain) is used which is a rapid method of malaria detection consisting of two solutions: the first one is the mixture of methylene blue, potassium, and sulfuric acid and the second one is the eosin dissolved in water. The optical microscope Leica DM1000 was used to capture images size of 640×480 pixel.
2. The second step is to eliminate the unwanted information from the image, that is, preprocessing. In image preprocessing, various techniques are available and these can be applied to preprocess the image. These techniques used various types of filters, noise removal techniques, and many more.
3. The third step is feature extraction. More than 60 samples of images are used in which each sample has normal and abnormal cells. The three main objects that are extracted from these samples are parasites, white blood cells, and artifacts. The calculated features are histogram features and shape-based measurement features. Some first-order statistical features are mean, variance skewness, and kurtosis and the features such as area, convexity, form factor, and compactness are the shape measurements which are used to check the change in size.
4. The last step is the classification. Support vector machine classification technique to classify parasite as healthy or infected (malaria).

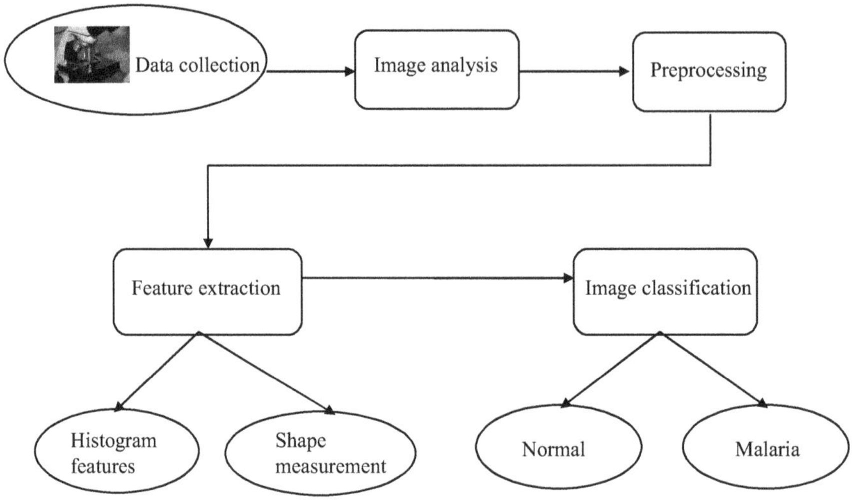

Figure 5.7 Framework of Image-based IoT system of malarial detection from blood sample images

5.3.5 IoT-enabled plant disease detection

Economic growth of any country is highly dependent on agriculture. Many farmers in the country have difficulty in choosing the right seed and right pesticide drug for them. So the selection of right seeds and right pesticides can increase agriculture significantly, which leads to increased economic growth. There are many diseases in plants that cause the crops to get worse which may affect the economic growth of the country. Identification of disease at the right time prevents the crop from spoiling. Diseases in plants are identified manually by experts, for whom they have to work a lot to find the correct disease and this method involves more time. But now the time has changed and the technology of identifying diseases in plants also has changed. These techniques identify the right disease in a short time. Image processing with IoT has proved to be a boon to identify the disease of plants in today's time. The image processing and IoT techniques work on the images of plants leaf, root, stem, and fruit to identify the type of disease. This technique analyzes the symptom by using IoT with the image processing technique and then classifies the category of disease. There are lots of methods for plant disease detection is shown in Figure 5.8 such as the direct method, indirect method, and by using sensor and image-based measurement system.

Direct method is used where there is a need to analyze a large number of samples and in this method, the disease that is caused by viruses and fungi can be detected. In the indirect method, disease identification is based on some parameters like climate change, organic chemicals, and water movement from root to other parts of the plant like leaves, stem, and buds. Various types of biosensors are available to measure the chemical transformation and can also identify some biological parameters like protein amount, enzymes, and microorganisms that help in

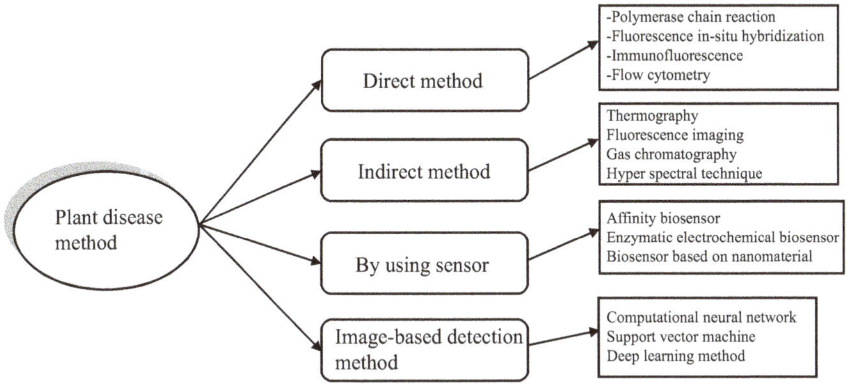

Figure 5.8 Plant disease detection methods

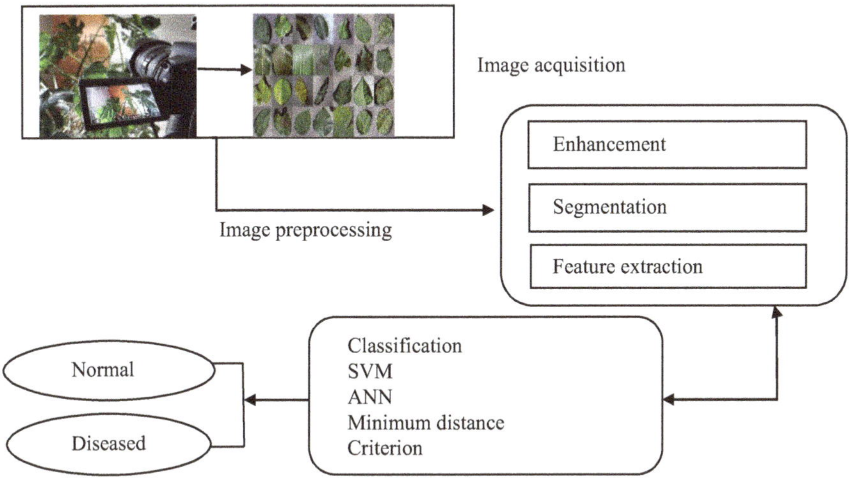

Figure 5.9 Image-based IoT system for plant disease detection method

identifying the type of disease in plants. These sensors are fiber-optic, piezo-electric, and biocatalytic. The image processing is a very fast and accurate method of disease detection by analyzing the plant leaf.

An image-based IoT system of plant disease detection method uses some powerful IoT-enabled image processing techniques that analyze the sample data for symptoms of various different disease and by analyzing these symptoms this method can detect correct disease in the plant [31–33].

Some steps to be followed for disease detection and for classification into specific diseases as shown in Figure 5.9.

1. Prepare images sample data set: The preparation of the sample data set is the first step (image acquisition). The image is captured by using a digital camera

and by using other imaging modalities that can capture a clear image of the plant leaf. In the data set, two types of images are included: healthy leaf images (normal) and unhealthy leaf images (diseased infected image). All images in the data set are RGB (red, green, and blue) color formed images.

2. Image enhancement: On this step, the image is processed to get only the desired portion from the whole image. The image is enhanced to improve the quality by removing noise and artifacts that appear at the captured time due to imaging capturing device and light. At this step only the affected area of the image can be extracted to make the further processing easy and less time consuming.
3. Image segmentation: Partitioning of the image into different parts is image segmentation. By using the segmentation method only ROI of affected portion can be segregated from the whole image. In the image processing method, a lot of segmentation techniques are available: edge and boundary detection algorithm, clustering algorithm, and region-based method. Here the change in the color of the image is one of the common symptoms of the disease, so the main focus is on the green color. The calculation of green color can be done by using the threshold method.
4. Feature calculation using color co-occurrence method: By using the color co-occurrence method, the unique features of an image like color and texture are calculated that represent the image. Before extracting the image features, first, convert the RGB color images of leaves into HIS color space. The texture features such as homogeneity, contrast, energy, entropy, and cluster shade of the image are calculated.
5. Classification: Classification is the last step of the process that analyzes all the image features and on which the result of the algorithm is dependent. The extracted features value is stored into database and then this data value is classified by classification techniques. For classification support, vector machine and minimum distance criteria technique is used. The success of the classification method is done by the classification gain method.

5.4 Fundamental steps of image-based IoT measurement system

The advanced technology of image processing and IoT plays a very important role in the identification of disease in plants, humans, and animals because the traditional way in which human is involved, and disease identification is very time consuming as well as costly and cannot diagnose disease 100% correct. The use of advanced IoT-powered computer technology with the help of experts provides a satisfactory result within seconds. Image-based IoT measurement system for disease diagnoses is fully depending on the technology of image processing. Image-based IoT system as a diagnostic tool follows some steps shown in Figure 5.10.

1. Image acquisition: This is the first step of image processing and the remaining steps depend on it. To capture the real-world data and transformed into

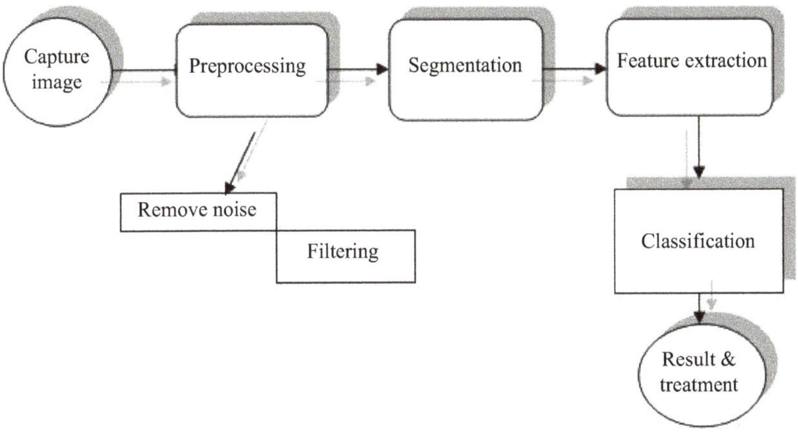

Figure 5.10 IoT-enabled image analysis system

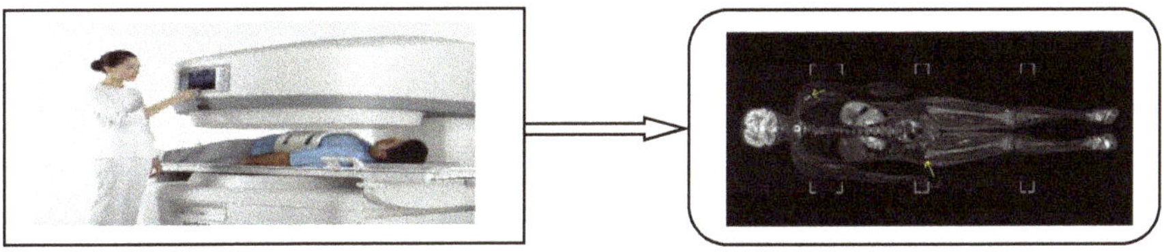

Figure 5.11 Image captures by MRI machine

numerical form which is understandable by the computer system is the process of image acquisition. In medical image processing, the various imaging techniques are available that are used to acquire the image of the internal body parts of humans. To acquire the image, MRI, CT scan, X-ray, fluorodeoxyglucose positron emission tomography, and single-photon emission computed tomography can be used but it depends on the condition for which image is required. In an image-based IoT measurement system, the defined imagining techniques can be used for disease diagnose of plants, humans, and animals as shown in Figure 5.11. For the detection of disease, the image of the affected part is needed for the analysis process.

2. Preprocessing

   Preprocessing in which input is image and output is also image but with some improvement. The preprocessing process removes the unwanted area which is not useful to make the image clear so the process can be completed in less time with less effort and the result would also correct. If the image of skin is used to detect any skin-related disease than the unwanted tissue like the hair is considered as a noise in an image and also creates a difficulty in the detection of correct disease. To remove noise from the image, various filters are available in image processing. In preprocessing, we can not only remove noise but we can add anything to make sharp and smooth the image. The result of preprocessing process is like as shown in below given Figure 5.12.

3. Segmentation

   The segmentation is done after the enhancement process, the division of the entire image into multiple parts so that the analysis process can be faster and easier. By segregation, the process can extract only the target area on which further processing is done as shown in Figure 5.13. The different types of segmentation algorithms can be used such as region growing, clustering approach, neural network, watershed segmentation, and threshold-based segmentation techniques.

4. Feature extraction

   The image features are something that differentiates the object with the other objects, boundary, and with its surrounding within the image. There may a lot of objects inside a single image and every object has some different feature which represents the pattern of the image. Too many features are available in image processing as shown in Figure 5.14 which can be color-based features, texture features, and shape-based features.

5. Classification

   Classification is the process in which objects are recognized and then categorized into the predefined group of class. The process of classification by visual perception is easy for human but this is difficult for computer machines.

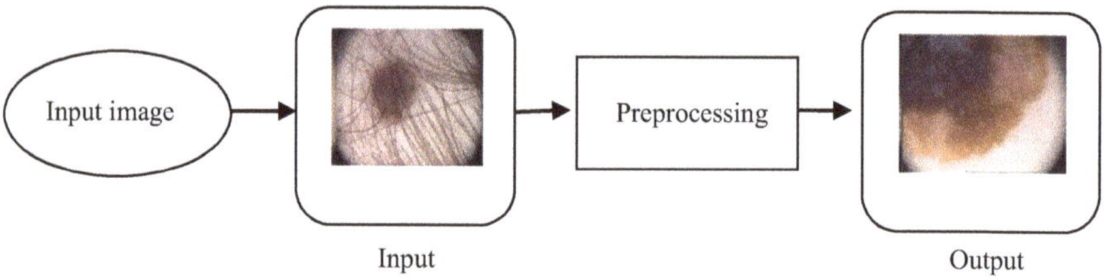

Figure 5.12 Image enhancement of captured noise image by removing unwanted area

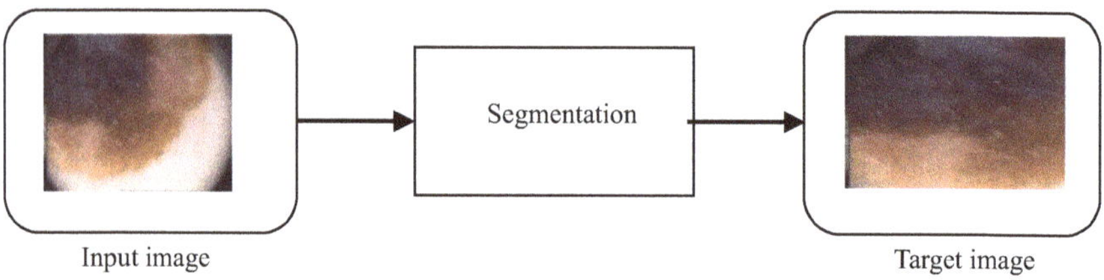

Figure 5.13 Extraction of the region of interest

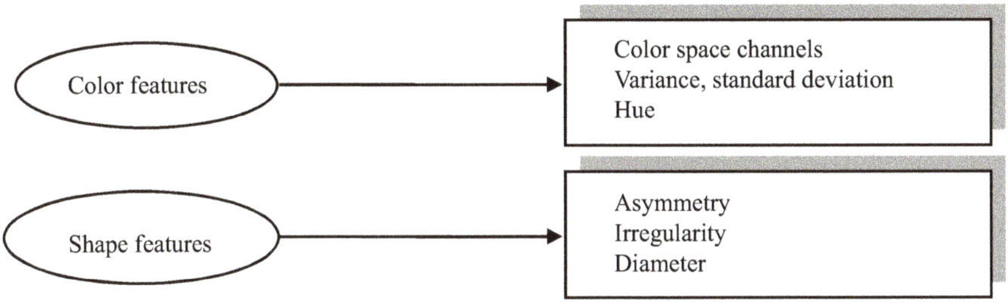

Figure 5.14 Different types of image features

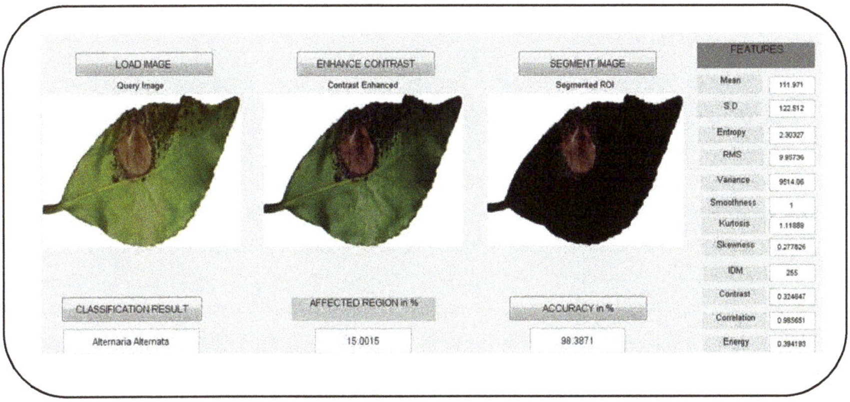

Figure 5.15 Result of the classification process

This process takes the help from feature extraction step so that it can read the pattern of the data and can put data into a particular class. At the time of skin disease detection, classification helps in finding the type of disease by analyzing the input data value or it divides the tissue into normal (healthy) and abnormal (unhealthy) (disease-affected tissue). The various classification techniques: K-nearest neighbor (KNN), Radial basis function, black propagation network, artificial neural network, neural network, support vector machine, and fuzzy measure are used in IoT-powered image processing for disease detection in plants, human and in animals. This process is a mathematical solver which takes a data as a problem and calculates the solution for this data, for example, for plant disease detection, first extract the features of leaves and then submit these features for next process which is classification. In the next phase, classification is done by using ANN*, and these extracted features are considered as neurons in ANN on which basis ANN works and it gives the result that what type of disease is there. Figure 5.15 shows the result of leaf disease.

*Artificial neural network is a supervised learning model inspired by biological nervous system is used here.

6. Result and discussion

   After getting the result of the classification process, it is clear that which disease is present in plants, humans, and animals. On this result, medical experts can take a decision that which treatment is right for the disease.

All the fundamental steps in disease diagnosis system play an equally important role before completion of the first step, we cannot process the next step.

5.5 IoT-based smartphone applications for disease detection

5.5.1 Leaf Doctor: an IoT-based expert system for plant disease detection

The plant disease becomes the reason for less production of food and food security. Plant parameters are used to detect the disease and its types. An attractive IoT-based mobile application Leaf Doctor used for pathometry produced by Cornell University with the collaboration of the University of Hawaii at Manoa, College of Tropical Agriculture and Human Resources. It is an iOS-based application that can be used on iPhone, iPad, and iPod and the android version is under implementation. The traditional method of plant disease severity detection is to see the leaf tissue of plants by the experts which is expensive, time consuming, and due to low efficiency it also slows down the development in agriculture. The automatic disease detection model is being used more due to the advanced technology in computer vision. IoT-based expert system Leaf Doctor is a new development in the field of plant disease detection which is an image-based IoT technique that uses leaf of plant for detection of disease type [34]. In this application, the user takes a photo of leaves and submits these leaves for calculation of disease percentage. The following steps are included in the Leaf Doctor application.

1. First download the app from any iOS-based iPhone, iPad, and iPod.
2. The instruction about how to use the application, its function, and about algorithm all are available under the "about" heading of app.
3. For disease detection, the user can take real time images of leaves and can use already saved images. But there are some important points to capture images: capture images without flash, use an umbrella to avoid sunlight which helps in capturing the natural colors that are useful in image analysis step and can differentiate the healthy and diseased affected area, and background should be black.
4. The image assessment process works on the user's touch ability and in this process for healthy tissue representation, the user can allocate values for eight different colors. The app algorithm works on these values which are assigned for a different color of healthy tissue and then give a status for healthy tissue or disease-affected tissue. This process checks every pixel in the image and then by analyzing the process, it is clear which pixel is healthy and which pixel is disease affected pixel, that is, the blue color is affected tissue. The threshold

slider bar is used to adjust the maximum distance from healthy tissue. If user is not satisfied with the result then the color for healthy tissue can change the color for healthy tissue and if satisfied than the percentage of disease is calculated.

Like Leaf Doctor, there are various IoT-based applications that can be used for different disease detection in plants: LeafScan, LeafAnalyzer, Black spot, Limani, BioLeaf, Plantix Skye, and many more application are available. Figure 5.16 is the process of Leaf Doctor mobile application for plant disease detection.

5.5.2 IoT-based Skin Vision app for skin disease detection

In western countries, the number of serious disease cancer patients is increasing day by day and the detection of cancer in the early stage is also very complicated and costly also. In this disease, the abnormal cell grows and spreads all over the body which can disturb the function that helps the body functioning well. The different types of cancer are prostate cancer, lung cancer, skin cancer, carcinoma cancer, and melanoma. If the cancer is not detected at the early stage then it can be difficult to cure it [35–38]. A technology where an individual can check any abnormal sign in his skin at home without skin specialist, for example, the change in size, color, and unusual growth in the mole. This technology is an IoT-enabled mobile phone application which was developed by an expert team of Romania's University of Bucharest called as "Skin Vision" in 2011. This IoT-based mobile app helps the people in checking any change in existing mole and can detect melanoma cancer at an early stage with 73% accuracy. The technology of mathematics branch "fractal geometry" is the base of "Skin Vision" mobile app which is used to analyze the tissue growth in a captured skin image. This algorithm detects the symptoms related to the abnormal mole on which basis it can detect the cancer. By using Skin Vision app, anybody can check their skin at home by following some simple steps:

1. Take an image of a body part that has a suspicious spot.
2. The captured image is analyzed by "fractal geometry" algorithm of "Skin Vision" and within 30 s, it gives the indication whether it is normal or abnormal.

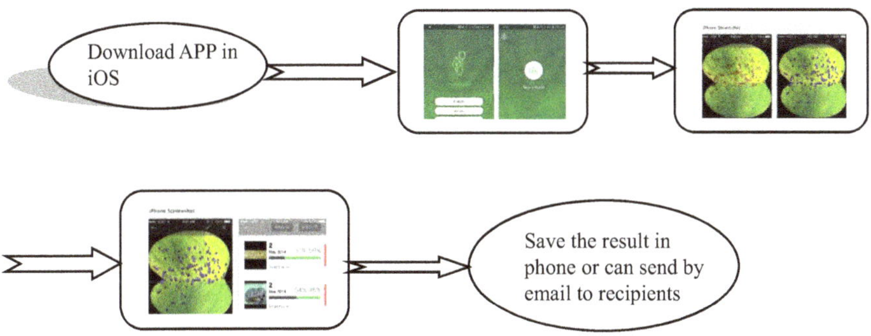

Figure 5.16 Working flow of an IoT-based expert system Leaf Doctor

This application was tested on melanoma patients and it is found out that out of 100 cases of melanoma, it identified 73 patients correctly. This mobile application can be downloaded from iTunes or Google play store. Figure 5.17 clearly shows that how the Skin Vision application works.

5.5.3 A smart way of anemia detection without taking blood sample

The blood condition "Anemia" is the result of the inadequacy of healthy red blood cells or can be called "hemoglobin" in the body. The low quantity of red blood cell or hemoglobin deficiency leads to many diseases like heart problems, pregnancy-related issues, and reduce oxygen level into the body. Hemoglobin is responsible for sending oxygen into the body cell for generating energy. Due to the lack of hemoglobin, the exact amount of oxygen does not reach the cell, due to which the body feels tired. The first symptom of anemia is body fatigue. According to the published report on anemia globally by WHO (World Health Organization), it affects more than 1.62 billion people in which the highest prevalence is in the pre-school children and pregnant women, and lowest prevalence in the men. The various types of anemia are iron deficiency, vitamin deficiency, aplastic anemia, and sickle cell anemia. For the detection of anemia, a blood sample is required every month by using an external invasive tool, so it is full of hassle for the patient.

An IoT-based smartphone app in which anemia is detected by without taking a blood sample developed by researcher Wilbur A. Lam and Mannio at Emory University and Georgia Institute of Technology, USA [39]. This application checks the hemoglobin level by just taking an image of fingernails and analyzes the color of fingernails [40–43]. The following steps are followed by the smartphone app at the time of hemoglobin checkup:

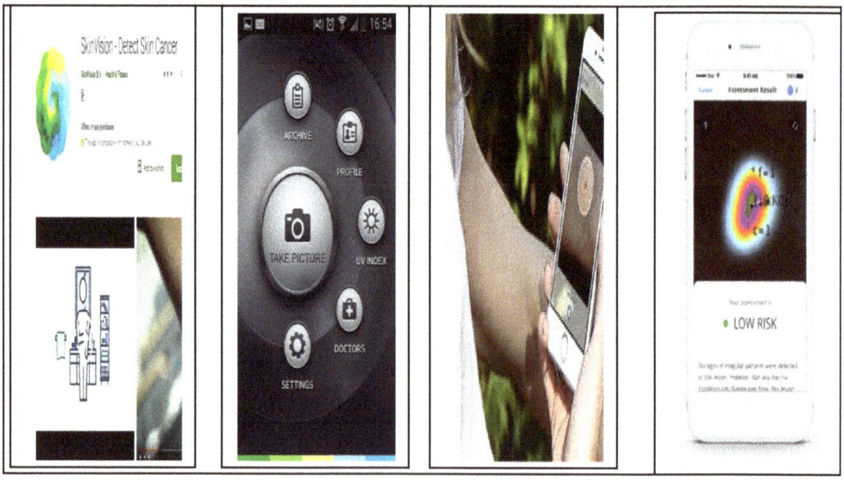

Figure 5.17 Skin Vision mobile app of skin cancer detection

1. Download application on your smartphone.
2. Open app and take an image of your fingernails. This app also tells that any white spot on fingernails is flashlight or leukonychia.
3. The app algorithms analyze the nails color and then give the HgB measurement level.

Figure 5.18 shows how the image and IoT-based anemia detection smartphone app works.

5.5.4 E-health monitoring system: iCare

In the medical field, the use of IoT applications is rapidly growing due to its accuracy, time-consuming process, and cost-saving process in disease diagnosis. In the mHealth, one application is available which we can call "ALL IN ONE" app and that provides healthcare services. The ALL IN ONE app is an IoT-enabled iCare health monitoring application that helps the people to check blood pressure, heart rate measurement, pedometer, lung capacity measurement, and workout. Everything is available in one app so no other device is needed to check the physiological parameters. This IoT-based health monitoring app through image processing measures the heart rate, oxygen level in the blood, and blood pressure by collecting photoelectric pulse wave signal stability [44–47]. Just by using your finger, you can measure all physiological parameters. The working process of this app is shown in given Figure 5.19. Some points to be followed for measuring these parameters:

1. First, download IoT-based expert iCare Health Monitor on your android or iOS phone.
2. Pressing the phone screen by using your finger.
3. Fully cover the rear camera by just placing your indexed finger.
4. Don't raise your finger from the camera before completing the measurement.

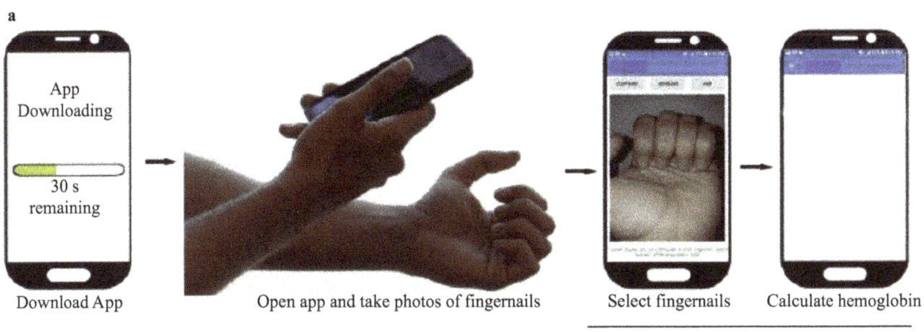

Figure 5.18 Process of an IoT-based mobile app of anemia detection by using fingernails

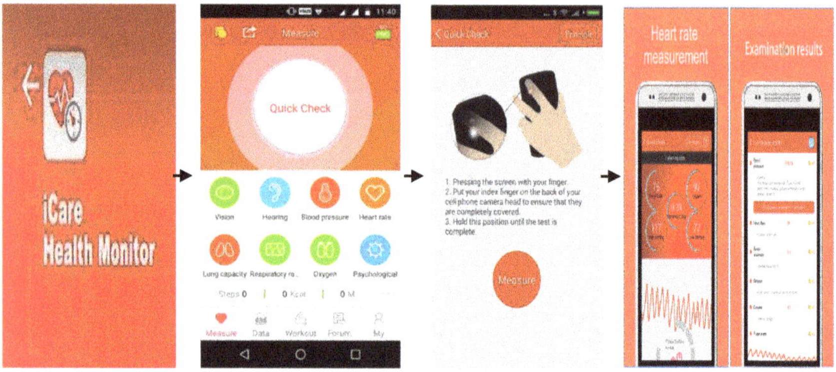

Figure 5.19 E-Health monitoring system iCare mobile app working process

Some important point that needs to follow for improving the assessment result:

1. Don't press your finger very hard, otherwise, blood circulation will be altered.
2. Don't use flash light and dark background. Capture image in daylight which works best in the measurement.
3. Stay calm and do not do any movement.

5.5.5 Cancer detection by using IoT: DERMA/CARE

The advanced technology of IoT through image processing in the medical field for detection and diagnosing of disease is becoming more helpful for medical experts. Researcher develops a smart mobile application, that is, depending on IoT technology which works on the algorithm of image processing for detection of various diseases such as skin disease, heart disease, and hypertension in human, plant, and animal. In skin diseases, the most serious disease is cancer which is caused by the growth of abnormal cells and it spreads all over the body. The most common cancer is "Basal Cell Carcinoma," the second most common is "Squamous Cell Carcinoma," and the most dangerous cancer is "melanoma." A skin cancer melanoma arises from the melanocytes (pigment cells) that become dangerous and spread all over the body. The report was given by the National Cancer Institute on melanoma cancer, in 2017, and around 87,110 new melanoma cancer cases were diagnosed [48].

The treatment of melanoma cancer is depending on the stage of cancer which means how far it has spread over the body [49]. Figure 5.20 shows the different melanoma cancer stages.

The researchers Alexandros Karargyris, Orestis Karargyris, and Alexandros Pantelopoulos developed a mobile application that makes the smartphone a powerful tool with the help of image processing and IoT which helps the skin disease patients so that they can check the suspicious change in an existing mole on their skin without medical expertise help. The mobile application DERMA/care is an IoT expert system integrated [50] with an inexpensive microscope for screening the

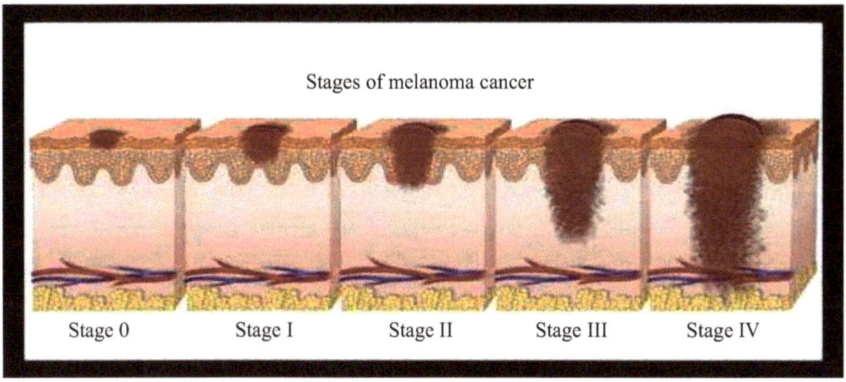

Figure 5.20 Stages of melanoma cancer

Table 5.2 Effect of melanoma skin cancer

Stages	Description
Stage 0	Melanoma in suit (present on the outermost layer of skin)
Stage 1	2 mm thick
Stage 2	May be thicker than 4 mm (at least 1.01 mm thick)
Stage 3	Spread over the nearby lymphatic channel and thicker than 4 mm
Stage 4	Spread to body organs like brain, liver, and lungs

skin and the use of image processing application in DERMA/care to achieve something important as follows:

1. Identification of disease affected area in captured skin image.
2. Image processing techniques that help in analyzing the skin disease affected area to find the types of disease.
3. Result classification.
4. Save results for comparison and further processing.

The following steps are included in image-based IoT system DERMA/care for melanoma cancer detection:

1. At the time of skin cancer melanoma detection, a microscope is mounted on a phone camera to acquire a clear and high-resolution image of the affected area on which the image processing algorithm works.
2. From the acquired image extract the ROI of the affected area, but the captured image is a colored image (Red-Green-Blue) and ROI should be redder or browner so here there is a need to convert the RGB image into some understandable color. Hue saturation value (HSV) is used here that represent the exact color of ROI.
3. On the next step, texture and color features of ROI are calculated. The geometrical features of ROI such as area, perimeters, and diameter are calculated so that the user can easily monitor change in suspicious spots on the skin.

Superficial spreading melanoma	***Nodular melanoma***	***Lentigo maligna melanoma***	***Acral lentiginous melanoma***
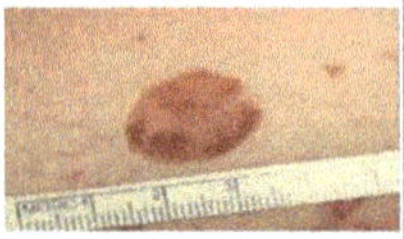	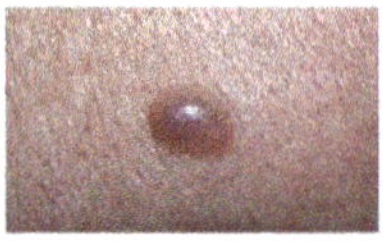	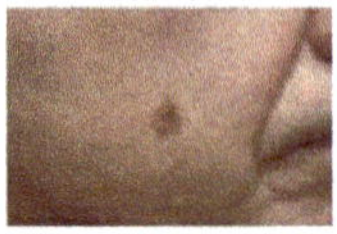	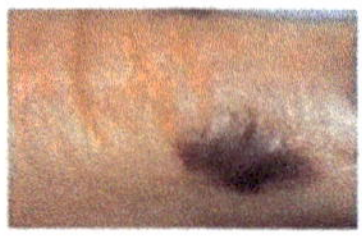
-Most common -Flat and irregular in shape and color -Shades of black and brown	-Usually starts as a raised area -Dark black/blue or bluish/red -Some are not colored	-Usually occurs in old skin types -Commonly on face, neck, arms, etc. -Abnormal skin areas usually large, flat, and tan with areas of brown	-Least common -Usually found on palms, soles, and even under fingernails

Figure 5.21 Different types of melanoma cancer

4. For data classification purposes, a supervised method support vector machine is used.

The E-Health smart mobile app DERMA/care working snapshots are shown in Figure 5.22.

5.6 Smart E-health monitoring medical imaging modalities

The visual representation of the tissue of plants, animals, and humans is important for clinical analysis. Here smart medical imaging modality helps in generating the visual representation of organs, muscle tissues, bones, and nerves which are hidden by skin and bones and it can also show the blood flow. These techniques help the physicians to look inside the human body, plants, and animals for diagnosis purposes. The various invasive and noninvasive modalities are available in the medical system: magnetic resonance imaging, fluorescent markers, computed tomography, endoscopy, bioluminescence, positron emission tomography, ultrasonography, positron emission tomography (PET), quantum dots, and single-photon emission tomography (SPECT). The medical imaging techniques can be invasive and noninvasive as shown in Figure 5.23. The invasive technique means which can be inserted inside the body; some invasive techniques are catheter venography and intravascular ultrasound. The noninvasive technique means which cannot be inserted inside the body. The noninvasive techniques are magnetic resonance imaging (MRI), CT scan, and X-ray.

5.6.1 Magnetic resonance imaging

A noninvasive procedure uses radioactive effects to produce a detailed high-resolution view of the internal organs of the body. It is a pain-free and safe technique that can study every part of the body like soft and hard tissue, heart, bones, brain, blood vessels, and spinal cord to check internal abnormality for the treatment planning of disease. The first MRI scan was done by American scientist

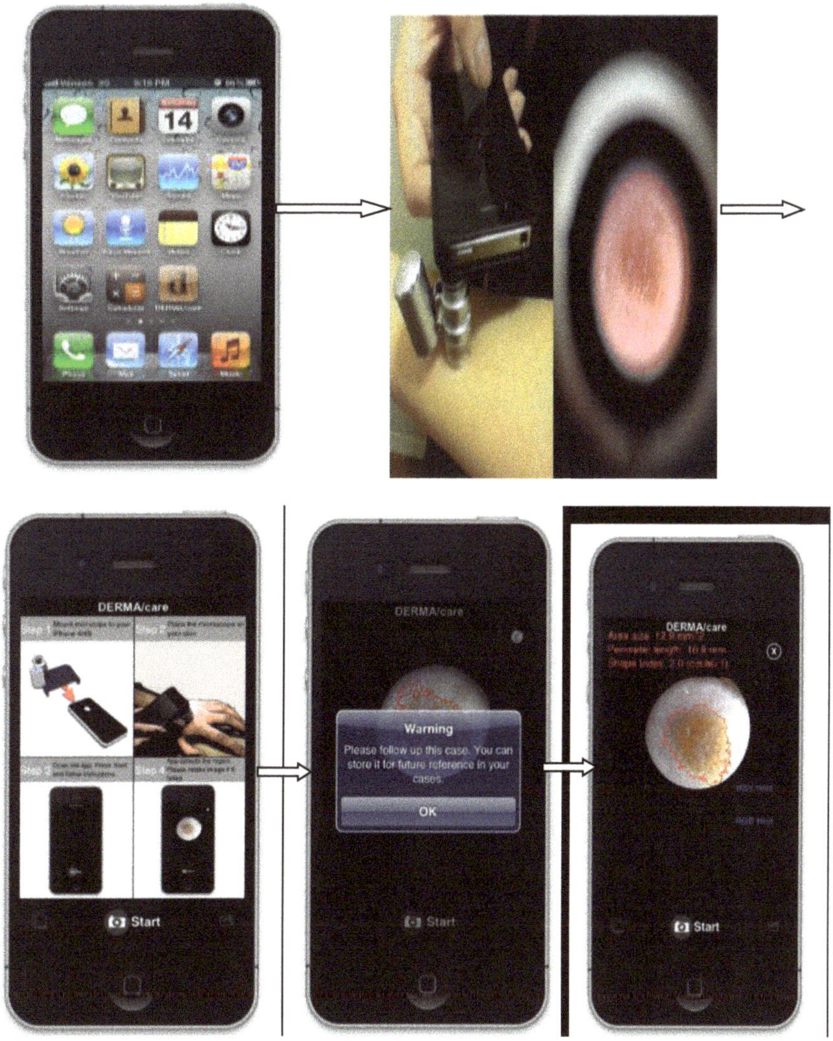

Figure 5.22 DERMA/care mobile app working process

Dr. R. Damadian in 1977 of human for the detection of cancer. Magnetic field and radio waves are the working principles of MRI scanner which helps in generating the information about the internal body. A report on the human body is presented by H. H. Mitchell who shows that the human body is totally made of water in which heart and lungs contain 73:83 percent water and skin have 64%, bones are composed of 31%, and muscles and kidney have 79% water. Hydrogen and oxygen atoms are two main components in water molecules and in each hydrogen center, very small particles protons like tiny magnet are available which are sensible to magnetic. At the scan time, radio waves are sent to body parts to active protons by

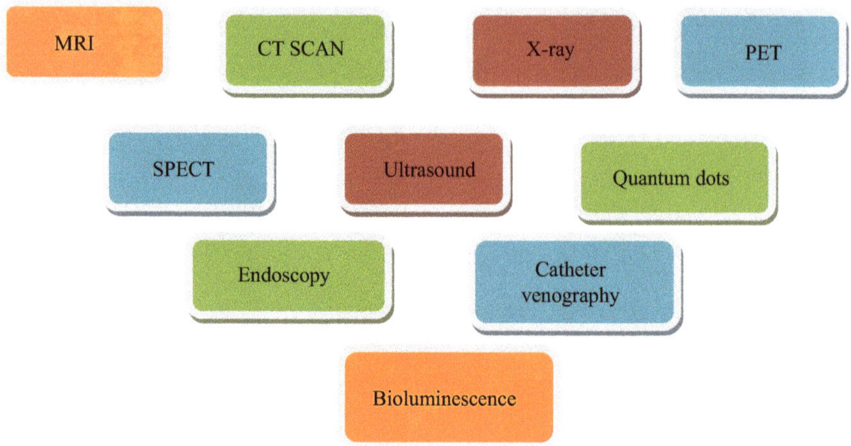

Figure 5.23 Different types of smart IoT-enabled imaging modality

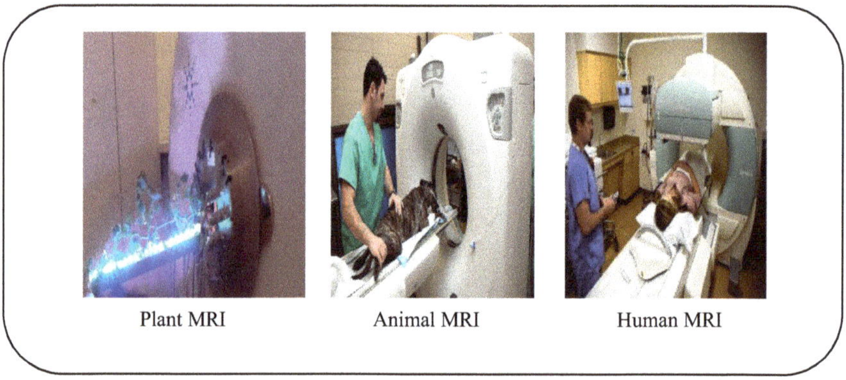

Figure 5.24 A use of MRI for plants, animals, and humans

which protons can spin out of equilibrium and when radioactive off then protons generate energy and this energy gives the information about the location of protons picked by MRI scanner and helps the physician to produce the information about the tissue. This MRI principle can work in the same for humans, plants, and animals as shown in Figure 5.24. MRI is an expensive tool but safe also because it does not use any ionizing radiations which are very dangerous.

MRI of humans is done for abnormality detection in the body such as tumors, brain strokes, any type of infection, neuro disorder dementia, blood vessels blockage, spinal injury, and many more diseases. Cardiac MRI helps in to diagnose different heart abnormalities such as tissue damage, artery blockage, and any damage due to heart attack [51,52].

MRI is also done for plants on the same working principle that works on humans. The two- or three-dimensional MRI of intact plants is analyzed to check the water level in different parts of plants such as stem, leaves, and roots, to check plants' growth by MRI analysis of seed germination, and also to check different stages of seed development. This technique also helps in checking the root growth by providing the three-dimensional view of root not only in liquid but also in sand or soil. Nuclear MRI of leaves, plant parts inside the soil, and roots are required to diagnose the disease of plants.

MRI, a common diagnostic tool, is used for humans and plants and the same can be used for animals to visualize the internal structure. This technique works on animals in the same way as it works on humans but the difference is that anesthesia is compulsory for animals [53].

This technique helps in diagnosing the disturbance in the brain and spinal cord due to the growth of abnormal tissue. It can detect brain cancer, infarcts, and abscesses. The spinal disorders such as herniated discs, spinal tumors, and nerve root impingement can be detected with the help of using MRI [54].

5.6.2 X-ray

X-ray or radiography introduced by Wurzburg University Prof. Wilhelm Conrad Roentgen is an imaging technique that uses radiation to capture multiple images from different angles inside your body. This imaging modality can generate images of hard and soft tissue in which hard tissue like bone absorb radiation in high quantity so it looks white on X-ray film and soft tissue is opposite to this, as soft tissue like muscles absorb radiation in low quantity so it appears dark in color on X-ray film. This IoT-powered imaging technique for humans can be used to detect different types of diseases such as bone fracture, a dental problem, and

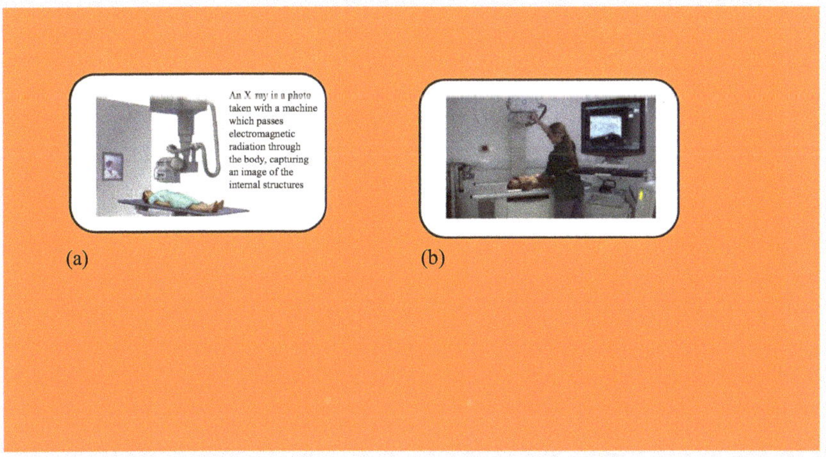

Figure 5.25 (a) Human lie on X-ray machine and (b) animal X-ray machine

chest X-ray to detect lung problems such as any cysts, cancer, and asthma. This technique is not only helpful for human but also it generates an image of animal's body as shown in Figure 5.25. For animals, an X-ray is used to check problems inside the body parts such as bones and organs by applying radiation on the target area. Like human X-rays, animals absorb different amounts of radiation due to hard and soft tissue. In animals, an X-ray imaging technique is used to check different problems such as dental problems (fracture and abscesses), stone, cysts, and reason for painful abdomen [55–57]. X-rays can be used to study plant parameters, root development, and different types of diseases in plants. In comparison to MRI, X-ray takes less time for imaging different parts of plants and too much affected with soil.

5.6.3 Ultrasound

The real-time, painless, and safe imaging technology uses high-frequency sound waves to generate images of internal body parts. This method of imaging depends on the advanced technology of IoT which can be used to diagnose the cause of swelling and infection in any body parts but especially it is used in pregnancy to monitor the fetus development. After a heart attack, ultrasound helps in the biopsy. This imaging process does not use any ionizing radiation so it is safe for the body and creates real time image of internal working organs as well as it can capture blood flow through vessels. Doppler ultrasound is helpful in checking how blood is moving through arteries to body parts such as leg, arms, neck, and brain, and physicians can also check blockage in blood flow [58,59].

5.6.4 Computed tomography

The "tomography" word originated from the Greek word "tomos graphein" where tomos means "a cut or a slice" and graphein means "to record" and first used in 1971 for humans. Computed tomography is an imaging procedure that uses a computerized X-ray machine to create an internal view of the body. This imaging technique can produce a detailed cross-sectional image of the internal body parts such as organs, bones, and hard and soft tissue by applying an X-ray beam. CT imaging is used to detect abnormal tissue growth and helps in to find out the presence, type, stage, and location of cancer and also helps in biopsy surgical treatment [60,61].

Before CT-scan, "dye" a contrast agent is used to get clear images of inside the body area which can be injected into a vein or by mouth [62]. This scan process can also be used in veterinary for pets to scan tissues of brain, bones, and spine. The same scan process that works for humans is also working for animals but little difference is that anesthesia is required in some critical situations for animals so the animal can remain calm during the scan procedure (see Figure 5.26(a)). At the scan time, the CT-scan table on which animal is lying is done inside the machine that executes the scan, that is, gantry. During the scanning process, the X-ray beam moves around the patient to capture the images into different angles and these captured images are recorded into the computer for analysis purposes. As the same of humans, CT-scan can also show the location and size of tumors, reason of

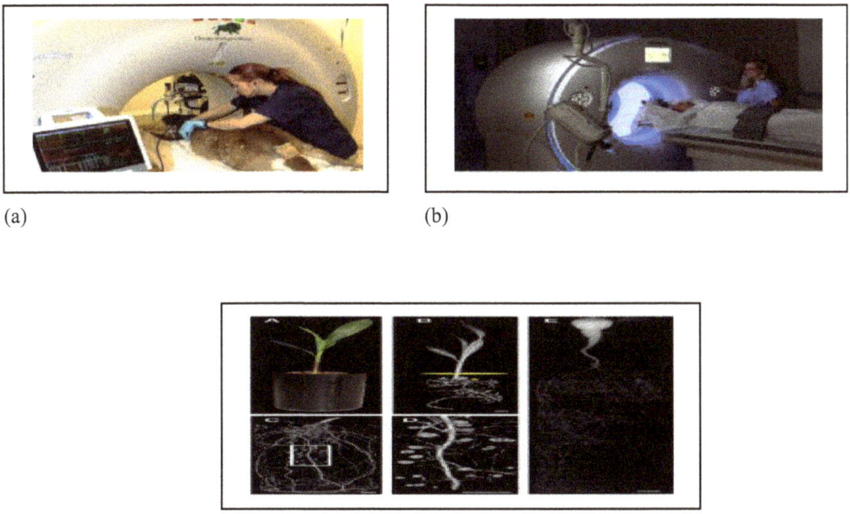

(a) (b) (c)

Figure 5.26 (a) Animal CT-scan, (b) human CT-scan, and (c) plant CT-scan

swelling, and cause of fracture of animal [63,64]. The plant development depends on the soil, so CT-scan also helps to study the soil structure and water content level by capturing the image of heterogeneous material and the development can also be checked by scanning the root architecture from different angles.

5.6.5 Nuclear medicine

In nuclear medicine, a special radioactive isotope which is made of carrier molecules is inserted in the body to access different internal body parts. It produced energy in the form of gamma rays when inserted into examining the area and these gamma rays are detected by a special camera to generate images of the particular area. In nuclear imaging, two special imaging techniques are used: single-photon emission computed tomography (SPECT) and positron emission tomography (PET). This smart imaging modality is comparatively better than other imaging techniques because it gives special information that cannot be achieved by using other imaging modalities and also helps in detecting the disease at the initial stage [65].

SPECT: A special imaging tool that is used to check how blood flows in the brain and also helps in diagnosing infection, stress, and seizure. This imaging process uses computed tomography and radioactive tracer in which radioactive is injected into a vein to produce gamma rays that are easily recognized by IoT-powered CT machine to capture the image of body parts. The study also shows that this technique is more sensitive to brain abnormalities rather than other scanning techniques such as MRI and CT machines [66].

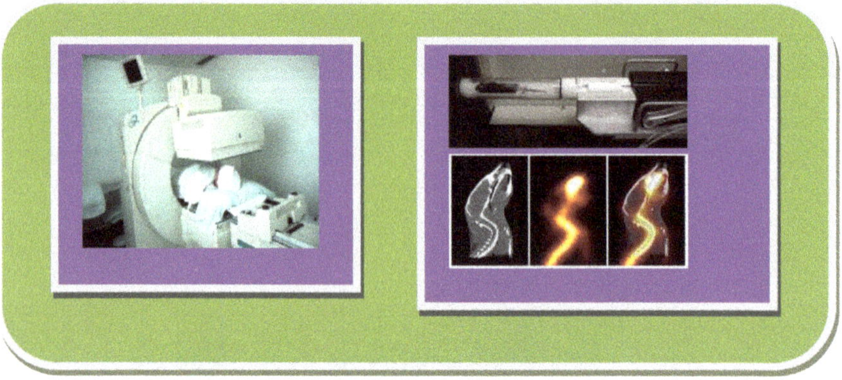

Figure 5.27 PET and SPECT images of human and animals

PET: The positron emission tomography imaging techniques used with the combination of MRI and CT scan to study how the body tissue and organs are functioning. The disease related to heart (blood flow), cancer, and brain (tumor) can be detected with the help of the PET nuclear imaging technique. The radiotracer used in PET generates a small particle positron which is detected by computer to produce images as shown in Figure 5.27.

5.7 Image-based IoT smart image analysis system

Image analysis system in the medical image system is used to diagnose different diseases in humans, animals, and plants. Image analysis can be defined as taking out some quantitative information from captured images and videos by using analysis software. As per the customer need, the use of image analysis is not limited but it used in different fields such as defense, material science, medicine, robotics, security, and microscopy. In the astronomy field, image analysis is used to check the size of the planet and in security service to check the person's physical characteristics such as hair and eye color. This image analysis process is very useful in the detection of disease in humans, plants, and animals based on image measurement. The analysis software read the scanning image to detect the type, size, and cancer place in body parts. The IoT-enabled image analysis software is different for humans, plants, and animals. Here we discuss some important analysis software (Wikipedia) (try to enter steps involved in it):

1. WinRHIZO (image analysis system specifically designed for root measurement) and
2. *IMAFISH_ML* (image analysis software for assessing fish disease detection).

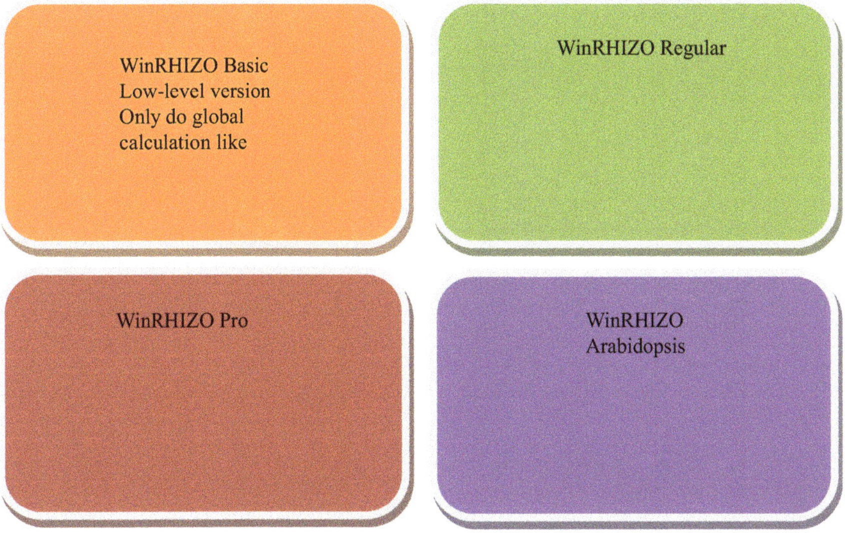

Figure 5.28 Four versions of smart WinRHIZO system

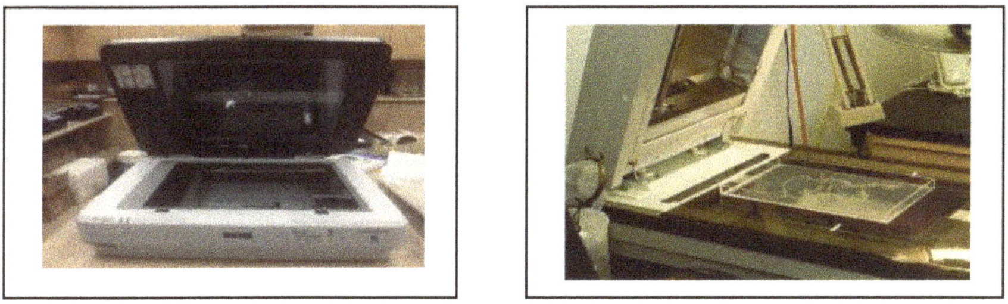

Figure 5.29 Regent scanner

5.7.1 IoT-based smart plant root measurement: WinRHIZO system

The plant root analysis software WinRHIZO, which is IoT-enabled system, is used to measure root in different forms which is an integrated program of image capturing device and computer programming. This integrated program analysis system can measure different parameters of roots such as area, length, color, and topology but works only on washed roots [67]. The four different versions are available of WinRHIZO which is shown in Figure 5.28.

The washed root analysis process by WinRHIZO can be completed in just one click. This process includes some important steps which are as follows:

1. Root placing on scanner: In this step, put the washed plant roots on a waterproof glass tray of the scanner. Here Regent scanner is used which has some

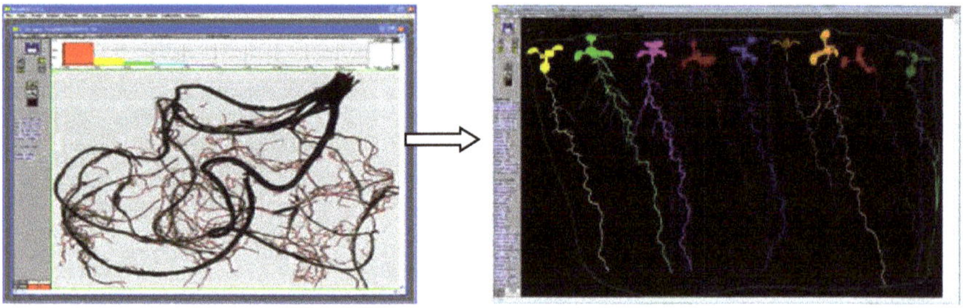

Figure 5.30 Digitized plant roots

unique characteristics like the use of a special lighting system helps in avoiding extra lighting like a shadow and also has a manual that helps the user at the scanning time. This same scanner can be used for document scanning. This scanner is available into different size standard area and large area sizes where 22×30 cm is the standard size area and 30×42 cm is the large area size. The regent scanner is shown in Figure 5.29.

2. Image acquit ion: Image can be digitized within a few seconds by just clicking the scanner icon (physical roots to imaging root). WinRHIZO is a "TWAIN" compatible because for analysis it can control scanner directly and can take root images from different cameras and can also analyze previously stored tiff and JPEG format images files. The root image in the WinRHIZO system is shown in Figure 5.30.
3. The root analysis: After getting the digitized image, the root image for WinRHIZO is displayed in different colors because root parameters such as length, area, and volume are coded as a different color for understanding purposes.
4. Data saving: In this process, the data that come from the root analysis process are saved for future use, reanalysis, and validation. This process is automatically done by WinRHIZO because, after completion of root analysis, it automatically saves the data into ASCII format which can be easily readable by other programs like "excel."

The two models of WinRHIZO available are WinRHIZO Tron and WinRHIZO Tron MF. The analysis of one frame at a time is done by WinRHIZO Tron and the analysis of multiple frames at a time is done by WinRHIZO Tron MF (multiple frames). The different analysis products are also available for leaf, seed, and tree rings. The WinFOLIA is used to analyze the leaf and WinSEEDLE analyzes the seeds. All the products work on the Window operating system including Windows 10.

5.7.2 Smart aquaculture IMAFISH system: real-time IoT-based smart system for fish disease identification

In today's time, the awareness about food quality is increasing day by day because the person has become very cautious about his diet and he wants to have more protein in food. The major source of protein is available in marine origin in the form of fish. According to the United Nation Food and Agriculture Organization to meet the growing demand for food, it is necessary to increase the production. In marine origin, the three main species of fishes are Gilthead sea bream, meager, and red porgy are the biggest part of eating habits to complete the protein. To provide good food, the aquaculture company needs to increase the production and quality of its food. So, to increase the production it would be better to measure the quality before delivery of the product. A noninvasive measurement system can be used to measure the morph metric and meristic characteristic of fish where morph metrics are measurable traits such as shape and size and meristic are countable traits such as fin rays, spines, and vertebrae slits. Fully automated and IoT-based image analysis system software is developed that measures the morph metric characteristic of three commercial fish species and helps the farmer and aquaculture company to measure their fish production. The name of the automated system software is IMAFISH_ML developed with the MATLAB® programming software [68]. The steps followed by IMAFISH_ML to measure the morphometric traits are as follows:

1. Image capture: In a small dark room, the digital camera with two fluorescent tubes and a sign of fish position is used to capture the fish image for further analysis. The two images, one is lateral whose background is red in color and other is a dorsal image whose background is white in color are captured of each fish in the dark room without using the flashlight. The lying position and head on the left side of the fish give the lateral image. The vertical position and head on the right side of the fish give the dorsal image.

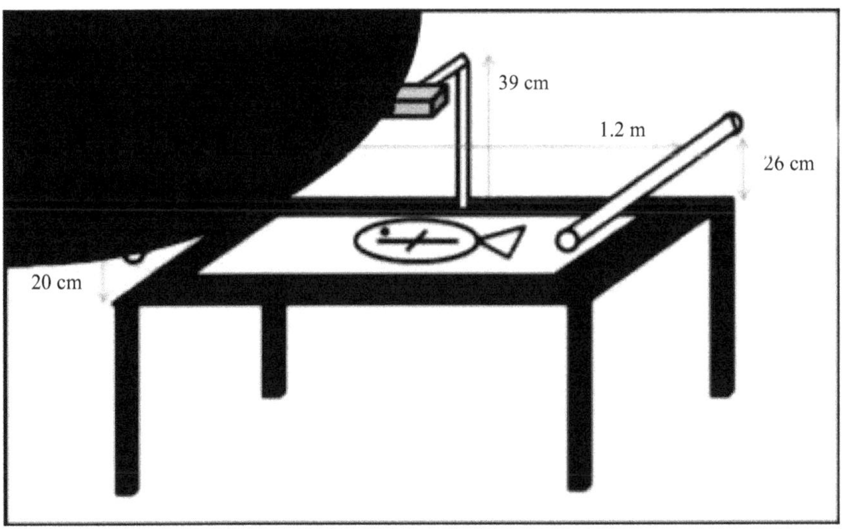

Figure 5.31 Fish image capturing process

Figure 5.31 shows the setting done for capturing the fish image.

2. Image analysis

   The advanced IoT-based image analysis system IMAFISH_ML software analyzes the image of three different species: gilthead seabream, meager, and a red porgy and in this software more than 27 morphometric characteristics per fish are included because each fish have different morphometric traits and this process takes the following steps to complete the analysis process:

   1. Image size calibration: An Otsu's binarization threshold is used for horizontal and vertical image size calibration of three different images.
   2. Image selection: Image is just entered into the programmed and the program read the image and then select it.
   3. K-means clustering segmentation is used to segregate the fish by the use of RGB space.
   4. The fish head size and the caudal fin is detected to measure the body length and then fish alignment is done.
   5. In the last, fish measurement is done. The fish measurement is done by the different morphometric traits of the lateral and dorsal view of the fish image.

Table 5.3 shows the morphometric traits obtained from the lateral and dorsal views.

IMAFISH_ML software calculates the minimum and maximum threshold value of each trait and for each fish species shown in Table 5.4.

Table 5.3 Morphometric traits of fish

Lateral view morphometric traits	Dorsal view morphometric traits
Total lateral area and length	Total dorsal area and length
Fish maximum height	Maximum width
Fillet area, area %, the maximum length	Five equidistant fish widths
Standard length	Fillet volume
Caudal peduncle height	
Fish and head eccentricity	
Tail-excluded length	

Table 5.4 Threshold value set for traits

	Fish species					
	Gilthead seabream		Meager		Red porgy	
Threshold	Minimum 13 cm	Maximum 36 cm	Minimum 15 cm	Maximum 55 cm	Minimum 15 cm	Maximum 30 cm

The IMAFISH_ML software is IoT-powered used for the smart aquaculture company to check the quality of fish and at that analysis time if the morphometric trait value of any image is less or greater than the minimum and maximum value of threshold given in Table 5.4 then the system generates an error message. So the aquaculture company can check their product quality to improve the production by providing good products to the customer.

References

[1] Pujari J. D., Yakkundimath R., and Byadgi A. S. "Grading and classification of anthracnose fungal disease of fruits based on statistical texture features." International Journal of Advanced Science and Technology. 2013; 52: 121–132.

[2] Deshpande T., Sengupta S., and Raghuvanshi K.S. "Grading and identification of disease in pomegranate leaf and fruit." IJCSIT. 2014; 5(3): 4638–4645.

[3] Gavhale K. R., Gawande U., and Hajari K. O. "Unhealthy region of citrus leaf detection using image processing techniques." IEEE International Conference on Convergence of Technology (I2CT), Pune, 2014, pp 1–6.

[4] Okuboyejo D. A., Olugbara O. O., and Odunaike S. A. "Automating skin disease diagnosis using image classification." Proceedings of the World Congress on Engineering and Computer Science 2013, Volume II, San Francisco, USA.

[5] Kaur D., and Sandhu P. "Human skin texture analysis using image processing techniques." International Journal of Science and Research (IJSR), India, 2012, ISSN: 2319-7064, pp. 17–20.

[6] Okuboyejo D. A., Olugbara O. O., and Odunaike S. A. "Automating skin disease diagnosis using image classification." Proceedings of the World Congress on Engineering and Computer Science 2013, Vol II WCECS 2013, October 23–25, 2013, San Francisco, USA.

[7] Al-Turjman F., Ever Y. K., Ever E., Nguyen H., and Deebak D. "Seamless key agreement framework for mobile-sink in IoT based cloud-centric secure public safety networks." IEEE Access. 2017; 5(1): 24617–24631.

[8] Aruta C. L., Calaguas C. R., Gameng J. K., Prudentino M. V., Chestel A. A., and Lubaton J. "Mobile-based medical assistance for diagnosing different types of skin diseases using case-based reasoning with image processing." International Journal of Conceptions on Computing and Information Technology. 2015; 3(3): 2345–9808.

[9] Duan Y., Fu Z., and Li D. "Toward developing and using web-based tele-diagnosis in aquaculture." Expert System with Applications. 2003; 25: 247–254.

[10] Al-Turjman F., Hasan M. Z., and Al-Rizzo H. "Task scheduling in cloud-based survivability applications using swarm optimization in IoT." Transactions on Emerging Telecommunications. 2019; 30: e3539.

[11] Zhang X., Fu Z., and Wang R. "Development of the ESFDD: Expert system for fish disease diagnosis." Proceedings of the Oceans '04 MTS/IEEE Techno-Ocean '04. Kobe, Japan, November 2004, pp. 482487.
[12] Li D., Zhu W., Duan Y., and Fu Z. "Toward developing a tele-diagnosis system on fish disease." M. Bramer (ed.), *Artificial intelligence in theory and practice*. Book Series IFIP International Federation for Information Processing, Vol. 217. Boston: Springer, 2006, pp. 445–454.
[13] Campioni F., Choudhury S., and Al-Turjman F. "Scheduling RFID networks in the IoT and smart health era." Journal of Ambient Intelligence and Humanized Computing. 2019. doi: 10.1007/s12652-019-01221-5.
[14] Park J.-S., Oh M.-J., Han S., and Eco Aquafarm Research Center, Chonnam National University, Korea. "Fish disease diagnosis system based on image processing of pathogens' microscopic images." IEEE, 2007.
[15] Al-Turjman F., and Alturjman S. "Context-sensitive access in Industrial Internet of Things (IIoT) healthcare applications." IEEE Transactions on Industrial Informatics. 2018; 14(6): 2736–2744.
[16] See https://www.aad.org/about/burden-of-skin-disease
[17] See https://www.healthline.com/health/skin-disorders#causes
[18] See https://www.medicalnewstoday.com/articles/316622.php#Permanent%20conditions
[19] See https://www.skinsite.com/index_dermatology_diseases.htm
[20] Yadav N., Narang V. K., and Shrivastava U. "Skin diseases detection models using image processing: A survey." International Journal of Computer Applications. 2016; 137(12): 34–39.
[21] Okuboyejo D. A., Olugbara O. O., and Odunaike S. A. "Automating skin disease diagnosis using image classification." *Proceedings of the World Congress on Engineering and Computer Science 2013*, Vol II WCECS 2013, October 23–25, 2013, San Francisco, USA.
[22] Ambad P. S., and Shirsat A. S. "A image analysis system to detect skin diseases." IOSR Journal of VLSI and Signal Processing. 2016; 6(5): 17–25.
[23] Harville M., Baker H., Bhatti N., and Süsstrunk S. "Consistent image-based measurement and classification of skin color." *IEEE International Conference on Image Processing (ICIP)*, September 11–14, 2005, Genoa, Italy.
[24] Tsumura N., and Ojima N. "Image-based skin color and texture analysis/synthesis by extracting hemoglobin and melanin information in the skin." ACM Transactions on Graphics. 2003; 22(3): 770–779.
[25] Chakraborty S., and Mali K. "Image based skin disease detection using hybrid neural network coupled bag-of-features." IEEE, 2017.
[26] See https://www.frontiersin.org/articles/10.3389/fpls.2017.01852/full
[27] Al-Turjman F., Zahmatkesh H., and Mostarda L., "Quantifying uncertainty in Internet of Medical Things and big-data services using intelligence and deep learning." IEEE Access, 2019. 10.1109/ACCESS.2019.2931637

[28] See https://en.wikipedia.org/wiki/Jaswant_Singh%E2%80%93Bhattacharji_(JSB)_stain

[29] See https://www.mayoclinic.org/diseases-conditions/malaria/symptoms-causes/syc-20351184

[30] See https://www.who.int/news-room/fact-sheets/detail/malaria

[31] Fang Y., and Ramasamy R. P. "Current and prospective methods for plant disease detection." *Biosensors.* 2015; 4: 537–561. doi: 10.3390/bios5030537.

[32] Wang G., Sun Y., and Wang J. "Automatic Image-based plant disease severity estimation using deep learning." Computational Intelligence and Neuroscience. 2017, Article ID 2917536.

[33] Khirade S. D. "Plant disease detection using image processing." IEEE, 2015. doi 10.1109/ICCUBEA.2015.153.

[34] See https://www.plant-image-analysis.org/software/leaf-doctor

[35] See https://www.businessinsider.in/This-App-Can-Detect-Skin-Cancer-In-7-Out-Of-10-Cases-Heres-How-It-Works/articleshow/45124113.cms

[36] See http://techroast.me/skinvision-the-skin-cancer-detector-mobile-app

[37] See https://www.skinvision.com/service

[38] See https://healthcareweekly.com/skinvision-and-central-launch-skin-cancer-screening-app/

[39] See https://www.who.int/vmnis/anaemia/prevalence/summary/anaemia_data_status_t2/en/

[40] See https://thenextweb.com/apps/2018/12/05/a-new-mobile-app-can-detect-anemia-without-a-blood-test/

[41] See https://www.healio.com/hematology-oncology/hematology/news/online/%7B752652a0-cfd5-4192-8464-88158426a7e8%7D/smartphone-app-uses-fingernail-photos-to-

[42] See https://www.webmd.com/a-to-z-guides/understanding-anemia-basics#1

[43] See https://www.k4health.org/toolkits/anemia-prevention/anemia-causes-prevalence-impact

[44] See https://steemit.com/steemhunt/@syarrf/icare-health-monitor-app

[45] See https://www.socialapphub.com/app/icare-health-monitor-bp-hr

[46] See https://www.amazon.co.uk/%E5%8C%97%E4%BA%AC%E5%98%89%E5%98%89%E5%BA%B7%E5%BA%B7%E7%A7%91%E6%8A%80%E6%9C%89%E9%99%90%E5%85%AC%E5%8F%B8-iCare-Health-Monitor/dp/B0196G4MT8

[47] See https://freeappsforme.com/apps-to-measure-blood-pressure/

[48] See https://www.pinterest.com/pin/445223113156887660

[49] See http://studymedicalphotos.blogspot.com/2017/05/malignant-melanoma.html

[50] Karargyris A., Karargyris O., and Pantelopoulos A. "DERMA/care: An advanced image-processing mobile application for monitoring skin cancer." IEEE, 2012.

[51] See www.healthline.com/health/heart-mri#uses

[52] See https://www.nibib.nih.gov/science-education/science-topics/magnetic-resonance-imaging-mri

[53] See http://www.ivghospitals.com/service/neurology/magnetic-resonance-imaging-mri/
[54] See https://animalwellnessmagazine.com/does-he-need-an-mri/
[55] See https://www.animaltrust.org.uk/our-services/pet-x-rays/
[56] See https://www.msah.com/services/blog/learn-differences-between-human-and-animal-x-rays
[57] See https://www.perintonvet.com/veterinary-services/diagnostics.html
[58] See https://www.radiologyinfo.org/en/info.cfm?pg=genus
[59] See https://www.betterhealth.vic.gov.au/health/conditionsandtreatments/ultrasound-scan
[60] See https://www.vetmed.ucdavis.edu/hospital/diagnostic-imaging/small-animal/ct
[61] See https://www.radiologyinfo.org/en/submenu.cfm?pg=ctscan
[62] See https://www.cancer.gov/about-cancer/diagnosis-staging/ct-scans-fact-sheet
[63] See https://www.petcoach.co/article/computed-tomography-ct-scans-is-used-in-veterinary-medicine/
[64] See https://www.vetmed.ucdavis.edu/hospital/diagnostic-imaging/small-animal/ct
[65] See https://www.nibib.nih.gov/science-education/science-topics/nuclear-medicine
[66] See https://mayfieldclinic.com/pe-spect.htm
[67] WinRHIZOTM Image Analysis for Plant Science. http://www.regentinstruments.com/assets/winrhizo_software.html.
[68] Navarroa A., and Lee-Montero I. "IMAFISH_ML: A fully-automated image analysis software for assessing fish morphometric traits on gilthead sea bream (Sparus aurata L.), meagre (Argyrosomus regius) and red porgy (Pagrus pagrus)." Computers and Electronics in Agriculture. February 2016.

Chapter 6

The development of a blood infusion warmer device: a new device

Auny El Jundi[1], Omran Alkhaldi[1], Mohamad Elamin[1], Sharmain Dube[1], Fadi Al-Turjman[2,3,4], Ilker Ozsahin[1,4] and Dilber Uzun Ozsahin[1,4]

During the blood transfusion process, even the smallest mistake can result in life-threatening problems. Maintaining the temperature of the blood delivered to the patient is very important because when blood temperature drops it can result in hypothermia and combining this with the two other risks of coagulation of blood and acidosis, this forms what is known as the triad of death. The purpose of this project is to lower the health risk that patients are faced with due to blood loss, hypothermia, transportation of the patient, or a sudden blood loss in the surgical operation unit. The main concept of this project is to develop, create, and design a better device that is more compact and more reliable. So cold refrigerated blood that is stored in the room is to be heated up using a dry-heat plate blood warmer made in this project. And one of the main objectives is to have a stable blood temperature and a flow rate. This process will reduce the possibility of the patient having hypothermia. Blood infusion warmers can save lives and increase the chance of the patient to live. The blood infusion warmer has a heater that warms up the blood and then the blood will be safe for transfusion. Blood warmers use different mechanisms to heat up the blood including water bath, intravenous tube warmer, dry-heat plate warmer, and forced air warmer.

6.1 Introduction

Approximately 5 million people die from traumatic injuries each year which make it one of the leading causes of death. Early trauma-related death is associated with hemorrhage in approximately 30% of cases. In cases where a lot of blood has been

[1]Department of Biomedical Engineering, Near East University, Nicosia / TRNC, Mersin-10, Turkey
[2]Department of Artificial Intelligence, Near East University, Nicosia / TRNC, Mersin-10, Turkey
[3]Research Center for AI and IoT, Near East University, Nicosia / TRNC, Mersin-10, Turkey
[4]DESAM Institute, Near East University, Nicosia / TRNC, Mersin-10, Turkey

lost and the loss is posing a threat in the injured people's lives, early transfusion of blood products is a critical component of treatment. During this process, there is a high risk of coagulopathy which can be due to certain key factors namely, injury severity, hypothermia, hypocalcemia, acidosis, and continued bleeding [1–3]. Coagulopathy, also known as bleeding disorder, is a condition which usually affects blood platelets thus impairing the body's ability to clot. This leads to excessive bleeding which can be life threatening to the patient as they can bleed to death. In cases of serious injuries or trauma, it is important to prevent coagulopathy. In short, these factors can be simply put as the trauma triad of death: coagulopathy, acidosis, and hypothermia [1,4–6].

Hypothermia can be reduced by using blood warmers to warm up the blood before infusion. There are different types of blood warmers used and these include water bath warmers, dry-heat warmers and intravenous fluid tube warmers, and forced-air blood warmers. Blood warmers are devices designed to prevent heat loss, not to reverse it therefore they warm up the blood before it is infused [7]. In-line type blood warmers are usually mounted on an IV pole between the infusion device and the patient line. They are easily understood by their in-line method of heat transfer which they use to use warm incoming blood and most of the blood warmers used are the in-line type [8,9].

In-line, the emerging Internet of Things (IoT) and blood warmers are important because they prevent hypothermia by preventing loss of heat from the blood before an infusion takes place [10–14]. With reduced chances of hypothermia, the chances of survival for the individual are higher as the infusion process is likely to be a success [15]. Blood infusion warmers can be used in most parts of the hospital where the process of transfusion could be necessary. It is used in emergency rooms, operating rooms, and intensive care units and they carry the same purpose of keeping blood in a desirable temperature before any blood transfusion so as to prevent hypothermia. The aim of this study is to make a dry-heat plate warmer which can be used to warm up blood during transfusion [16–19].

6.2 Related work

In-line type blood warmers are divided into different groups through their mean of heat transfer and because of this each blood warmer type possesses different characteristics which make them unique and also affects the efficiency of each type.

6.2.1 Water bath blood warmers

The water bath warmer system warms up the blood or the intravenous fluid with prewarmed water [20,21]. The maximum temperature that can be reached is 38 °C. The main advantage of using the water bath warmers is that they are very cheap; however, the disadvantages are that they are inefficient at high infusion rates [22], and also the fact that in order to ensure even warming of the blood it needs to be agitated so as to spread the temperature throughout [23]. An example of a water bath blood warmer is the infusion warmer constant temperature water bath (CTWB) (see

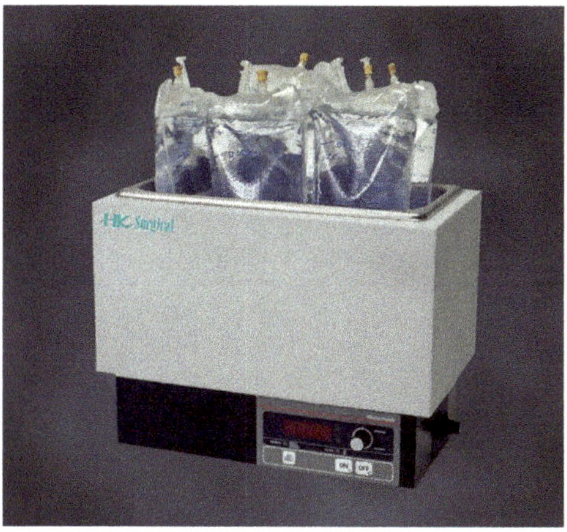

Figure 6.1 CTWB infusion warmer

Figure 6.1). The device contains a water bath with a blood bag. The water is heated to a certain temperature so that blood bag is heated and infusion of the heated blood to the patient. The device can use more than one blood bag, four blood bags to serve four patients (four blood bags 1,000 ml). It also has an alarm indicator where a red light-emitting diode (LED) light will glimmer if the heating plate reaches or exceeds 47 °C, meanwhile the warmer gives an audible alarm. The alarm will release automatically once the temperature of heating plate gets back to normal.

6.2.2 Intravenous (IV) tube warmers

The IV tube warmer works by warming the IV fluid or blood in a unique tube by using water which forms an outer layer and circulates from one side to the reservoir. The tube is made to be long and heavy and each specific one requires a specifically designed IV tube warmer coil. The rate at the blood is warmed depends on the power of the heating device. As the temperature of the fluid rises it is displayed on the display screen which also shows infusion rate so as to avoid overheating (above 41 °C). Blood warmers require regular servicing and maintenance because if ever there is a malfunction, for example, hemolysis and air embolism, it can result in a lot of complications during the infusion process which can be fatal to the patient being treated. The intravenous fluid tube warmer is more efficient at lower flow rates for the maintenance of euthermia, but this can also be a limitation [24–26].

An example of an IV tube warmer is the Hotline HL-90 Level 1 (see Figure 6.2), which requires a specially designed IV tube warmer coil (REF L-70). It warms up IV fluids in a specially designed tube which has a central lumen (internal diameter of 3.0 mm) and is surrounded by an outer layer through which warm water circulates downside and then up to the reservoir. The highest temperature setting is 40 °C–42 °C [20,27].

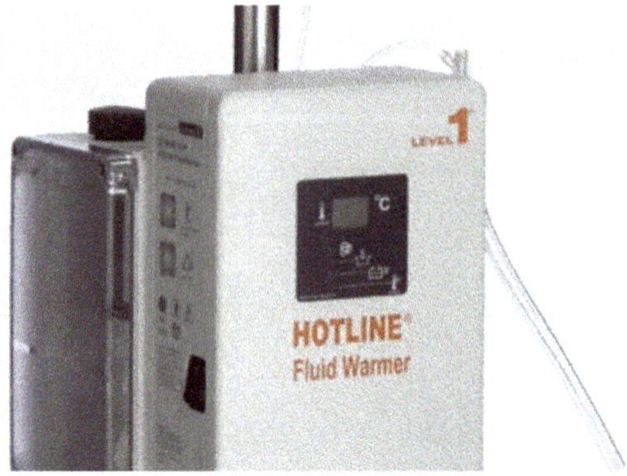

Figure 6.2 Hotline HL-90 Level 1. Adapted from [20]

6.2.3 Forced-air blood warmers

Force air or active air blood warming devices rely on convection heating to increase the fluid temperature. It warms the IV fluid in a duct of warmed air to the blanket. The type of blood warming system has vented airflow releasing air at up to 43 °C. This provides extra thermal energy which increases temperature gradients that can delay or prevent laminar flow. The limitation of the forced air blood warmers is that they have been shown to cause disruption of laminar flow and increase the risk of site contamination in cases of surgical operations. It has been shown that potentially pathogenic microorganisms can be detected in the tubes of these warming devices [20,28–30]. An example of the forced air blood warmer is the Bair Hugger 241 Blood/fluid warming system (see Figure 6.3). The system can heat up IV fluid to the temperatures of about 42 °C–43 °C [20,31].

6.2.4 Dry-heat plate blood warmer

Dry-heat plate warmer system warms up the intravenous fluid or blood in a cassette between the heat-plates. The advantage of using this warming device is that it increases heat transfer capability of the material and also allows temperature increase of up to 41 °C or even higher temperature [23]. It is also good for maintaining euthermia in IV fluid [24]. The disadvantage is that higher flow rates are necessary for it to work more efficiently. A good example of a dry-heat plate blood warmer is the infusion warmer BIW 204 (see Figure 6.4). The device has ability to deliver large quantities of blood effectively and maintain smooth blood flow. Blood can be heated without the risk of pollution, in addition to its work easy and quick to use. Self-exposure and self-testing can be introduced and operated permanently. The device is easy to design, simple to form and flexible, and it moves blood easily to reach the scene [32] (Table 6.1).

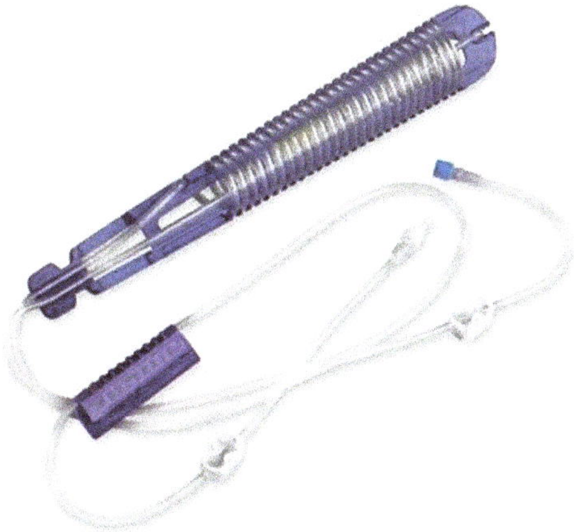

Figure 6.3 Bair Hugger 241 blood warming system. Adapted from [20]

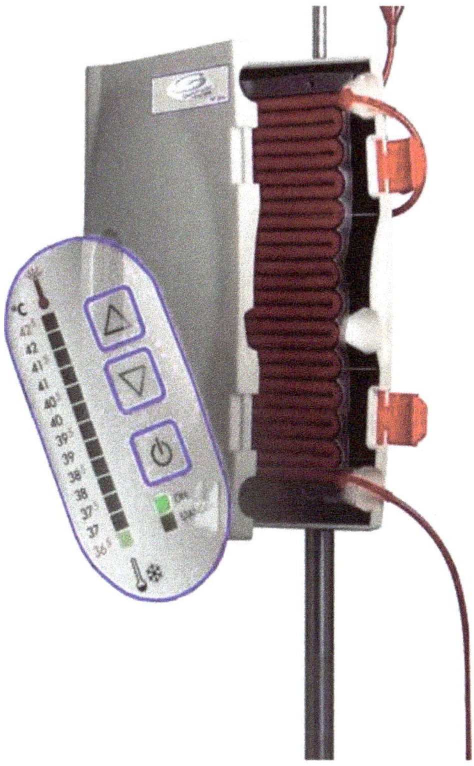

Figure 6.4 Infusion warmer BIW 204

Table 6.1 Summary of related work

Blood warmer type	**Design factors**								
	Safe	**Cost $**	**Size**	**Usability**	**Temp. °C**	**Flow rate**	**Sensing technology**	**Communication technology**	**Refs.**
Water bath	Yes	300	Small-large	Easy	38	Up to 4,000 per use	Yes	Yes	[20,23]
IV tube warmer	Yes	250	Small	Medium	42	500 mL/h	Yes	Yes	[20,21,26]
Forced air	No	120	Small-medium	Medium	43	3,000 mL/h	Yes	Yes	[19,20,30]
Dry-heat plate	Yes	80	Small-medium	Easy	41	Up to 4,000 mL per use	Yes	Yes	[20,22,32]

6.3 Methodology

6.3.1 Functionality

The process of designing and making a dry-heat plate blood warmer requires a number of components that come together to form an effective system. Each component of the device has specific properties and a purpose which contributes to the efficacy of the device.

6.3.1.1 Architecture of the dry-heat plate blood warmer

For warm blood to be delivered to the patient during transfusion, the dry-heat plate blood warmer components have to be connected properly (as shown in Figure 6.5). The power supply which could be electric or from a battery is connected to the breadboard which allows connections of almost all the other components. The peristaltic pump is connected to the IV set which is connected to the patient because it monitors the mass flow rate of the blood on its way to the patient's body. The temperature and any other signals or results obtained are analyzed by the sensory and communication components and then displayed through the display screen. Another connection to the IV set is the bubble detector which detects any bubbles available in the IV tube which moves the warmed blood from the reservoir to the patient. If any bubbles are detected in the IV tube the bubble detector will send a message to the system and blood flow will be instantly stopped from the IV tube to the patient to avoid any air bubbles entering the patient's system.

6.3.2 Components of the in-line IV tube warmer

1. Battery: It is used to provide power in order for the device to work without the main power supply (see Figure 6.6).

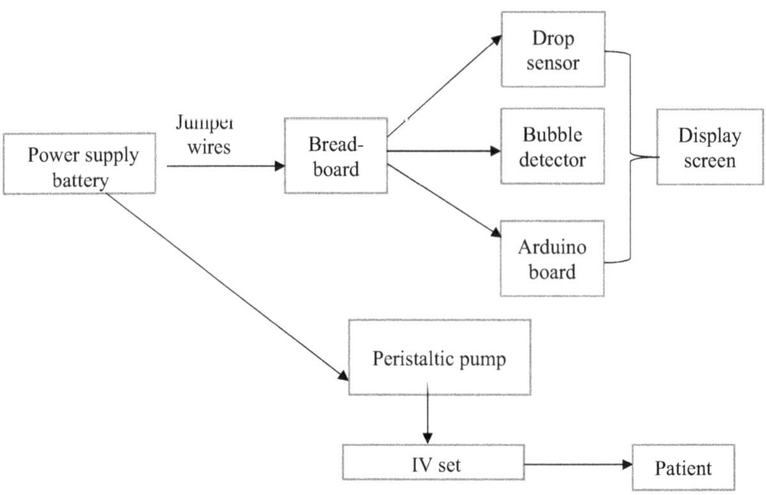

Figure 6.5 Architecture of dry-heat plate blood warmers

Figure 6.6 Battery for the project's part

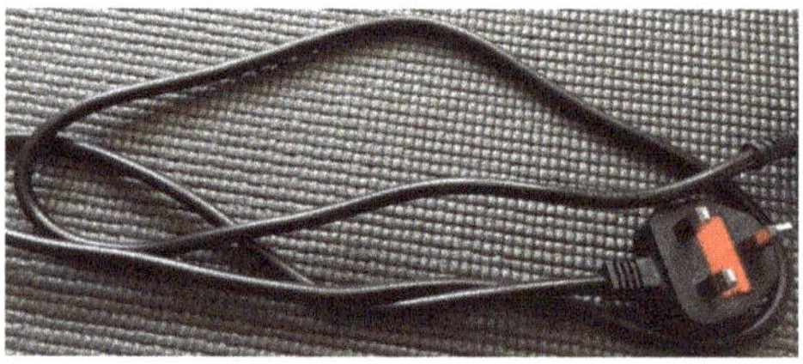

Figure 6.7 Power cable for powering the warmer

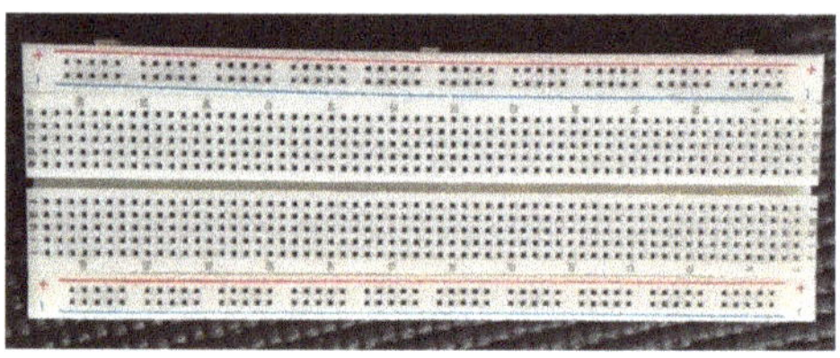

Figure 6.8 Breadboard for connecting the sensors

2. Power supply: Power supply is needed to give electricity to the boards and to the battery (see Figure 6.7).
3. Breadboard: This component is used for connecting and powering the parts of the project, such as a bubble detector and drop sensor (Figure 6.8).
4. Jumper wires: These wires are used to connect and power the sensors that are in the breadboard (see Figure 6.9). Jumper wires are usually used with breadboards and other prototyping materials to make it easy to change a circuit whenever necessary.
5. Arduino uno: Arduino uno is used for codding, processing, and to order the devices what to do and it receives and sends information to the rest of the device. This software runs on Mac, Windows, and Linux. It is used for a lot of projects of different types because it is easy to use for beginners, but it is also

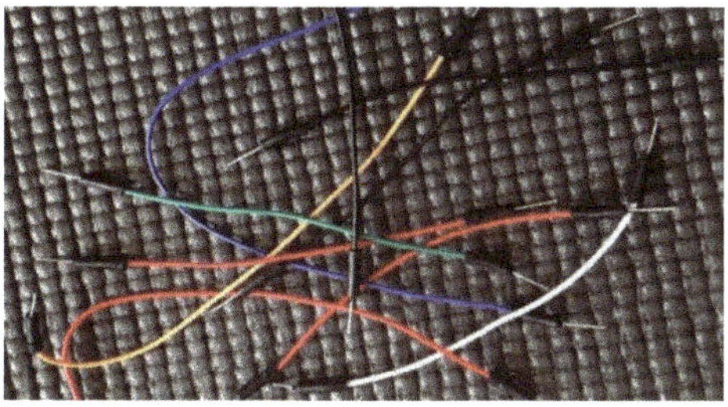

Figure 6.9 Cables and jumper are for connections

Figure 6.10 Arduino uno for programing the sensors

flexible enough for advanced users. It is more cost-effective than other microcontroller platforms. Figure 6.10 shows the Arduino uno.

6. Peristaltic pump: The peristaltic pump is a pump that is used mostly in the medical field, and it is used to obtain infusion and blood moves toward the patient in a constant rate, in other words, the peristaltic pump controls mass flow rate of the IV fluid. In place of the peristaltic pump, a dialysis pump can be used to achieve higher flow rates (300–600 mL/min) (see Figure 6.11).
7. Bubble detector: The bubble detector is used to see and detect if there are any air bubbles in the blood tube. This is important because if any bubbles escape to the blood tube can enter the blood system which can cause complications in the blood transfusion process. In case bubbles are detected, the sensor will send an information to the Arduino uno [33]. The bubble sensor is shown in Figure 6.12.
8. Drop sensor: This part of the device is used to check and detect if the drip chamber that is found in the blood sack is empty, full, or normal (see Figure 6.13). If the drip chamber is full this means that there is a blood clot in the tube and then the device will stop, and if the drip chamber is empty this

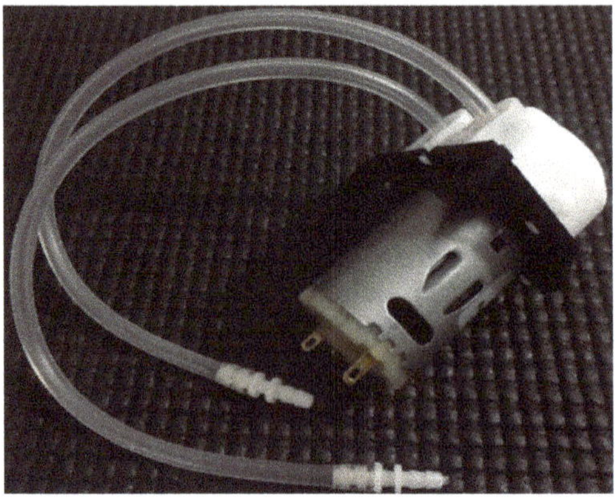

Figure 6.11 Peristaltic pump for infusion

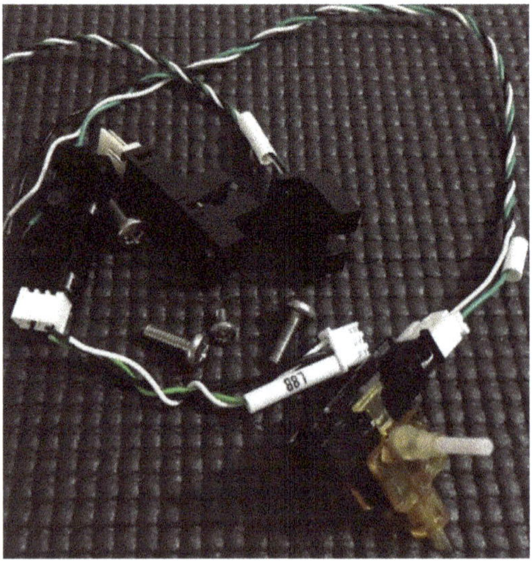

Figure 6.12 Bubble sensor for detecting the air bubble in the tube

means that all of the blood/fluid are infused to the patient and the device will stop automatically just in case the nurse forgot to turn the device off.

9. Heating mechanism: The heating system will heat the blood to approximately between 37 °C and 38.5 °C (see Figure 6.14). By heating the blood, this prevents hypothermia to the patient.
10. Display screen: The screen will display the current temperature and it will also show the infusion speed rate and it can also show if there is an error (see Figure 6.15). A normal Arduino LCD display will be used here.

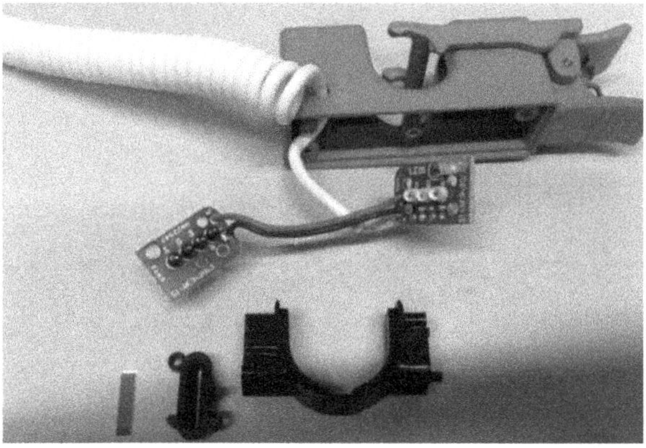

Figure 6.13 Drop sensor for detecting the drip chamber

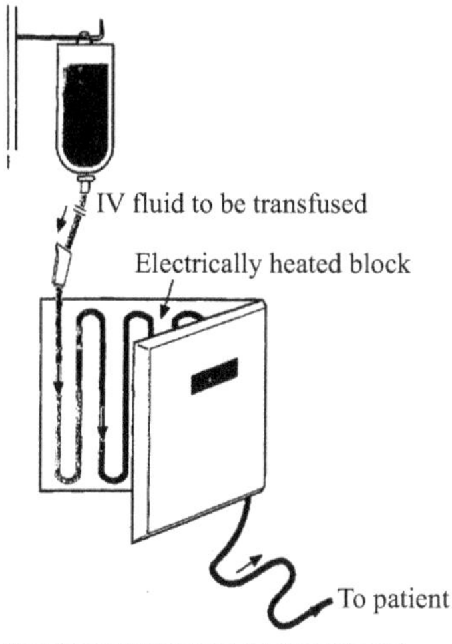

Figure 6.14 Heating mechanism

11. IV set: The IV set will include drip chamber, blood sack, tubes, and roller flow control clamp (see Figure 6.16).

6.4 Discussions

The dry-heat plate warmers are very important intravenous warming devices whose effectiveness is measured by the flow rate as well as the delivered temperatures of the blood to be warmed. The dry-heat plate blood warmer is also very efficient when it comes to handling thermal load. A research shows that the Fenwal blood warmer which is a dry-heat plate type of warmer varied from 95% efficient at the lowest

Figure 6.15 LCD screen for viewing the setup

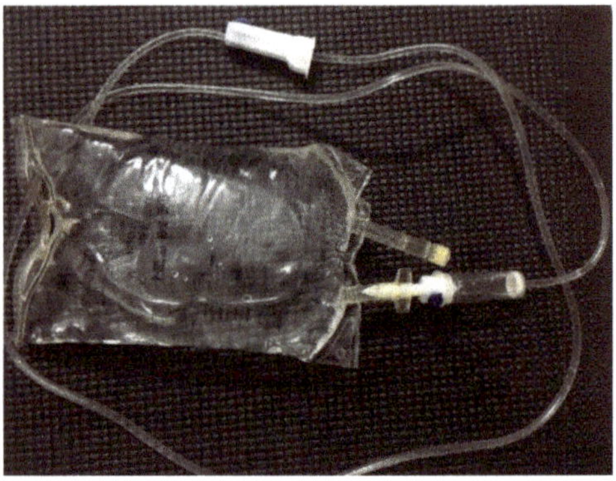

Figure 6.16 IV set for the device

thermal load to about 82% at the highest thermal load [7] which was highest compared to other three water bath type warmers. This dry-heat plate blood warmer achieved a maximum temperature of 40 °C. The amount of time that was required by the blood warmer to warm up was about 30–40 s. They are effective in low flow rates, as well as in higher flow rates. The delivered temperature of the blood is usually adequate as it is monitored not to pass 41 °C. As the flow rates are usually low for IV tube warmers and water bath blood warmers, if the flow rate tends to be high, a device using dry-heat plate is advised to be used as metal is a better conductor of heat [23,32]. In this project, the dry-heat plate blood warmer that was made resulted in a maximum flow rate of 120–150 mL/min. This is because a peristaltic pump was used in the design, if a dialysis pump had been used, a flow rate of 300–600 mL/min could have been achieved.

Since dry-heat plate warmers require no water baths for their use this is an added advantage because there are no water-related risks as well as risk of infection. The presence of the bubble detector improves the efficacy of the blood warmer because the bubble detector works by detecting if there any bubbles present in the IV tube which is carrying the blood being transfused to the patient. If there is bubble present

Figure 6.17 Dry-heat plate blood warmer

in the IV tube, the bubble detector sends a message to the Arduino uno and blood flow from the IV tube to the patient is stopped instantly. This reduces the chances of air embolism which can result in complications that can be life threatening.

The overall cost of the equipment required to make this dry-heat plate blood warmer is $250 which is cost-efficient considering the factors of reaching desired temperatures for transfusion and also the flow rate which was effectively high. From this work, the dry-heat blood water that was made has an advantage which makes it different from most conventional blood warming devices. This blood warmer can instantaneously warm up the blood to be transfused to the patient whilst the transfusion process is taking place, which speeds up the process and can greatly reduce the risks of hypothermia.

Complications can be faced when using the dry-heat plate warmers as they are devices which cannot be 100%. One of the biggest risks that can be faced is that of blood overheating which can lead to hemolysis of blood which then will cause life-threatening complications to the patient. The opposite of this case can occur when blood delivered has a lower temperature than favored as it may lead to hypothermia, a very dangerous factor that can result in fatal problems (see Figure 6.17 for the final result of the dry-heat plate blood warmer).

6.5 Conclusions

All blood warming techniques and devices have advantages and disadvantages of using them as much as they serve the same purpose. There are a number of factors to consider when choosing a certain warming technique, which include flow rate, temperature delivery, whether the use of water as a way of transmitting heat to the intravenous fluids is required or not, how much power it requires, and so on. The dry-heat plate warmer models are usually more expensive than the other blood warmer types, but in most cases, they display a lot of characteristics which are proof of its efficiency. It is important to choose a warmer according to its characteristics as well as its performance so as to get the most successful results when it comes to transfusion [20,23,32].

References

[1] Milligan, J., Lee, A., Gill, M., Weatherall, A., Tetlow, C., and Garner, A. A. "Performance comparison of improvised prehospital blood warming techniques and a commercial blood warmer." *Injury*. 2016; 47(8): 1824–1827.

[2] Polderman, K., Lundbye, J., Nichol, G., and Le May, M. "Therapeutic hypothermia in postcardiac arrest." *Therapeutic Hypothermia and Temperature Management*. 2019; 9(2): 102–107.

[3] Bullock, R., Foreman, M., and Conterato, M. "Temperature and trauma in accidental hypothermia." *Therapeutic Hypothermia and Temperature Management*. 2011; 1(4): 179–183.

[4] Kushimoto, S., Kudo, D., and Kawazoe, Y. "Coagulation abnormality in the acute phase of trauma: Acute traumatic coagulopathy and trauma-induced coagulopathy." *Japanese Journal of Thrombosis and Hemostasis*. 2016; 27(4): 399–407.

[5] Mitra, B., Tullio, F., Cameron, P., and Fitzgerald, M. "Trauma patients with the 'triad of death," *Emergency Medicine Journal*. 2011; 29(8): 622–625.

[6] Keane, M. "Triad of death: The importance of temperature monitoring in trauma patients." *Emergency Nurse*. 2016; 24(5): 19–23.

[7] Harrison, M. J., and Healy, T.E.J. "A comparison of four blood warmers." *Anaesthesia*. 1975; 30: 651–655.

[8] McMahon, M., and Shereen, S. (ed.). 2003–2020 Conjecture Corporation. 7 May 2020.

[9] Weatherall, A., Gill, M., and Milligan, J., *et al.* Comparison of portable blood-warming devices under simulated pre-hospital conditions: A randomised in-vitro blood circuit study. *Anaesthesia*. 2019.

[10] Al-Turjman, F., Zahmatkesh, H., and Mostarda, L. "Quantifying uncertainty in Internet of Medical Things and big-data services using intelligence and deep learning." *IEEE Access*. 2019; 7(1): 115749–115759.

[11] Al-Turjman, F., Ulusar, U., and Nawaz, M. "Intelligence in the Internet of Medical Things era: A systematic review of current and future trends." *Elsevier Computer Communications Journal*. 2020; 150(15): 644–660.

[12] Al-Turjman, F., and Alturjman, S. "Context-sensitive access in Industrial Internet of Things (IIoT) healthcare applications." *IEEE Transactions on Industrial Informatics*. 2018; 14(6): 2736–2744.

[13] Al-Turjman, F. "Intelligence and security in big 5G-oriented IoNT: An overview." *Elsevier Future Generation Computer Systems*. 2020; 102(1): 357–368.

[14] Al-Turjman, F. "A rational data delivery framework for disaster-inspired Internet of Nano-Things (IoNT) in practice." *Springer Cluster Computing*. 2019; 22(1): 1751–1763.

[15] Al-Turjman, F. "A cognitive routing protocol for bio-inspired networking in the Internet of Nano-Things (IoNT)." *Springer Mobile Networks and Applications*. 2017. doi: 10.1007/s11036-017-0940-8.

[16] Crossley, B. "EKG signals, blanket warmers, and fluid warmers." *Biomedical Instrumentation & Technology*. 2012; 46(1): 72.

[17] Erkin, Y., Tasdogen, A., and Gonnullu, E. "Is there a risk of emboli during infusion using line type liquid warmers." *Revista Brasileira de Anestesiologia*. 2013; 63(5): 389–392.

[18] Hendrarsakti, J., and Ichsan, Y. "Experimental study of isothermal plate uniformity for blood warmer development using geothermal energy." *IOP Conference Series: Earth and Environmental Science*. 2016; 42: 012018.

[19] Kumawat, V., and Aribandi, A. "Air bubbles produced during rapid blood warming with inline blood warmer leading to panic of air embolism." *Indian Journal of Hematology and Blood Transfusion*. 2016; 33(2): 281–282.

[20] Ohtsuka, N., Yamakage, M., Chen, X., *et al*. "Evaluation of four techniques of warming intravenous fluids. *Journal of Anesthesia*. 2002; 16: 145–149.

[21] Pashmakova, M., Barr, J., and Bishop, M. "Stability of hemostatic proteins in canine fresh-frozen plasma thawed with a modified commercial microwave warmer or warm water bath." *American Journal of Veterinary Research*. 2015; 76(5): 420–425.

[22] Milligan, J., Lee, A., Gill, M., Weatherall, A., Tetlow, C., and Garner, A. "Performance comparison of improvised prehospital blood warming techniques and a commercial blood warmer." *Injury*. 2016; 47(8): 1824–1827.

[23] Nickson, C. "Blood warmer, life in the fast lane." April 22, 2019. Accessed at https://litfl.com/blood-warmer/

[24] Sud, S., Dwivedi, D., Sawhney, S., Bhatia, J.S., Davis, J., and Dudeja, P. "Comparison of three different fluid warming techniques used to maintain euthermia in patients who underwent cesarean section: A retrospective audit." Journal of Obstetric Anaesthesia and Critical Care. 2019; 9: 35–39.

[25] Kaul, N., Nair, R., and Khan, R. "Cisatracurium degradation: Intravenous fluid warmer the culprit?" *Indian Journal of Anaesthesia*. 2015; 59(5): 323.

[26] Alexander, M. Blood and fluid warmers. *Journal of Infusion Nursing*. January–February 2006; 29(1): S35.

[27] Woon, S., and Talke, P. "Possible risk of air emboli using hotline HL-90 fluid warmer." *Anesthesiology*. 1998; 89(Supplement): 1204A.

[28] Tjoakarfa, C., David, V., Ko, A., and Hau, R. "Reflective blankets are as effective as forced air warmers in maintaining patient normothermia during hip and knee arthroplasty surgery." *The Journal of Arthroplasty*. 2017; 32 (2): 624–627.

[29] Crossley, B. "EKG signals, blanket warmers, and fluid warmers." *Biomedical Instrumentation & Technology*. 2012; 46(1): 72.

[30] Erdoğan, H., Işıl, C., Türk, H., Ergen, G., and Oba, S. "Comparison of forced-air warming systems and intravenous fluid warmers in the prevention of pediatric perioperative hypothermia." *Medical Bulletin of Haseki*. 2019; 57(3): 225–231.

[31] Trentman, T., Weinmeister, K., Hentz, J., Laney, M., and Simula, D. "Randomized non-inferiority trial of the vitalHEAT™ temperature management system vs the Bair Hugger® warmer during total knee arthroplasty." *Canadian Journal of Anesthesia/Journal canadien d'anesthésie*. 2009; 56 (12): 914–920.
[32] Chiang, V., Hopper, K., and Mellema, M. "In vitro evaluation of the efficacy of a veterinary dry heat fluid warmer." *Journal of Veterinary Emergency and Critical Care*. 2011.
[33] Fidan, U., Yilmaz, E., and Ziğarli, H. "The implementation of optical and capacitive bubble detector for intravenous fluid therapy." *Afyon Kocatepe University Journal of Sciences and Engineering*. 2016; 16(2): 285–291.

Chapter 7

Wireless sensor devices in medical applications: an overview

Samuel Nii Tackie[1], Kamil Dimililer[1,2] and Fadi Al-Turjman[2,3]

Application of wireless body area networks (WBAN) in medicine over the last decade has witnessed a rapid increase. This can largely be attributed to improvements made in wireless sensor technologies which have led to significant improvements in medical delivery especially with the introduction of low-cost medical-based wireless sensors. In this chapter, a review of wireless sensor devices and networks in medical application is investigated. The focus of this review is to examine the various wireless sensor devices and networks with respect to security, unobtrusiveness, reliable communication, and interoperability. This is done by reviewing past but current publications and suggesting future trends.

7.1 Introduction

Modern health facilities employ a variety of new tools in the delivery of well-organized healthcare, and this is possible due to improvements in communication structures and medical devices such as wireless sensors technologies. These physiological sensing devices should have the following features: integrated circuits requiring low power, intelligent or smart monitoring devices, lightweight, and minute size while the communicating structures are shot-range wireless systems [1,2]. Wireless sensor networks (WSNs) are composed of two important parts: wireless sensor devices or technologies and wireless communication networks. The importance of WSN makes it suitable for application in health, military, education, firefighting and prevention, and psychology. WSNs have the capability to perform a multitude of tasks such as monitoring, evaluation, and comprehension of a phenomena [3]. WSN in healthcare application is able to minimize mortality rate in

[1]Electrical and Electronic Engineering Department, Faculty of Engineering, Near East University, Nicosia, Mersin 10, Turkey
[2]Research Centre for AI and IoT, Near East University, Nicosia, Mersin 10, Turkey
[3]Artificial Intelligence Engineering Department, Faculty of Engineering, Near East University, Nicosia, Mersin 10, Turkey

cases of cardiac arrest. The survival rate in cardiac arrest in the first 720 s (12 min) is 48%–75% as reported by the American Heart Association (AHA). A detailed description of the WSN application is shown in Figure 7.1. There are several wireless sensor technologies which are regarded as an offshoot of WSN and the commonest is wireless body area network (WBAN). WBANs which are also referred to as BSNs (body sensor networks) or BANs (body area networks) are fashioned with thin, small lightweight sensors dispersed around, on, and in a human body to function as a monitoring device for the body and its immediate environment [5]. The WBAN term was initially introduced by [6], and it gradually gained momentum as the preferred wireless monitoring device in the health sector. Wireless BAN applications are found in the following areas: vital signs monitoring, daily activities, gait model, and senior patient monitoring such as Parkinson's patients, etc. [7].

Basically, WBAN functions as a monitoring, data detection and collection, and wireless data transfer system. Usually, personal digital assistant (PDA) and smart mobile phones are used to transfer the data to the health professional through a main wireless system. WSBN uses sensor nodes to detect and measure the data of selected phenomena such as heartbeat, temperature, sound, vibration humidity, etc. The measured data are either processed or transferred in its raw state through a single gateway or multigateway. The sensor nodes are commonly made-up of sensing component, processing component, communicating component, and a power unit. Together they sense, collect, process, and transmit data wirelessly to a central receiver. The following features are important for the most reliable and efficient sensor nodes: low cost, power efficient, wireless capabilities, multihop

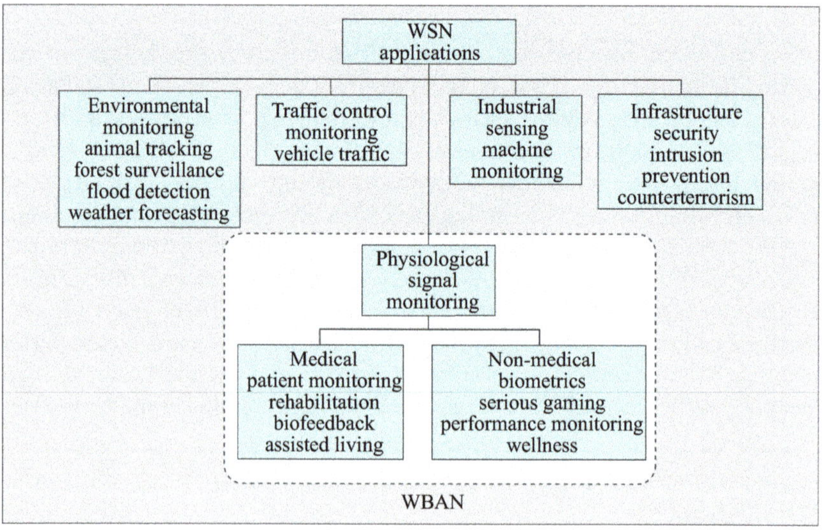

Figure 7.1 WSN applications. Adapted from [4]

data routing, and decentralized processing [8]. The lifespan of sensor nodes is mostly affected by the quality and duration of the power source, that is, the battery. Previous studies have tried to increase the lifespan of sensor nodes by balancing or duty cycling the load among the sensor node. However, recent studies seek to prolong the node's lifespan by integrating renewable energy in the sensor structure [9–12]. Deployment of sensor nodes is done in multitudes because of their low and small size. Although WBAN is referred to as an offshoot of WSN, there are several differences between these two systems, and these parameters of differences are presented in Figure 7.2. Energy efficiency and reliability are very important parameters in both systems. The other parameters are more important in WBAN than in WSN, for instance, maximum security is required in WBAN because of the sensitive nature of the data being transmitted (Table 7.1).

A review of WBAN is investigated in this research. The focus of this comprehensive review is a selection of important WBAN features and providing detailed analysis with more emphasis on the following:

- Medical applications
- System architectures
- WBAN nodes
- WBAN standards
- WBAN layers
- Wireless communication technologies in WBAN

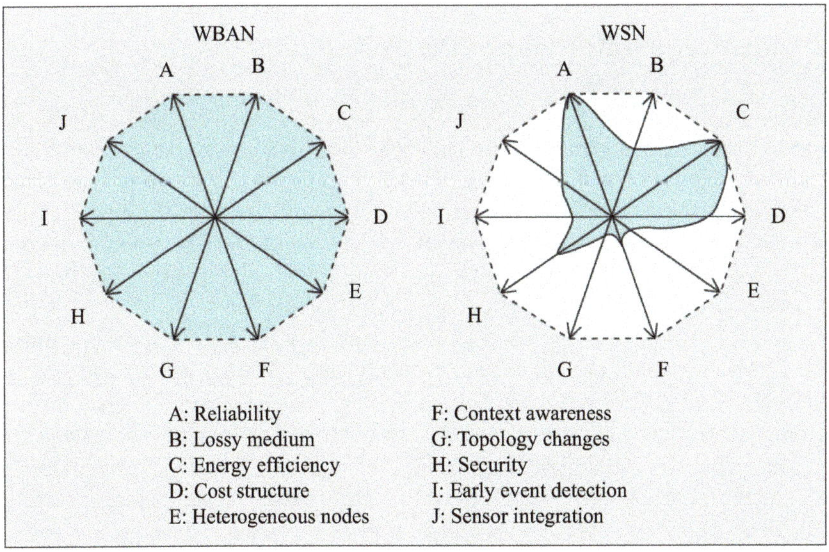

Figure 7.2 WBAN and WSN challenges. Adapted from [13]

Table 7.1. Abbreviations and acronyms used in this chapter

Abbreviations	Definitions
AAL	Ambient assisted living
ADC	Analog to digital converter
AHA	American Heart Association
A/F	Amplifier and a filter
APs	Access points
BAN	Body area network
BLE	Bluetooth low energy
ECG	Electrocardiogram
EEG	Electroencephalogram
EMG	Electromyography
HBC	Human body communication
IEEE	Institute of Electrical and Electronics Engineer
HME	Hub management entity
MAC	Medium access control
MWBAN	Medical WBAN
mW	Milli-Watts
PDA	Personal digital assistant
NB	Narrow band
PAN	Personal area network
PHY	Physical
SFD	Start frame delimiter
TDMA	Time division multiple access
UWB	Ultra-wide band
WSN	Wireless sensor networks
WBAN	Wireless body area network

7.2 Medical applications of the WBAN

WBANs are applied in a variety of fields such as sports, military, health, and gaming. Basically, if a physiological signal or vital signal monitoring is required, in both medical and nonmedical fields, WBAN is the appropriate technology to use. Medical application of WBAN has the potential to transform healthcare delivery by real-time monitoring and diagnosing of critical diseases. The goal of medical WBAN is to achieve lifelong monitoring without restricting the daily activities of the wearer [14]. In medical monitoring, WBAN is referred to as Medical WBAN (MWBAN) [15] and they are used to keep track of vital signals such as temperature, blood pressure, electromyogram, heartbeat, glucose levels, and electrocardiogram, and these signals are critical in diagnosing diseases and also useful in deciding appropriate treatments. WBAN in medical applications are categorized into four parts: distant patient monitoring [16], assisted living [17], rehabilitation [18], and biofeedback [19]. Examples of MWBAN-based physiological sensors are ECG, EEG, and EMG. WBANs are also used as therapeutic devices; it transmits electrical pulses to targeted areas for specific functions, for example, the pacemakers [20]. Two common devices, actuators and sensors, are used in achieving

Table 7.2. WBAN medical applications

WBAN medical applications		
Wearable WBAN	**Implanted WBAN**	**RC of medical devices**
Evaluating the battle readiness and fatigue of a soldier	Detection of cancer	AAL (ambient assisted living)
Assisting amateurs and professionals in sports training	Cardiovascular diseases	Monitoring of patient
Health monitoring		Tele-medicine
Stage sleeping		

these functions. The actuators are used to implement or perform specific functions such as insulin administering while sensors are used to monitor and measure specific signals [11]. Table 7.2 shows the medical application of WBAN based on wearable and implanted WBANs.

7.3 WBAN architecture

Generally, WBAN structure is composed of three different parts: these are intra wireless BAN, extra wireless BAN, and beyond wireless BAN, that is, there are three different tiers of architecture which constitute the WBAN topology [21]. In the first tier (intra-WBAN), communication occurs between sensor nodes and coordinator nodes, the sensors are variable and used to transmit physiological data to a PS (personal server), and the range of transmission is about 2 m. The sensor nodes should be able to perform any of the following: physiological measurements, actuator functionality, and electrical pulse conductivity. However, communication in the second tier (extra WBAN) occurs between coordinator nodes and single or multiple APs (access points), and the second tier communication aims to bridge data transmission between WBAN and other network devices [22]. Finally, communication between the coordinator nodes and medical servers constitutes the third tier (beyond WBAN), where PDA is applied in interconnecting the second and third tiers. The third tier is mostly employed in metropolitan areas [23]. A fourth tier system is proposed by [24] where a black-box made up of data collection, filtering, analyzing, and decision system is embedded between the second and third tiers (Figure 7.3).

The design of WBAN requires considering the following specifications to produce a robust, reliable, and secure system:

- **Wearability:** The size of the sensor node should be minute and its weight should be thin. These factors actually depend on the battery's weight and size, this is to say that the WBAN size and weight are dependent on the size of the battery and a battery's volume is directly proportional to its size. A lightweight small size WBAN is a wearable WBAN.

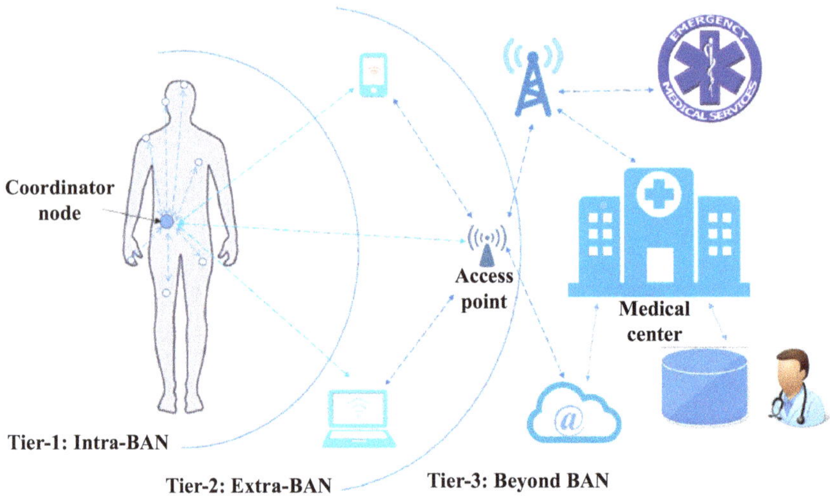

Figure 7.3 WBAN system

- **Security:** The applied communication technology should be well-secured to provide the best protection to sensitive data of clients. Also all levels of communication should have the best protection. This will eliminate the intrusion of confidential, private, and sensitive personal medical data of clients.
- **Communication:** The chosen communication technology should be reliable, and should have the desired capacity and the required speed. Different wireless connections are available for WBAN communication, for example, ZigBee, WLAN (wireless local area network), etc.
- **Interoperability:** Assembling of a powerful WBAN sensor should be done in the shortest possible duration and should meet the health specification of the client. WBAN sensor interoperability will promote market competition thereby driving the system cost down.
- **Latency:** Transmission of data in real-time devoid of late response is of critical importance in medical applications especially in cases of emergency. Hence, it's important to eliminate long or late response in data transmission.

7.4 Sensor nodes

The term ***node*** as applied in WBAN systems is a self-sufficient device which has communication functionality. Categorization of WBAN nodes is based on the following three specifications: (1) network role, (2) functionality, and (3) implementation. Based on network role classifications, nodes are grouped in relays, coordinators, and end nodes. Based on functionality classifications, nodes are

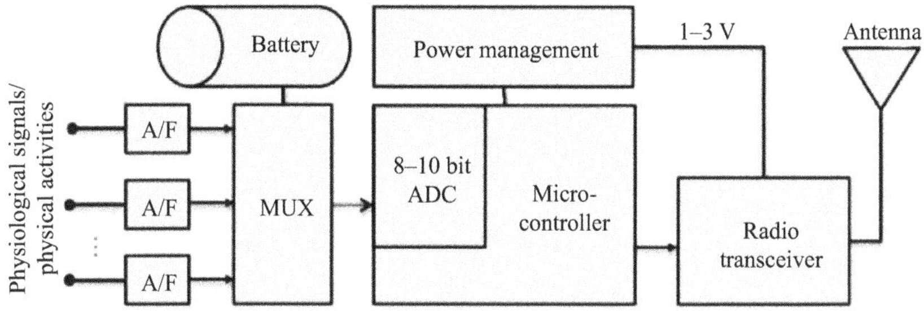

Figure 7.4 WBAN node architecture

grouped into the following three groups of actuators [25]: sensors [26,27] and personal devices [28] and finally, based on implementation mechanisms, classification of nodes are implant nodes, external nodes, and body surface nodes [29,30]. The majority of the sensor nodes located on the human body are power by batteries. High power consumption is a major drawback of these types of sensor nodes. A typical WBAN node architecture is illustrated in Figure 7.4. The physiological human data measured are weak and contain noise when an amplifier and a filter (A/F) are required to condition the data to the desired quality. Switching between multiple sensors is done by a multiplexer after which analog to digital converter (ADC) is employed for further signal conditioning. Signal coding for protection is done by the microcontroller, and this signal is then ready for transmission through the appropriate channel. There are two types of sensor nodes when they are classified based on the medium of transmission, and these are ultra-wide-band sensors and narrow-band sensors. Narrow-band frequencies are the first and most popular wireless communication technology which is commonly used by sensor nodes communication, and it operates at 2.4 GHz bandwidth. Sensor nodes are mostly implanted in everyday devices such as eyeglasses, fingerings, and earpieces because they have to unobtrusive [31–34].

7.5 Standards of WBAN

WBAN is an emerging technology and due to its vast areas of application has limited standards. The little standards available now were introduced recently and do not support all areas of WBAN applications. In some WBAN application areas, there are no set standards, and nonetheless, gradual steps are being taken to develop guidelines to meet the following standards of noninterference, low power consumption, reliability, and QoS (quality of service) [35]. The working group of IEEE has established a few standards for medical-based WBANs. These standards provided guidelines for wireless short-range communication of sensors within the parameter (inside or around) of the human body. In 2012, IEEE 802.15.6 was introduced to meet the standards mentioned above [36]. Most importantly some regulations were provided to help minimize interferences such as channel hopping, beacon shifting, and superframe interleaving [36]. Some existing IEEE standards are the IEEE 802.11 which sets the standards for high-speed communication, IEEE 802.15.1 which is responsible for the range of PAN, and IEEE 802.15.4 which is responsible for low power consumption and proximate range of the WBAN devices [37]. Table 7.3 shows the specifications of WBAN using the current IEEE 802.15.6 standard.

7.6 WBAN layers

The nodes in WBAN are divided into two layers: the PHY (physical) layer and the MAC (medium access control) layer. The goal of IEEE 802.15.6 is to set guidelines for PHY and MAC layer communications where lower power devices are used, and these layers of WBAN provide high QoS such as highly reliable, low complexity, short-range wireless communication, and extremely low power consumption. Also the guidelines permit reciprocity of information between PHY, MAC, and other layers and network management for NME (node management entity) and HME (hub management entity). IEEE 802.15.6 PHY is responsible for the following function activation of radio transceiver, deactivation of radio transceiver, CCA (clear channel assessment) inside the present channel, data reception, and data transmission. The type of WBAN determines the selection of PHY layer for both medical and nonmedical applications for on-body or in-body implementations. There are three different PHY layers set out by IEEE 802.15.6 and these layers are UWB (ultra-wide band), HBC (human body communication), and NB (narrow band). Communication between on-body and off-body devices and also communication between multiple on-body devices is enabled by the ultra-wide band layer. The prerequisites of EFC (electrostatic field communication) which envelopes packet structure and modulation SFD (start frame delimiter) are provided by PHY human body communication layer. The narrow band PHY layer oversees data reception and transmission of data, activation of radio transceiver, deactivation of radio transceiver, and CCA [20]. On the one hand, the MAC layer is designed above the physical layer to provide regulation of the channel access. The previous

Table 7.3 Specification of WBAN

Parameter	**WBAN**
IEEE standard	802.15.6
Node size	Mini is required
Density of node	Less 64
Node type	Heterogeneous
Node range	2 m–5 m
Node rate	1 Kbps–10 Mbps
Node power	30 mW
Technology	Wireless communication

MAC IEEE 802.15.4 version had some limitations such as absence of frequency hopping, undependable, restricted bandwidth, and limitless latency. However, an improved type referred to as MAC IEEE 802.15.4e was introduced in 2012 [38]. In the MAC layer, the coordinator segments the total channel into a series of superframe for time reference resource distribution. The coordinator is also responsible for allocating equal bacon periods and also correcting periods of beacon offsets [39]. Basically, the MAC layer should provide the following merits if properly implemented: minimum delay time, maximum channel efficiency, highly reliable, and flexible transmission system [40]. Three categories of MAC layers exist and these are polling, TDMA (time division multiple access), and contention-hinged protocols. Organization of the channel access which is the responsibility of the coordinator (hub) is achieved by any of the three solutions provided by [20].

7.7 Wireless connection

The current wireless communication technologies which are able to support the quick rollout of WBAN systems are reviewed in this section. There are multiple tiers in the WBAN architecture, and each tier requires different wireless communication technology because of factors such as security, range, and power consumption.

7.7.1 Bluetooth

Bluetooth technology is a WPAN system [41] which was developed for wireless communication in relatively short distances and it provides maximum security. Bluetooth technology enables concurrent communication of a single device with seven devices, that is, the single device functions as the master while the other seven devices function as the slaves, and the communication lasts for the period of the piconet lifespan. Bluetooth supports communications in all types of Bluetooth-enabled devices and also these devices do not require line-of-sight for connection purposes. A better version of Bluetooth known as BLE (Bluetooth Low Energy) [42,43] was designed in 2006 by Nokia to reduce significantly the power in idle state, easy device location, and maximum dependability in data transfer. Both technologies are suitable for WBAN application; however, Bluetooth low energy is most suitable especially in the first tier system because of low power. Interference may be a limitation in Bluetooth technology because it uses the 2.4 GHz bandwidth.

7.7.2 Zigbee and IEEE 802.15.4

Zigbee technology is one of the most commonly used wireless communication networks in WBAN systems together with Bluetooth. It's an improved version of IEEE 802.15.4 which allocates extra protocol layers such as security [44], network, and extra layers located above MAC layers. It possesses the following advantages: minimum data rate, increased battery life, well-protected network, and a longer period of operation. However, Zigbee has interference limitation because it shares the same 2.4 GHz bandwidth with WLAN which uses high power for transmission, and also the low data rate makes it unsuitable for transmitting large volumes of data. There are two types of Zigbee technology: Zigbee alliance and IEEE 802.15.4 [45]. There are other low power technologies such as ANT [46], RuBee [47], Sensium [48], Zarlink [49,50], Z-Wave [51], Insteon [52], Wavenis [53], BodyLAN [54], Dash7 [55,56], ONE-NET [57], and Enocean [28].

7.7.3 Wi-Fi

The Wi-Fi technology which is also referred to as WLAN (wireless LAN) is a data transfer or communication systems for terrestrial networks. WLAN also use 2.4 GHz bandwidth and its hinged on IEEE 802.11 criterion and variety of version such as 802.11a/b/g/n are available. The coverage areas of the 802.11g and 802.11b versions are 50 m for indoor applications and 100 m for outdoor applications [58] (Table 7.4).

7.8 Data delivery and intelligence in WBAN

There are three levels of wires communication WBAN systems: in the first level of wireless communication, nanodevices are interconnected via millimeter wireless communication; in the second level, sensors and nano devices are connected together to the gateway using wireless communication technologies such as Bluetooth, BLE, ZigBee, etc., while in the third level, the gateway is connected to permanent storage systems as clouds [59]. However, data delivery or transfer between wearables and third-party systems in the WBAN architecture are done in two forms: these are wearable data delivery and warehouse data delivery, and the latter form of data transfer has some limitations when compared to a wearable form of data transfer. Accessing data from these modes of data transfer is grouped into indirect and direct data access [60]. Several data delivery approaches have been implemented or reported in the literature, and an example is an energy-efficient algorithm which resolves the limitation of data transfer in Internet of Nano Things (IoNT) and also it guarantees higher levels of QoS. In WSN routing protocols (RPs), it has been proven by research that multipath RP is superior to single-path

Table 7.4. Medical sensor market globally

Sensor types	Sensor location	Sensor application	Final user
Pressure sensors	Implantable sensors	Wellness and fitness	Home care
Flow sensors	Wearable		Hospitals
Motion sensors	Wireless sensors	Therapeutics	Rehabilitation centers
Biosensors		Monitoring	Clinics
Temperature sensors		Diagnostics	Diagnostic centers
Image sensors			

RP [61,62]. Some examples of these RPs are SAR [63], ReInForM [64], GAF [65], GEAR [66], etc. Artificial intelligence provides smart ways of doing things thereby improving efficiency, with emphasis on WBAN systems, and an overview of machine learning-based WBAN intelligent system is reviewed in [61].

7.9 Conclusion

WBAN is one the fastest emerging technologies which when applied in healthcare service improves healthcare delivery thereby increasing the lifespan of patients. As an emerging technology, IEEE 802.15.6 provides the guidelines for the WBAN standards: some of which are low power consumption, non-interference, data rate, security, wireless communication, and quality of service. In this research, a review of WBAN with respect to medical applications, architectures, WBAN nodes, WBAN standards, WBAN layers, and acceptable wireless communication technologies are investigated. Some drawbacks of WBAN are power consumption, security, sensor deployment period, expandability, and sensor density. Improving the above WBAN limitations will drastically revolutionize healthcare experiences.

References

[1] Jain P. C. "Wireless body area network for medical healthcare." IETE Technical Review. 2011; 28(4): 362–371.

[2] Akyildiz I. F., Su W., Sankarasubramaniam Y., and Cargici E. "Wireless sensor networks: A survey." Computer Networks. 2002; 38(4): 393–422.

[3] Stephan T., Al-Turjman F., Joseph K., Balusamy B., and Srivastava S. "Artificial intelligence inspired energy and spectrum aware cluster based routing protocol for cognitive radio sensor networks." Journal of Parallel and Distributed Computing. 2020; 142(1): 90–105.

[4] Tobón D. P., Falk T. H., and Maier M. "Context awareness in WBANs: A survey on medical and non-medical applications." IEEE Wireless Communications. 2013; 20(4): 30–37.

[5] Vallejos de Schatz, C. H., Medeiros, H. P., Schneider, F. K., and Abatti, P. J. "Wireless medical sensor networks: Design requirements and enabling technologies." Telemedicine and e-Health. 2012; 18(5): 394–399.

[6] Dam K. Van, Pitchers S., and Barnard M. "Body area networks: Towards a wearable future," *Proceedings of WWRF Kick Off Meeting*, Munich, Germany, March 6–7, 2001.

[7] Poon C. C. Y., Lo B. P. L., Yuce M. R., Alomainy A., and Hao Y. "Body sensor networks: In the era of big data and beyond." IEEE Reviews in Biomedical Engineering. 2015; 8: 4–16.

[8] Viera M. A. M., Choelo C. N. J., da Silva D. C. J., and da Mata J. M. "Survey on wireless sensor networks devices." *Proceedings of ETFA-03*, Lisbon, Portugal, September 16–19 2003.

[9] Park C., and Chou P. "Ambimax: Autonomous energy harvesting platform for multi-supply wireless sensor nodes." SECON. 2006; 168–177.

[10] Jiang X., Polastre J., and Culler D. "Perpetual environmentally powered sensor networks." IPSN. 2005; 463–468.

[11] Dutta P., Hui J., Jeong J., et al. "Enabling sustainable and scalable outdoor wireless sensor network deployments." IPSN, 2006; 407–415.

[12] Lin K., Yu J., Hsu J., et al. "Enabling long-lived sensor networks through solar energy harvesting." *Proceedings of the 3rd ACM Conference on Embedded Networked Sensor Systems (SenSys)*. 2005, 309.

[13] Latrc B., Braem B., Moerman J., Blondia C., and Demeester P. "A survey on wireless body area networks." Wireless Network. 2011; 17: 1–18.

[14] Bradai N., Chaari L., and Kamoun L. "A comprehensive overview of wireless body area networks (WBAN)." International Journal of E-Health and Medical Communications. 2011. 2: 1–30.

[15] Sodagari S., Bozorgchami B., and Aghvami H. "Technologies and challenges for cognitive radio enabled medical wireless body area networks." IEEE Access. 2018; 6: 29567–29586.

[16] Ammari M. "Wireless sensor networks: Current status and future trends." *Review of applications of wireless sensor networks*. Boca Raton, Florida: CRC Press; 2013. pp. 3–31, Ch. 1.

[17] Wood A. D., Stankovic J. A., Virone G., Selavo L., He Z., and Cao Q. "Context-aware wireless sensor networks for assisted living and residential monitoring." IEEE Network. 2008; 22(4); 26–33.

[18] Hadjidj A., Souil M., Bouabdallah A., Challal Y., and Owen H. "Wireless sensor networks for rehabilitation applications: Challenges and opportunities." Journal of Network and Computer Applications. 2013; 36(1).

[19] van den Broek E. L. and Westerink J. H. D. M. "Biofeedback systems for stress reduction: Towards a bright future for a revitalized field." *Proceedings of International Conference on Medical and Health Informatics*. February 2012, pp. 499–504.

[20] Haddad O., and Khalighi M. A. "Enabling communication technologies for medical wireless body-area networks." *Global LIFI Congress (GLC)*. Paris, France, June 2019.

[21] Movassaghi S., Abolhasan M., Lipman J., Smith D., and Jamalipour A. "Wireless body area networks: A survey." IEEE Communications Surveys & Tutorials. 2014; 16(3): 1658–1686.

[22] Chen M., Gonzalez S., Vasilakos A., Cao H., and Leung V. "Body area networks: A survey." Mobile Networks and Applications. 2011; 16: 171–193.

[23] Latré B., Braem B., Moerman I., Blondia C., and Demeester P. "A survey on wireless body area networks." Wireless Network. 2011; 17: 1–18.

[24] Ghamari A., Janko B., Sherratt S., Harwin W., Piechockic R., and Soltanpur C. "A survey on wireless body area networks for eHealthcare systems in residential environments." Sensors. 2016; 16(16): 831.

[25] Wang S. and Park J.-T. "Modeling and analysis of multi-type failures in wireless body area networks with semi-Markov model." Communications Letters. 2010; 14: 6–8.

[26] Latr´e B., Braem B., Moerman I., Blondia C., and Demeester P. "A survey on wireless body area networks." Wireless Network. 2011; 17: 1–18.

[27] Hanson M., Powell H., Barth A., et al. "Body area sensor networks: Challenges and opportunities." Computer. 2009; 42: 58–65.

[28] EnOcean 2015. Available from https://www.enocean.com/en/home/ [Accessed December 2, 2019].

[29] Xing J. and Zhu Y. "A survey on body area network." *5th International Conference on Wireless Communications, Networking and Mobile Computing (WiCom '09)*. September 2009, pp. 1–4.

[30] Yazdandoost K. Y. and Sayrafian-Pour K. "Channel model for body area network (BAN)." Networks. 2009; 91.

[31] Yang B. -H. and Rhee S. "Development of the ring sensor for healthcare automation." Robotics and Autonomous Systems. 2000; 30: 273–281.

[32] Asada H. H., Shaltis P., Reisner A., Rhee S., and Hutchinson R. C. "Mobile monitoring with wearable photoplethysmographic biosensors." IEEE Engineering in Medicine and Biology Magazine. 2003; 22(3): 28–40.

[33] Atallah L., Aziz O., Gray E., Lo B., and Yang G. Z. "An ear-worn sensor for the detection of gait impairment after abdominal surgery." Surgical Innovation. 2013; 20: 86–94.

[34] Zheng Y., Leung B., Sy S., Zhang Y., and Poon C. C. Y. "A clip free eyeglasses-based wearable monitoring device for measuring photoplethysmographic signals." *Proceedings of the Annual International Conference of the IEEE Engineering in Medicine and Biology Society*. 2012, pp. 5022–5025.

[35] Salehi S. A., Razzaque M., Tomeo-Reyes I., and Hussain N. "IEEE 802.15.6 standard in wireless body area networks from a healthcare point of view." *Proceedings of the 22nd Asia-Pacific Conference on Communications (APCC)*, 2016, pp. 523–528.

[36] Velusamy B., and Pushpan S. C. "An enhanced channel access method to mitigate the effect of interference among body sensor networks for smart healthcare." IEEE Sensors Journal. 2019; 19(16): 7082–7088.

[37] Lal C., Laxmi V., and Gaur M. S. "Video streaming over MANETs: Testing and analysis using real-time emulation." IEEE 2013 19th Asia-Pacific Conference on Communications (APCC), 2013, pp. 190–195.

[38] De Guglielmo D., Anastasi G., and Seghetti A. "From IEEE 802.15.4 to IEEE 802.15.4e: A step towards the Internet of Things." *Advances in intelligent systems and computing.* Cham, Switzerland: Springer; 2014. pp. 135–152.

[39] Kwak K., Ullah S., and Ullah N. "An overview of IEEE 802.15.6 standard." *3rd International Symposium on Applied Sciences in Biomedical and Communication Technologies (ISABEL)*. November 2010, pp. 1–6.

[40] Ullah S., Higgins H., Braem B., et al. "A comprehensive survey of wireless body area networks on PHY, MAC, and network layers solutions." Journal of Medical Systems. 2012; 36: 1065–1094.
[41] Bluetooth. 2015. Available from http://www.bluetooth.com/Pages/Bluetooth-Home.aspx [Accessed December 2, 2019].
[42] Bluetooth SIG. "Bluetooth low energy." 2015. Available from http://www.bluetooth.com/Pages/lowenergy-tech-info.aspx [Accessed June 4, 2016].
[43] Alam, M. M., and Hamida, E. B. "Strategies for optimal mac parameters tuning in IEEE 802.15. 6 wearable wireless sensor networks." Journal of Medical Systems. 2015; 39(9): 1–16.
[44] Al-Turjman F. "Intelligence and security in big 5G-oriented IoNT: An overview." Elsevier Future Generation Computer Systems. 2020; 102(1): 357–368.
[45] Acampora, G., Cook, D. J., Rashidi, P., and Vasilakos, A. V. A survey on ambient intelligence in healthcare. *Proceedings of the IEEE*. 2013; 101(12): 2470–2494.
[46] ANT+. 2015. Available from http://www.thisisant.com/ [Accessed December 2, 2019].
[47] RuBee. 2015. Available from http://www.rubee.com/Techno/index.html [Accessed December 2, 2019].
[48] Toumaz Group. 2015. Available from http://www.toumaz.com/ [Accessed December 2, 2019].
[49] Microsemi. 2015. Available from http://www.microsemi.com/ [Accessed December 2, 2019].
[50] Zarlink Semiconductor. ZL70101 Medical Implantable RF Transceiver. 2009. Available from http://ulp.zarlink.com/zarlink/zweb-zl70101-datasheet-dec09.pdf [Accessed December 2, 2019].
[51] Z-Wave. 2015. Available from http://www.z-wave.com [Accessed December 2, 2019].
[52] Insteon. 2015. Available from http://www.insteon.com/ [Accessed December 2, 2019].
[53] Elster. Wavenis Technology. 2015. Available from http://www.elstermetering.com/en/wavenistechnology [Accessed December 2, 2019].
[54] FitLinxx. BodyLANWireless Protocol. 2015. Available from http://www.fitlinxx.net/bodylan-wirelessprotocol.htm [Accessed December 2, 2019].
[55] DASH7 Alliance.2015. Available from http://www.dash7-alliance.org/ [Accessed December 2, 2019].
[56] Weyn, M., Ergeerts, G., Wante, L., Vercauteren, C., and Hellinckx, P. "Survey of the DASH7 alliance protocol for 433 MHz wireless sensor communication." International Journal of Distributed Sensor Networks. 2013.
[57] One-Net. 2015. Available from http://www.one-net.info/ [Accessed December 2, 2019].
[58] IEEE standard 802.11-2007. "Wireless local area network medium access control (MAC) and physical layer (PHY) specifications." Available from

http://standrards.ieee.org/getieee802/download/802.11-2007.pdf [Accessed June 12, 2007].

[59] Al-Turjman F., and Baali I. "Machine learning for wearable IoT-based applications: A survey." Transactions on Emerging Telecommunications Technologies. 2019.

[60] de Arriba Pérez F., Caeiro Rodríguez M., and Santos Gago J. M. "Collection and processing of data from wrist wearable devices in heterogeneous and multiple-user scenarios." Sensors. 2016; 16(9): 1538.

[61] Al-Turjman F. "A cognitive routing protocol for bio-inspired networking in the internet of Nano-Things (IoNT)." Mobile Networks and Applications. 2017; 1–15.

[62] Al-Turjman F. "A rational data delivery framework for disaster-inspired internet of Nano-Things (IoNT) in practice." Springer Cluster Computing. 2019; 22(Suppl. 1): 1751–1763.

[63] Sohrabi K., Gao J., Ailawadhi V., and Pottie G. J. "Protocols for self-organization of a wireless sensor network." IEEE Personal Communications. 2000; 7: 16–27.

[64] Deb B., Bhatnagar S., and Nath B. "ReInForM: Reliable information forwarding using multiple paths in sensor networks." *Proceedings of IEEE Conference on Local Computer Networks.* Germany: Bonn/Konigswinter; 2003. pp. 406–415

[65] Hasan M., Al-Rizzo H., and Al-Turjman F. "A survey on multipath routing protocols for QoS Assurances in real-time multimedia wireless sensor networks." IEEE Communications Surveys and Tutorials. 2017. https://doi. org/ 10.1109/COMST.2017.2661201

[66] Singh S. K., Singh M. P., and Singh D. K. "Routing protocols in wireless sensor networks: A survey." International Journal of Computer Science and Engineering Survey (IJCSES). 2010; 1(2).

Chapter 8

Toward a smart hospital room and automated systems

Mohamad Bassl Alramli[1], Mohamad Dib[1], Mohammad Amrou Dib[1], Hussam Macha Alghazalat[1], Mubarak Mustapha[1,2], Fadi Al-Turjman[2,3,4], Ilker Ozsahin[1,4] and Dilber Uzun Ozsahin[1,4]

The healthcare industry is at the cutting edge of voice-user interface (VUI) design and making great progress to improve patient care and comfort through developing a hospital room equipped with state of art automated systems aim at simplifying life, saving energy, and time. This study's objective is to design an automated hospital voice recognition system that can utilize the patient's voice to control room appliances such as fan, light, etc. The design components used include Arduino, microphone, fan, LEDs, lamp, relay, servo motor, 12 V DC battery, etc. Also, the system will provide a cost-effective approach to curb the rising number of personnel needed in the hospital. After a series of developmental testing, the prototype was deployed to a real-world testing and a successful recognition rate of the movement and the verification command were obtained. Finally, the system was designed and implemented in a cost-effective manner.

8.1 Introduction

The advantages of the voice-user interface (VUI) extend far beyond simple conveniences for patients; it has a profound impact on healthcare improvement [1,2]. Just like a person, a well-designed VUI can use the tone of voice and other elements in conversation to shape behaviors or calm nerves. Because of the high increase in the number of inpatients in the hospitals, there becomes an urgent need to engage numerous health personnel to help with the basic necessity and comfort in the various rooms [3]. This can be a tedious and capital-intensive endeavor for

[1]Department of Biomedical Engineering, Near East University, Nicosia / TRNC, Mersin-10, Turkey
[2]Department of Artificial Intelligence, Near East University, Nicosia / TRNC, Mersin-10, Turkey
[3]Research Center for AI and IoT, Near East University, Nicosia / TRNC, Mersin-10, Turkey
[4]DESAM Institute, Near East University, Nicosia / TRNC, Mersin-10, Turkey

both the patient and the hospital management, and one of the promising ways to provide comfort in patients' room at a cost-effective manner is the use of hospital automation system (HAS) [4].

Hospital room automation is gaining more attention both from academia and industry [5–9]. HAS gives a great research opportunity in creating a new field of computing and engineering [10–13]. HAS comprises automated lighting, security lock, fan, bed, etc., a control system that aims at providing comfort to patients and reduces work for hospital personnel [10,12,13]. HAS is becoming prevalent and the market for such a system is increasing. Thus, end-users (patients) find this system to be very effective and efficient. Presently, computerization and automation in all fields are vital. A hospital room or home equipped with state of art automated systems aims at simplifying life. Apparently, the aim of improving technologies is to provide and protect patients with the necessary comfort. Thus, incorporating intelligent equipment into the hospital room simplifies life and saves energy and time. An intelligent hospital room is equipped with several power management systems and various monitoring modes [14,15]. This study aims at designing an automated hospital voice recognition system that can utilize the voice of the patient in controlling the room appliances such as fan, light, etc.

Nowadays, some patients find it difficult to control the appliances (fan, light, etc.) and bed in their rooms. These are a few of the challenges facing engineers most especially the biomedical engineers. However, HAS offers a present-day approach in which a sick person controls the hospital room via an automated design system [16]. The importance of HAS is becoming vital especially when dealing with patients who are physically challenged. In most cases, technology improvement is aimed at providing and protecting people with all the necessary comfort. To this end, this study's main objective is to design a system that will control the hospital room appliances that include light, fan, and bed with the aim of easing utilization using voice recognition rather than hand control.

8.2 Literature review

Several types of research have been conducted on the design and implementation of home/hospital automation system. The authors in [17] developed a home automation system (HAS) that utilizes Bluetooth, remote controller, and microcontroller. They used the Bluetooth module in providing a wireless-based control and the remote controller was used to send the signal to the wireless control. Also, the authors in [18] designed a system that utilized PC in converting voice commands into text and sending the text to mobile phone via a cellular network. The text is read by a microcontroller at the receiving end via another phone. Similarly, authors in [19] developed an automated system that can control devices from distance. Wi-Fi is used to send the voice command after converting them into symbols. The voice recognition system was developed on a PC via Microsoft Visualbasic.net. The control circuit is used to transfer information when the command is given through the PC parallel port.

Furthermore, the authors in [20] reviewed the home automation system using voice recognition. The study highlighted various means by which home automation is done. Authors in [21] studied home automated voice recognition for paralyzed people. The authors designed and implemented a low-cost system for physically challenged persons suffering from paraplegia so that they can control home appliances. The system comprises Arduino, voice recognition module, relay, and adjustable bed. Similarly, authors in [4] designed voice recognition for the advanced patient room. The proposed system is designed in a way that it can assist patients in the hospital room without requiring hospital attendants.

8.3 Methodology

The design of the system was done analytically by selecting components based on their manufacturer's datasheet and theoretical background. Some of the components were chosen based on their design output and the system's general design compatibility. However, the design was done taking into consideration the system stability, application, and best efficiency attainable. Also, the design phase considered components that can be replaced easily with less cost.

8.3.1 Circuit design

The aim of this study is to design and implement a HAS. Servo motor, voice recognition module, microphone, Arduino, fan, power supply unit, lamp, voltage regulator, and relay were used in the design of the system. Properly connected components in a suitable and stable manner lead to output with continuous out produced. The voice recognition module is connected to the Arduino via pin 12 and 13 while the microphone is attached to the voice recognition module. The servo motor is connected to the Arduino via pin 9 and also to 5 V voltage source. The relay in the circuit as shown in Figure 8.1 is connected to the Arduino via pin 5. The voice recognition module gets its power supply from 5 V supply. LED 1 and LED 2 are connected to the Arduino via pin 7 and 6, respectively, while the lamp is connected to the Arduino via pin 8. The 12 V source in the circuit is connected to the fan.

In designing this system, the study was divided into three blocks, namely:

Voice recognition module
Arduino mega
Power supply circuit

8.4 System design

This section will present the study design that includes the voice recognition module, Arduino mega, and power supply circuit.

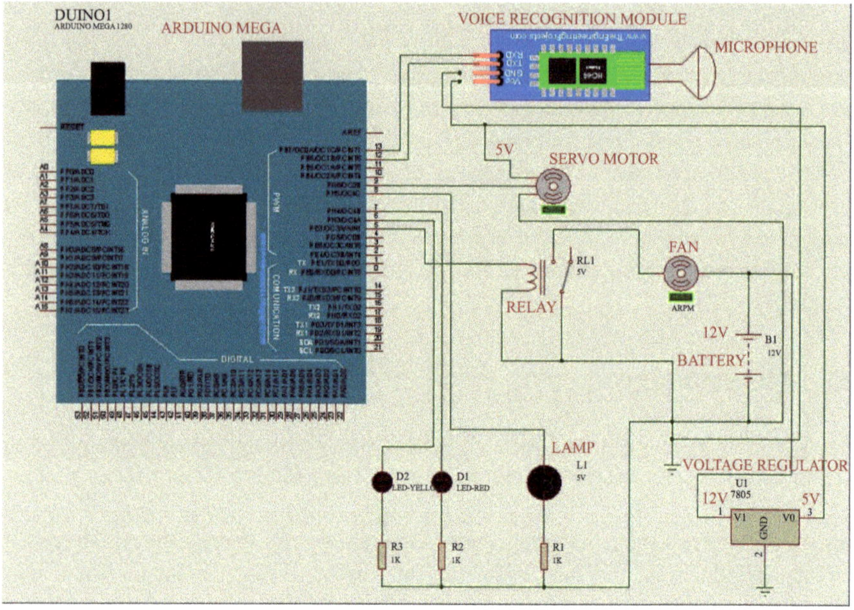

Figure 8.1 Circuit diagram

8.4.1 Voice recognition module

As shown in Figure 8.2, the train voice recognition module is presented. The voice command is transferred through the microphone in Figure 8.3 to the voice recognition module. For a voice to be actually recognized by the voice command, the voice recognition module has to be trained. The input speed from the microphone is given to the voice recognition module and then compared with the stored voice and if the voice matches then the control action via the circuit control is taken. Thus, this device can store about 80 commands but only 7 commands can be loaded into the recognizer for the process of recognition.

The train voice recognition module sends Sigtrain in order to train record 0 with the signature on when "speak now" appears on the serial monitor. This voice can be anything such as "on." Thus, when the "speak again" appears for the second time, the voice needs to be repeated again. If the two voices match, "success" is printed on the serial monitor and 0 is recorded. But when the voices are not matched, the voice has to be repeated until success. However, a signature describes the voice command [22].

If your 7 voice commands are "1, 2, 3, 4, 5, 6, 7," you could train in the following way:

Sigtrain 0 OnL, Sigtrain 1 OffL, Sigtrain 2 OnF, Sigtrain 3 OffF, Sigtrain 4 MB45, Sigtrain 5 MB90, and Sigtrain 6 MB180.

Where OnL means "on light," OffL means "off light," OnF means "on fan," OffF means "off fan," MB45 means "move bed to 45 degrees." MB90 means

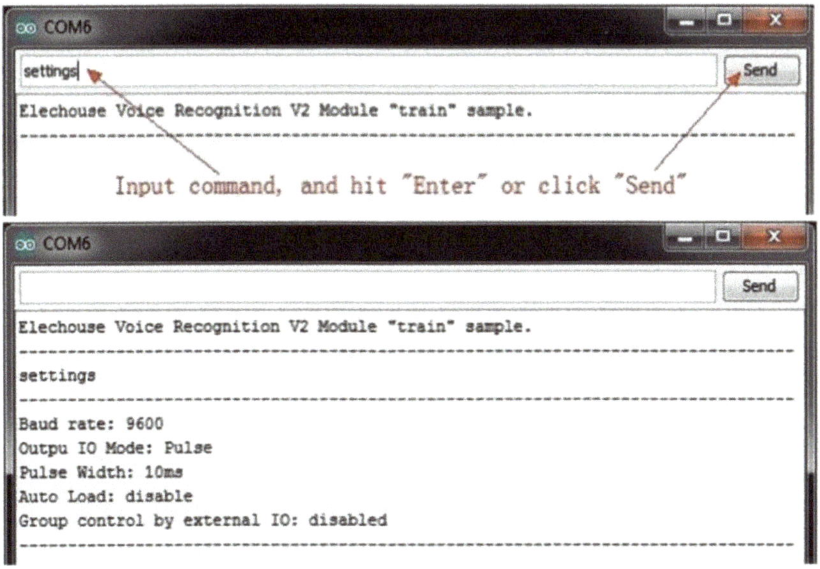

Figure 8.2 Train voice recognition module

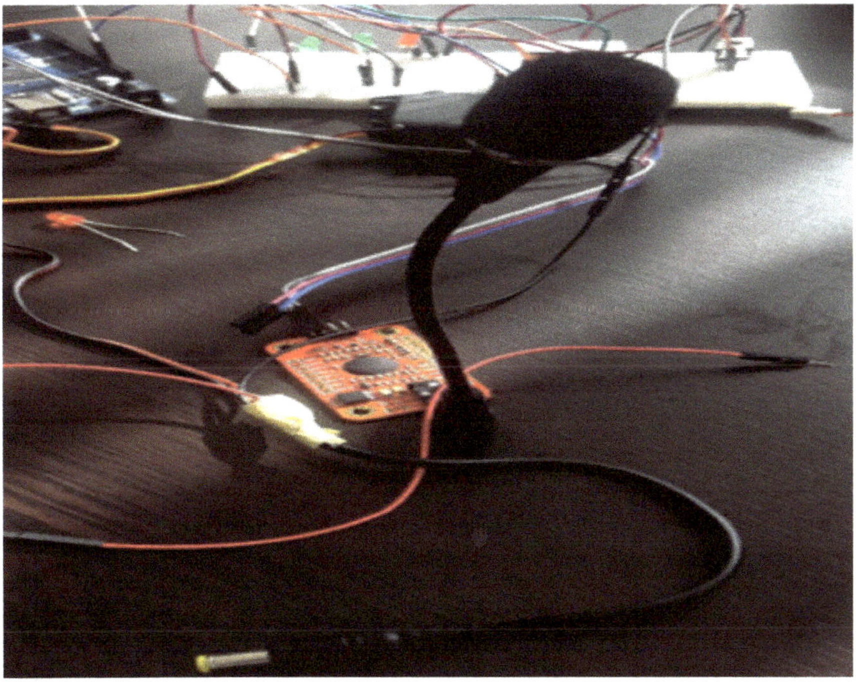

Figure 8.3 Construction showing the microphone

Figure 8.4 Construction showing the connection to the Arduino

"move bed 90 degrees," and MB180 means "move bed 180 degrees." If the command is called, the signature is displayed.

The two LEDs on the VRM can indicate the training process during training. The (yellow) SYS LED blinks to alert you to get ready. The voice command is set when the red (LED) light is turned on. As soon as the red light goes off, the recording process ends. When the SYS LED blinks, it indicates get ready for the next recording process. The status LED and SYS LED blink at the same time as soon as the training process successfully ends. And when training fails, the status LED AND SYS LED quickly blink at the same time.

8.4.2 Arduino mega

The proposed system controller is Arduino Uno microcontroller as shown in Figure 8.4. This device was chosen due to its cost-effectiveness and easy to use. The device has an integrated development environment (IDE) that can run on a computer.

8.4.3 Power supply circuit

The circuit needs a good power supply in order to give the required voltage. Low voltage (DC) is used as a power source in most of the electronic systems. The power unit gives the required voltage supply. Thus, the power unit needed to power the HAS in this study is a 12 V voltage supply.

8.5 Discussions

Designing an automated smart hospital room has not only made life comfortable for patients admitted in the hospital, it has also open ways for newer technology in ensuring that the best treatment can be provided. Decades ago, no one believed there will be an artificial/information age but here we are open possibilities beyond human thinking.

The system design in this study provides comfort on another level to the patient in the absence of a doctor, nurse, or other allied health workers. Patients can now adjust the shade of their room, switch off the bulb of the room, adjust the position of the bed, and so on without having to call anyone. This comfort psychologically aids the speedy recovery of the patient and in the end improves healthcare delivery. It is often argued by several professionals that this is going to cause unemployment at an unimaginable scale but what they do not understand is the newer jobs it is going to create.

Also, the result of the study if implemented will save the hospital the cost of recruiting and training new staff. The introduction of artificial intelligence into hospitals will also fear of losing a patient to resignation, holiday, or sick leave as machines only need maintenance and can function all year round. The money allocated for this can be spent in another area where it is much more needed.

The whole construction and testing of the system are in five phases with the aim of designing an automated hospital room system in a cost-effective manner. The second phase deals with buying the components and setting out the system construction. The third phase has to do with components arrangements on the breadboard and final soldering. The fourth phase is training for voice recognition while the final phase is about system test of each component for design specification and conformity.

After conducting a tedious market survey and purchase of the required components, construction began in stages described below.

8.5.1 Breadboard layout

This is used to arrange and test the circuit. The breadboard as shown in Figure 8.5 has several holes in which the circuit components can be inserted. It has metal strips that run underneath the board and connect the holes on the board. Thus, during the construction phase of this device, the components' legs were inserted in the hole with each set of the hole connected by metal strips underneath forming a node. The connections between different components are done by putting all the legs to a common node. Jumper wires are also used in connecting some of the components. Finally, with the necessary

Figure 8.5 Construction showing breadboard connections

connections, the circuit is designed and tested while the required responses were derived.

8.5.2 Soldering

A well-ventilated area was used during the soldering process of the system. The wires from each component were connected together and soldered. At the end of the soldering, the lips were cut in order to avoid a short circuit. The following factors were put into consideration when choosing the components used:

Sound mechanical support
Ease of installation
Electrical earthen capability
Thermal conductivity

After the tedious design, a final phase which deals with system test of each component for design specification and conformity. The system was ready as depicted in the following figures.

8.5.3 Testing

To test the continuity of the system, a multimeter as shown in Figure 8.6 was used [23,24]. Thus, each individual component was tested. It should be noted that the

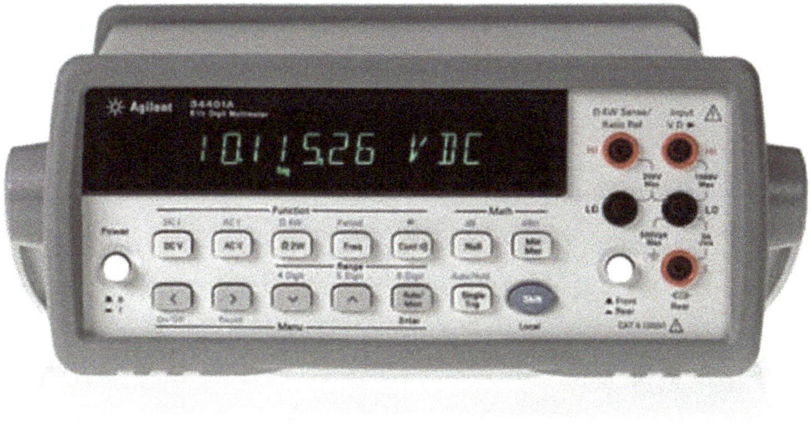

Figure 8.6 A digital multimeter

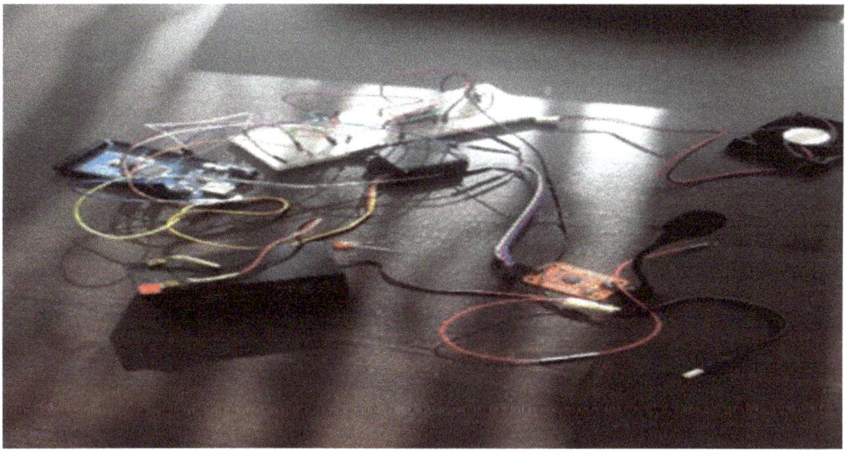

Figure 8.7 Construction showing all the components

test was done when the power supply unit was in the off-stage. The testing gives the status of all the components as shown in Figure 8.7 used in the study. Furthermore, the oscilloscope test was done when the power was supplying to the circuit. However, probes connections were done in a good manner and the pulse test was done at the output of the decade counter to evaluate the output sequences. After all these mentioned tests, the results were obtained and the circuit was assembled.

As presented in Table 8.1, the system cost implication shows that the automated hospital room system is cost-effective.

Table 8.1 Components cost implications

Components	Quantity	Price ($)
Voice recognition module	1	75
Arduino mega	1	30
Breadboard	1	5
DC fan	1	10
DC lamp	1	10
Servo motor	1	10
LEDs	2	2
Resistors	2	1
Battery	1	25
USB TTL	1	25
5 V DC relay	2	
5 V voltage regulator	1	
Total		

8.6 Conclusion

The main aim of this study is to design, construct, and test hospital room automated system. Although some problems were encountered during the design and construction phase, the device test shows that the hospital room automated system is effective and can be used to reduce the stress in operating the hospital room appliances and managing the hospital as a whole. If implemented, the result can be effective in reducing the cost of employing and training new staff, and running the hospital. This fund can be diverted into other areas of the hospital where they will be more effective. Also, the test results show that the device will reduce stress to both patients and hospital personnel. However, there are indications that some healthcare personnel with menial jobs might lose their jobs. This idea is not acceptable by the labor and will, therefore, be debated. In the end, the importance outweighs the disadvantages.

References

[1] Spyropoulos, B. "Smart hospital-room and modern photonics emerging clinical reality based on optical systems." *Optics*. 2018; 7(1): 18.

[2] Farag, W. "ClimaCon: An autonomous energy efficient climate control solution for smart buildings." *Asian Journal of Control*. 2016; 19(4): 13751391.

[3] Al-Refaie, A., Chen, T., and Judeh, M. "Optimal operating room scheduling for normal and unexpected events in a smart hospital." *Operational Research*. 2016; 18(3): 579–602.

[4] Singh, T., and Yadav, N. "Voice recognition based advance patient's room automation." *IJRET: International Journal of Research in Engineering and Technology*. 2015; 4: 308–310.

[5] Akaboshi, K. "S1-3. Present status of smart rehabilitation room in Shonan Keiiku hospital." *Clinical Neurophysiology*. 2019; 130(10): e188.

[6] Reddy, S. "Smart health for smart cities." *International Journal of Research Foundation of Hospital and Healthcare Administration*. 2016; 4(2): 0–0.

[7] Al-Turjman, F., Zahmatkesh, H., and Mostarda, L. "Quantifying uncertainty in Internet of Medical Things and big-data services using intelligence and deep learning." *IEEE Access*. 2019; 7(1): 115749–115759.

[8] Al-Turjman, F., Ulusar, U., and Nawaz, M. "Intelligence in the Internet of Medical Things era: A systematic review of current and future trends." *Elsevier Computer Communications Journal*. 2020; 150(15): 644–660.

[9] Al-Turjman, F., and Alturjman, S. "Context-sensitive access in Industrial Internet of Things (IIoT) healthcare applications." *IEEE Transactions on Industrial Informatics*. 2018; 14(6): 2736–2744.

[10] Koca, M., Gulhan, Y., and Yilmaz, S. In terms of hospital management, employee perception of hospital automation system. *PressAcademia*. 2017; 3(1): 770–782.

[11] Al-Turjman, F. "Intelligence and security in Big 5G-oriented IoNT: An overview." *Elsevier Future Generation Computer Systems*. 2020; 102(1): 357–368.

[12] Al-Turjman, F. "A rational data delivery framework for disaster-inspired Internet of Nano-Things (IoNT) in practice." *Springer Cluster Computing*. 2019; 22(Suppl. 1): 1751–1763.

[13] Al-Turjman, F. "A cognitive routing protocol for bio-inspired networking in the Internet of Nano-Things (IoNT)." *Springer Mobile Networks and Applications*. 2017. DOI: 10.1007/s11036-017-0940-8.

[14] Garbey, M., Joerger, G., Huang, A., *et al.* "An intelligent hospital operating room to improve patient health care." *Journal of Computational Surgery*. 2015; 2(1). https://doi.org/10.1186/s40244-015-0016-7

[15] Sanborn, M., and Cohen, T. "Get smart: Effective use of smart pump technology." *Hospital Pharmacy*. 2009; 44(4): 348–353.

[16] Mowad, M. A. E. L., Fathy, A., and Hafez, A. "Smart home automated control system using android application and micro-controller." *International Journal of Scientific & Engineering Research*. 2014; 5(5): 935–939.

[17] Jawarkar, N. P., Ahmed, V., and Thakare, R. D. (2007, February). "Remote control using mobile through spoken commands." *2007 International Conference on Signal Processing, Communications and Networking*. Chennai, 2007; 622–625. DOI: 10.1109/ICSCN.2007.350684.

[18] Mardiana, B., Hazura, H., Fauziyah, S., Zahariah, M., Hanim, A. R., and MK, N. S. "Homes appliances controlled using speech recognition in wireless network environment." *International Conference on Computer*

Technology and Development. Kota Kinabalu, 2009; 285–288. DOI: 10.1109/ICCTD.2009.178.

[19] Kamdar, H., Karkera, R., Khanna, A., Kulkarni, P., and Agrawal, S. "A review on home automation using voice recognition." *International Research Journal of Engineering and Technology*. 2017; 4(10): 1795–1799.

[20] Kumar, M., and Shimi, S. L. "Voice recognition based home automation system for paralyzed people." *International Journal of Advanced Research in Electronics and Communication Engineering (IJARECE)*. 2015; 4(10): 2278909.

[21] Shen, W. Voice recognition module V3. Speak to control (Arduino compatible). May 9, 2014.

[22] Newman, K., and Blei, M. "Evaluation of smart phones for remote control of a standard hospital room." *Wireless Personal Communications*. 2013; 75(2): 1005–1013.

[23] Agustin, E., Yunardi, R., and Firdaus, A. "Voice recognition system for controlling electrical appliances in smart hospital room." *TELKOMNIKA (Telecommunication Computing Electronics and Control)*. 2019; 17(2): 965.

[24] Zhang, P., Li, F., and Bhatt, N. "Next-generation monitoring, analysis, and control for the future smart control center." *IEEE Transactions on Smart Grid*. 2010; 1(2): 186–192.

Chapter 9

Security issues in wireless medical sensor networks

Pranshu Dhingra[1], *Gayathri Nagasubramanian*[2], *Rakesh Kumar Sakthivel*[2] *and Ramesh Chandran*[3]

9.1 Introduction

9.1.1 Emergence of WMSNs

The unfolding of various scope in wireless medical sensor networks (WMSNs) in not only the medical domain but also its use in smart buildings, transport and logistics, security and surveillance, etc. is growing in the industry expeditiously. It also had many pragmatic impacts including convenience in medical access, cost reduction, and timeliness of delivering service even to people in remote areas. It is expanding gradually and has emerged as a new paradigm for gathering information by the collaborative efforts of the various self-organized sensing nodes. Healthcare is the most promising field growing and expanding its scope at a high rate driven by the motivation to ameliorate the medical industry's efficiency and accuracy. It's increasing applications in wearable sensors (which can help to detect vital signs) and the location tags can help in tracking the health condition and the overall status of the patient along with his location in real-time. With an increase in expectancy of human life, the medical healthcare domain is expanding to provide all means of healthcare facilities to even the people residing in remote areas. With the increase in expectations of medical services and the lack of professionals in this field, the scope of expanding the concept of "eHealth" would be quite favorable as a whole.

This emerging technology of wireless sensor networks (WSNs) has the propensity to improve human health and hence his life expectancy. Thus, this technology of "wireless medical sensor networks" is considered as one of the primary and salient research areas providing significant information and such mechanisms of alerting against certain odd conditions, it is made for detecting so as to alert the

[1]Subject Matter Expert-Statistics, NCR Eduservices Pvt. Ltd., Noida, Uttar Pradesh, India
[2]School of Computing Science and Engineering, Galgotias University, Uttar Pradesh, India
[3]Bannari Amman Institute of Technology, Sathyamangalam, Tamil Nadu, India

health care-taker about any odds detected. This minimizes the need for the number of care-takers and helps in efficiently distributing time among various patients. It also helps in providing quality care to babies and children whose both parents are working and to the elderly too with real-time monitoring. Any emergency or unanticipated situation can be handled nimbly with the help of this transpiring technology [1].

While these networks help in real-time analysis and efficient monitoring of the patients, it is wholly based upon the collection of data from these sensors which are deployed in quite accessible areas which can increase the likelihood of eavesdropping by unauthorized and dangerous people. Thus, leveraging the potential of WSNs in the healthcare domain requires facing of multitude of technical challenges including limited network capacity, scarce energy reserves, processing constraints, memory constraints, power and communication constraints as well as privacy and security issues. Maintaining the privacy and security of the patient's data collected by the sensors is the main concern which needs to be resolved. The patient may have some personal concerns or some sensitive disease which he might be very conscious about. Thus, stringent requirements on the quality of service and system reliability are needed along with maintaining the privacy of the patient's data. These requirements are quite diverse and may range from pre-hospital, home-monitoring, ambulatory, or even in-hospital to long-term collection of data in the databases for analyzing the trends and patterns. The data for medical monitoring that is collected could prove dangerous and thus only authorized users should be able to query or monitor the ongoing networks which would help in preventing the malicious interaction that could tamper or leak the data [2].

The aim of this chapter is to ameliorate the privacy and security of the sensitive data of the patient that is collected since the success of any healthcare application relies on it, for not only ethical but also legal reasons.

9.1.2 Wireless medical sensor devices: current trends and future directions

The enhancement of healthcare services from in-hospital/out-hospital communications to environment monitoring can be achieved by proper integration of medical sensor networks. Security involves the protection of data at each layer of the communication protocol it uses, as our system is only as strong and as secure as our weakest link. Current sensor network applications are in diverse fields with diverse requirements and several requirements based on the type of problem/situation that exists. Their application in ambient intelligence range from monitoring diverse ecosystems and tracking of assets and people in the industrial processes to the maintenance of buildings and many more. WSNs bring out extensive applications and undoubtedly quite significant improvements in the healthcare domain but the biggest drawback of it being its deployment in the exposed environment what is needed to be taken care of [3].

Thus, a secure sensor network must have the following features:

- Trust set up with key source establishments
- Confidentiality of the data
- Maintaining privacy
- Integrity and authentication services
- Secure routing protocols
- Secure group management
- Aggregation of data

These features, their requirements, and processes shall be discussed in the further sections 9.2.1, 9.2.2.2, and 9.2.3.

Applying security enhancements in an environment full of constraints of sensor networks, network performance is also a key aspect in terms of its robustness, scalability, and service availability as well. Thus, it can be said that the strength of the security as well as network performance both are equally important in forging the security architecture of the "WMSNs." And this is the rudimentary challenge for ensuring the usability of WSNs in the healthcare applications.

WSNs are usually devices with low communication bandwidth, limited computational power, small memory capacity, and limited power supply. Furthermore, as it has been discussed earlier, they are deployed in areas easily accessible to the surrounding people and their physical environment which might lead to unwanted exposure which further causes security vulnerabilities in case the security system is not sound enough. These problems aggravate security challenges causing to impose further restrictions for maintaining the security threshold of the protocols.

9.1.3 Growing aspect of WMSNs in healthcare applications

There are numerous healthcare applications of sensor networks some of which include telemonitoring of human physiological data, diagnostic purposes, drug administration in hospitals, providing an interface for the child or elderly integrated patient monitoring and tracking, and monitoring doctors or patients inside a hospital [3].

- Drug administration in hospitals: The sensor nodes can be attached to the medication the patient is currently on, to minimize the chances of taking or prescribing the wrong medications. Thus, sensor nodes can identify the allergies in the patients and hence medications required according to their allergies.
- Wireless body area networks: The expansion of the scope of WSNs for the eHealth domain is increasingly turning certain technologies to body sensor networks (BSNs) which have biosensors in them with the capability to record electrocardiograms, body temperature or blood pressure, electromyographs, electrodermal activities, among other healthcare parameters.
- Telemonitoring of human physiological data: The data collected by the sensor networks may be stored in them for a long period of time which can later be used in detecting the pattern of symptoms of a particular disease or other medical investigations as and when needed. The behavior and the physical activities of a patient can also be tracked through these.

- Tracking and monitoring doctors or patients inside a hospital: Every patient under monitor has a sensor node attached to them which performs their own specific functions and other sensor node might be carried with the doctor to locate them within the hospital. The sensor node might be detecting the heart rate, blood pressure, etc.

9.2 Related work

9.2.1 *Privacy and security requirements: essential factor for use of WMSNs*

Collection of data from the sensors should be secure enough so that no modification or wiretapping during transmission could breach the data. The medical sensor networks usually have one or more base stations which serve as data sinks and often as gateways to IP networks. The base stations are considered as secure if it is protected physically or has a tamper-resistant hardware with multiple protocol layers which increase the privacy of the data. These security protocols shall be discussed in detail in the forthcoming section 9.2.3.

General security requirements of WSNs are availability of data, maintaining integrity, and authentication along with confidentiality. Some other requirements include self-organization, data freshness, and source localization. Sensor networks have to fulfill these requirements in order to be trusted as secure as these requirements give them protection against the attacks and malicious invaders to the information conveyed over the network.

- Data confidentiality: In sensor networks, data flows from many nodes and networks to the base node for transmission which proliferates the odds of data breaching. The disclosure of personal information containing medical records is highly dangerous. Hence, the authorized people must be able to use it by sending the encrypted data which can be deciphered or decrypted by the end-authenticated-user only. Furthermore, for the items related to the invasion of privacy, a function that allows for the selection of confidentiality information should be provided in which the patient's details exist [4].
- Data availability: Delay in the information and/or incomplete transmission of the information should not occur while processing of the data. The availability of the network is essential for its intended use even in case of some attacks like denial of service. It is important to compensate in case of degradation either by node compromise or benign node failures [5].
- Nonrepudiation: It accounts for the specific acts in the remote areas specifically. Precisely, the authentication certification is based on a public-key which helps secure reliability among various medical activities [4].
- Data integrity: The data received by the nodes should be received by the end-user as such without any alterations or modifications. The intruder may breach in and modify the data according to his needs thus forwarding it to the end

user. This can be prevented by encryption methods, authentication, and security channels making the previously stored data unmodified [4].

- Data authentication: It is used to ensure that the person using the data is an actual user or not before giving access to the data. Furthermore, it enables sensor nodes to detect any malicious packets injected into it by verifying the origin of the packet and ensuring data integrity. Although this method may prevent outsiders from injecting parody packets, it still does not solve the problem of the nodes that have been compromised, which also have the secret keys of legitimate nodes. And this can be solved by using some detection techniques intrusively to find the compromised node and revoke its keys of cryptography [6].
- Privacy: Protecting the sensed data from the eavesdroppers is important which can be done by using standard encryption functions like AES block cipher. It can also be achieved by sharing a password key between the two parties to maintain the secrecy. Along with this, the control policies to access the information at the base station needs to be enforced for preventing any misuse of information [6].
- Data freshness: Data freshness ensures that the old messages or the replicated data are not replayed by any of the nodes by the addition of some time-related counter. Thus, every message transferred over the network is fresh [7].
- Self-organization: This is precisely the greatest challenge of security in WSNs as they have no fixed infrastructure and each node is independently capable of adapting itself to different situations thus maintaining the organizing and the healing properties by itself [7].
- Source localization: Some nodes/applications use the information of location from the sink/base node before transmitting the data. Hence, providing security to the sink node is important so that no malicious signals could be sent from a "false" sink node. This could transfer all of the data to the erroneous place [7].

9.2.2 Major security challenges and threats

9.2.2.1 Security issues/challenges

Security issues prevailing in the eHealth monitoring applications have always been the main point of concern and a functional area of research in the recent years. Maintaining security and privacy has been an inseparable part of any system and is of utmost importance. The use of these wireless communication networks of sensor nodes in healthcare can pose serious threats to the social life of an individual using these devices. The illegal disclosure and improper use of the location tracing data or the EHRs can cause grave consequences in people's lives. Securing the eHealth care record requires anonymity and unlinkability which protects the individual from any indifferent behavior of the society [8].

The sensor nodes have been characterized as low-power, bandwidth, and computational capabilities because of their resource-constrained behavior which further sets several constraints on the security framework of the application. Thus, only a small fraction of the total available space can be used for cryptographic algorithms because of which the encryption coding needs to be very lightweight along with no delays in the time of execution. Another thing that is significant to note is that the overheads required for providing additional security due to low bandwidth issues, should not in any case, degrade the overall efficiency of the system security framework in use. The sensor networks are more susceptible to attacks by spoofed packets or malicious environments than any other ordinary networks. Another point to keep in mind is that there should be a minimum generation of key exchange or authentication messages as the nodes and their topology is subject to frequent changes. Although lightweight cryptographic algorithms must reside within the nodes, it should be made sure that they can be used efficiently from within the nodes. As the topology networks and the nodes are subject to frequent changes, the security architecture must be scalable for leveraging the use of scarce energy-constrained resources [9].

Various parameters and frameworks have been developed for securing and maintaining the privacy of the patient's data [10] which will be discussed in the forthcoming section 9.2.3 on the solutions for combating the security issues.

9.2.2.2 Security threats

Threats in "WMSNs" occur if certain requirements maintaining the security are not met up to the expectations or the threshold value, thus resulting in grave consequences [4].

- If access to sensing devices is permitted without the certification or authentication, then the credibility of the data or source node cannot be given.
- If control of sensor nodes and their accessing is not taken into consideration for accurate decision making by the nodes, it will be difficult to process the data.
- If the authentication of medical staff, who is the main controller of telemedicine or eHealth, regarding patients' data is not taken into account, it can lead to misuse of the eMedical services and facilities and no accountability of the medical data or practice can be given.

Due to the high chances of security breach in eHealth applications of WSNs, the data and system are prone to numerous threats and attacks. The threats faced by a biomedical sensor can be categorized into various classes. The attackers can modify the data or add their own bits of data, reply old packets, or might attack the radio transmission. A system is considered secure if it can handle all these types of threats and support all security properties. Even overwriting the memory of normally deployed nodes is possible by the deployment of malicious nodes with exactly the same properties and capabilities as that of the base/nominal node [11]:

- Routing-based
- Protocol layer-based
- Capability-based

Patient's vital medical records are sensed by different medical sensors which are supposed to transfer it via different wireless connections through Internet. This demands security at a high level so that intruding in the data by attackers through eavesdropping or modifying the patient's sensitive data is infeasible. The sensed data may contain information regarding the location of a person allowing the attacker to harm the person by any means physically.

Denial of service attack results in accidental failure of nodes in case of any kind of malicious activity that not only aims to disrupt or destroy a network but is also responsible for any event that diminishes the network's strength and capability reducing its overall power [11].

On the basis of routing, an attacker can steal the information or modify it while the transmission of data is taking place from the source node to the sink (base station). Some attacks while routing are explained as below [7,11]:

1. **Wormhole attacks:** In this type of attack, there are two or more than two malicious nodes present in the system at various different locations which receive the information transmitted by the sending node and forward it to another malicious node which further directs the information to its neighboring nodes. So, basically, it fools the sender and receiver node by convincing them that the distance between two actual nodes is one or two hops distance away when it actually is multiple hops away both of which are usually out of range. Usually, wormhole attack is combined with selective forwarding attack and Sybil attack which make detection of attack onerous.
2. **Sybil attack:** This attack involves one malicious node with multiple identities by deceiving the sensor network to consider it as being present in multiple locations. These nodes which are malicious are known as Sybil nodes and are used against distributed systems having redundancy mechanism like that of attacking different kinds of protocols. This makes it a serious concern to the protocols that have been based on location and which exchange information about location tracing for efficient routing.
3. **HELLO flood attack:** In order to detect the neighbors a source node has, it broadcasts a HELLO message. The receiver node in its data transmission range sends this sensed data to the broadcaster. The attacker, in order to infiltrate, broadcasts HELLO message with such a high power of transmission that the nodes receiving it sends the data packets to the node of the attacker. This way the attacker can modify or drop unwanted packets leading to energy wastage and futile congestion of networks. It is one of the least difficult attacks for attacking in WSNs.
4. **Sinkhole attack:** In this type of attack, a malicious node attracts the network traffic by fake routing information. WSNs are quite vulnerable in this attack because communication in this type of attack takes place in many forms. In the figure below we can see that the malicious node in this type of attack has more power than other nodes and connects to the base station in just a single hop. Single hop path is used by it to attract more traffic in one go as most of the algorithms using the routing protocols use the shortest distance path for data transmission.
5. **Selective forwarding attack:** This attack involves one or more than one malicious nodes intruding in the path of communication process. This malicious node(s) acts as black holes which either selectively forwards the data packets to its

neighboring nodes or it might drop all of the data packets. In the latter case, it becomes difficult to recognize that there is some loophole in the communication and the other nodes, thinking the data route to have failed, starts trying for other routes of communication. The situation becomes more complex when the data packets are selectively forwarded. The odds of detecting the malicious nodes become onerous in case the attacker included the nodes externally to the paths. Based on the drop of packets, it is categorized into two categories:

- Specified type drop packets and
- Specified nodes drop packets.

6. **Replay attack:** This type is one of the kind in which valid data transmission is either repeated or delayed deceivingly by the creator or the attacker who intrudes in the data retransmitting, which possibly occurs because of a masquerade attack by IP packet swapping. The attacker copies the forwarded data packets and resends them continuously to the victim in order to exhaust his power supplies or to degrade the sink nodes and the network connections. Thus, the replayed data packets can degrade the poorly designed system or exploit vulnerable holes.
7. **Attack on the information in transit:** While the sensors are performing their assigned actions and sending the data to the sink node, it may be altered or spoofed upon or damaged as the wireless network is highly prone to eavesdropping or any attack on the traffic flow while data transit and can get into action by modifying or fabricating the packets.

On the basis of capability, the levels of data access and the damage caused to it depends upon the following:

1. **Active versus passive attacks:** It depends on the level of damage done to the data or the level at which the attacker has easy access to the data. In passive attacks, there is no interruption in the actual communication and only the data can be accessed by the attacker and no modifications can be made. Thus, sensitive data that are collected by the attacker exploit the overall confidentiality of the patient's data. Examples of passive attacks include traffic monitoring, interception, and analysis of the data. This differs from the active attack in the way that data collected by the attacker can be modified and misleading information can be added in the data traffic network which thus affects the performance of the network [8].

   This differs from the active attack in the way that data collected by the attacker can be modified and misleading information can be added in the data traffic network which affects the overall performance of the network.
2. **Insider versus outsider attacks:** The outsider attacks are the type of attacks in which the attacker finds no as such access of the deployed nodes and yet wants to harm them. The attacker node which wishes to harm them authorizes themselves to harm the network even if they are not part of the network. These are known as external attacks. In the insider attacks, the malicious node is situated inside the network itself and is more dangerous as all the

critical information riddles are already known and it has all types of access rights.

3. **Mote-class versus laptop-class:** In mote-class attacks, the attacker has few nodes with same capability and processing strength as that of the normal node with at least one authorized node in order to steal away the unique key for opening up the scope for intrusion of more such malicious nodes. Whereas in laptop-class attack, the attacker has special access to more powerful devices having more processing capability, sensitive antenna, powerful radio transmission, etc. rather than the network as a whole. For example, a laptop or any workstation equivalent to it can be more harmful as it has the capacity to make the whole network stuck whereas a standard node just has the capacity to disturb its neighboring nodes or nearby networks.

On the basis of protocols, WSNs can be divided into different layers in which each layer has a different function to perform. The attacks [7] on the basis of various layers are described below:

1. **Physical layer:** This layer is used for the transmission of data over wired or wireless medium in the form of raw bits which can be easily jammed from the point of view of an attacker. The common attacks over this layer include eavesdropping, jamming of network, or tampering. Jamming usually occurs in the case of denial of service (DoS) attack when the interference with radio frequency signals is caused which completely changes the way of working of the network. In the eavesdropping attack, an unauthorized person can read the messages.
2. **Link layer:** This layer ensures the proper communication of the data on physical layers among the nodes. This layer is responsible for the detection of errors, preventing collision of data packets or repeated transmission of data, and so on. The threats involved in this layer include packet replay, interrogation, or collisions among packets of data. The number of collisions among the packets can be decreased by using correcting codes or error detection but it increases the overheads in the routing of that network. Another type of connection layer danger is in the denial-of-sleep attack. In this, the node cannot go to the inactive state when not in use hence decreasing the overall lifetime of a network.
3. **Network layer:** For data routing mechanisms to be effective from source nodes to sinks or from nodes to body sensors or vice versa, the network layer is responsible. Direct attack of the attackers on this layer can give them the control of interfering with the flow of data and network traffic between the source and destination. Thus, secure and powerful protocols are required for the effective transfer of data to manage any kind of security attacks or node failures. Some routing protocol attacks include selective forwarding, spoofing, black holes, wormholes attack, and so on.
4. **Transport layer:** This layer is used for building up a communication link that is used for sensor networks outside of the range joined via Internet and can be considered as the most complex issue in the WSNs. The attacks usually found

in this layer include desynchronization and flooding. In the former attack, the attacker node replicates the packets to at least end of the user via sequence numbers which are all different on the packets thus making the host user requests for retransmission of the data packets that have been missed. In the latter case, numerous requests are sent for the establishment of the connection which depletes the node's memory.

9.2.3 Solutions to breach in security

Research in the field of "WMSNs" is still at its infancy stage as much of the salient issues haven't been completely resolved yet. Stringent security frameworks and rules need to be expanded for multifencing security that could be rooted in every component and layer of the network. Various security mechanisms have been developed for avoiding unanticipated and menacing situations to occur as loss of confidentiality poses adverse impact on the security of the health record data whereas loss of integrity in data can even lead to threat to a patient's life. This section discusses various security mechanisms that have been used to protect the patient's data from any kind of threats, attacks, or breaches.

Reference [12] deals with developing a framework that leverages the use of cloud system and blockchain technology together to secure the eHealth records of the patients. The centralized storage of data is prone to cyberattacks and hence constant reviewing of the data of each patient is required, which is very difficult. To avoid this, a security framework capable enough to counter the attacks of the foreign invaders needs to be developed. This chapter uses the blockchain technology to maintain the integrity of the data and the cloud-based system for ensuring authentication. The secrecy of the digital signatures with the help of keyless signature infrastructure also ensures the base of legal authentication to the data being accessed. The blockchain technology strengthens the security framework by partially adding the data in blocks which further generates a timestamp for the data that is stored at the base or that has been updated. To use these data stored in different blocks, cloud users need to pay some amount for accessing the service. Thus, this combined system enables the cloud users to make use of blockchain technology for ensuring high-level security to eHealth records of the patients.

As chronic diseases are exponentially growing in number, the need for developing eHealth care facilities is something quite favorable for the government, people as well as the country as a whole. The current government resources and the healthcare systems are deploying at such a state that the government has to make expense of almost half of their budgets to healthcare systems. Thus, scrutinizing the situations and leveraging the use of IT development, eHealth monitoring is one of the best and most promising solutions. A decent reduction of chronic disease risks has been seen if technologies and prevention schemes are well implemented. Reference [13] proposes a framework based on SOA architecture and cloud-based environment. The integration of these technologies, services, and applications while considering the limited capabilities and power constraints can communicate vital data of the patient's health with the utmost privacy and security. The patient's data, which are collected through various kinds of sensors, are linked with the cloud which stores all this information. This information is then made available from the Cloud environment via SOA architecture which can be accessed by the physicians, the medical practitioners, or any of the authenticated persons easily. The SOA system, besides being an adequate system for integration of heterogeneous systems, is also a platform which maximizes data reliability and availability while reducing the cost of the infrastructure. Service-oriented e-health system along with the Cloud-based technology enables real time monitoring along with taking appropriate action to identify or counteract any risk factors. The main advantage of this framework is that the system can be integrated in an efficient manner with other healthcare systems as well, allowing high interoperability among various heterogeneous systems.

Reference [14] proposes an architecture for effective health care monitoring and mutual authentication protocol for preserving the anonymity of the patients. It uses the AVISPA tool to resist unknown attacks. Along with the mentioned tool, it uses the BAN logic model which confirms the proposed protocol for mutual authentication features. The proposed protocol is somewhat based on the AVISPA software in order to ensure the robustness to security attacks and unknown threats. This chapter also gives an informal cryptanalysis to ensure that all known attacks are resilient to the proposed system. It not only protects from common security attacks and threats but also ensures robust mutual authentication, user-friendly phases of password changing scenarios as well as efficient login systems.

The main concern, either way, is the unauthorized access of the patient's data. Reference [15] presents an efficient and strong authentication protocol which has a 5D approach in maintaining strong user authentication which is a must-have for large-scale deployment and hence the success of the "WMSNs." The proposed system of "E-SAP" has five major developments for efficient security maintenance. First, it provides mutual authentication between the medical practitioner or the professional staff and the medical sensor deployed for collecting the sensitive data of the patient. Second, it provides a two-factor key, that is, the professional authentication of the smartcard and password. Third, it provides the facility of symmetric encryption or decryption of messages (that could contain some sensitive information) for providing confidentiality in the messages being exchanged. Fourth, it looks to providing secure key of the session at the end of authentication. And lastly, it enables professionals to change their passwords maintaining its secrecy at the highest level. Furthermore, three messages are required for exchanges between the professional person at one end, gateway node, and the medical sensor node in the proposed protocol. This way it achieves efficiency by minimizing the cost and the computational power. Thus, through the performance, security, and formal analysis done on this protocol, it can be said that E-SAP achieves its set security goals and defense mechanism against various threats and is a protocol that is well-suited for clinical, hospital as well as applications in homecare in "WMSNs."

Reference [16] proposes a practical approach in preventing the insider attacks on the data by the use of multiple data servers for storing the patient's data. It basically helps in distributing the patients data across multiple data servers securely and further employing "Paillier and ElGamal cryptosystem" approach for performing the necessary analyses on the data of the patient without compromising on its security. Further steps have been proposed in this chapter for securing the communication between the data servers and the medical sensors. The use of lightweight encryption mechanism combined with MAC generation scheme based on SHA-3 helps to secure the communication. As the patient's data are stored in three data servers, so, as long as none of the data servers is compromised, the security of their data can be preserved. For performing analysis based on statistics on the received data, some new protocols, such as, for average, variance, correlation, and regression analysis have been proposed, where all of the three data servers on a friendly note, cooperate with each other for processing the patient's data without revealing their privacy and providing the end users with the analysis results. The performance analysis on the structure developed shows the practicality of the protocols proposed whereas the security analysis shows how well-fortified our system is.

Reference [17] proposes a "privacy-assured healthcare monitoring system via compressive sensing." It is based on a secure cloud framework which enables fast data acquisition and indexing in the health care monitoring systems which is well aware of the privacy. Compressive sensing technique has been adopted for efficient collection and maintenance of data and its easy sampling and recovery. An encrypted high-performance index has been proposed to handle the continuously generated medical data samples which can be built fast via concurrent insertion threads. In this chapter, privacy-aware healthcare system has been developed that supports fast indexing as well as serves as a secure platform in cloud environment for the encrypted medical data. Further improvements in the building performance for practical considerations have been done through the reduction in the bandwidth of secure retrieval of data, through caching or via exploring the relationship between the efficiency and accuracy of the data. Moreover, exposing the data to a public-based cloud environment would raise security issues and thus, for making the use of cloud environment as a success, some critical and fundamental challenges need to be addressed whose solution frameworks have been elaborated in this chapter.

Several "wireless body area networks" have emerged which serves in revolutionizing the health industry further. They are classified as Off-body, In-body, and On-body communication which opens a plethora of opportunities to monitor the health of a patient during normal day to day activities for prolonged periods. These devices make use of various wireless medical sensors, implants, and wearable sensors, set over whole of the body in a connected fashion in such a way that the overall health status can be updated to the physician with real time monitoring. The deployment of WBANs must ensure that stringent security requirements ranging from medical to nonmedical must be met. Reference [18] discusses various security modes for WBAN that include AES-CTR, AES-CCM, and AES_CBC_MAC nodes whose selection is controlled at the application layer of the security system protocol. Moreover, to reduce the extra overheads incurred during the transmission process, the use of a cipher stream for encryption would be beneficial because of the same size of the plain as well as cipher text.

The basic technique for securing the data in a network is cryptography which involves encrypting of messages from the original end which are decrypted at the receiver end to obtain original data. Various types of keys are used in this process. Some of the protocols for solving the issue of security is WMSNs proposed by different authors are explained below [7].

1. **TINYSEC:** It is that layer of the architecture which provides and strengthens the link layer of the application. Besides being a lightweight protocol, it supports various features of maintaining security by providing integrity, authentication, and confidentiality. This layer uses cipher-block chaining (CBC) mode with stealing of the cipher text for encryption to achieve confidentiality in the data. It doesn't use any counters and uses CBC-MAC for authentication. The fundamental thing in this architecture is that it has two security options out of which one is for encrypted and authenticated messages—TinySec-AE, and the other is for just the authenticated messages—TinySec-Auth. The former security option involves data payload encryption and the authentication of the received packet is done with a MAC, whereas in the latter one, the data received is authenticated with a MAC but the payload of the data is not in the encrypted format.
2. **LEAP:** Localized Encryption and Authentication Protocol is used for large-scale distribution of sensor networks with a key management scheme that is highly efficient security mechanisms and is generally supported for inside network processing like that of data aggregation. Multiple keys mechanism is used in this protocol for achieving confidentiality and authentication to the sensed data for transmission. The multiple keys used in this system are all symmetric and are known as pairwise, group, individual, and cluster keys. The pairwise key is shared with other sensor nodes, the group key is globally shared and hence used by all network nodes, the individual key is used for communication between the front-end and the back-end user nodes which is unique in every sense and lastly, and the cluster key is used for transfer of data packets between the main node and its neighboring nodes by broadcasting messages locally. LEAP is basically used for protecting the system against Sybil Attack, HELLO Floods Attack, and Wormhole Attack which have already been discussed.
3. **SPINs:** The Sensor Protocols for Information that works through negotiation is done in three steps. First, a node having the data advertises the ADV packet to all its neighboring nodes. It contains all the metadata. If the receiving node would be interested in receiving the data, it would send requests for the data using a packet, here, the REQ packet. Finally, after receiving the request from the receiver node, the advertiser node prepares the packet for transmission to the requestor node. This type of protocol works best in small networks. This is because of the latency properties that exist in it. A SPIN generally consists of two blocks which have a backhand in making it secure. They are uTESLA and SNEP. TESLA helps to authenticate the data using digital signatures whereas SNEP uses message authentication code (MAC) to authenticate the data and helps maintaining the integrity as well as confidentiality using encryption techniques.

4. **ZIGBEE:** It is a wireless technology typically used in applications such as monitoring the environment, military security, and home automation further supporting data confidentiality. This mechanism uses 128 bit keys and uses the standard of IEEE 802.15.4. It also involves a trust center which allows other networks to join it only after authentication. That trust center is basically the ZIGBEE coordinator. Thus, it can be said that the three different roles of ZIGBEE are that of the trust manager, the network manager, and the configuration manager. The devices send requests for the authentication and the trust manager, when allows the access to the request for connection to the main network, the network manager manages the network keys and thereby helps in the distribution of the network keys, and the configuration manager configures the mechanism of security for ensuring secure connections between the end users. The two modes which ZIGBEE works in are the residential mode and the commercial mode. Less security is needed in the residential mode and hence no keys are used while in the commercial mode high security is required and hence keys and counters need to be maintained.

Several applications for healthcare have been proposed which aims to take eHealth to another level of advancement.

CodeBlue [19] is a popular project based on healthcare research in a medical sensor network. Several medical sensor networks such as EMG, SpO2, EKG, pulse oximeter, and such sensor boards on the Mica2 notes are placed on the body of the patient. These sensors sense the patient's body data and transmit to the end user devices such as laptops, personal computers, or PDAs for analyses. The basic idea of CodeBlue [20] is based on the "publish" and "subscribe" architecture in which the particular data queries are put and the data are published to the specific channels. TinyADMR routing protocol facilitates multicast routing and minimal losses in the path along with RF-based localization which enables the monitoring user to locate a patient's location accurately. The above-discussed link layer protocol is also good for use in this project as it is based on symmetric encryption. The CodeBlue architecture has been proposed to be deployed in disaster response, for stroke patients, and for pre-hospital as well as in-hospital emergency cares as well [21]

Another project designed for the same purpose is that of a heterogeneous network architecture specifically designed for the patient health care monitoring named Alarm-Net [22]. It consists of the body as well as environmental sensor networks and three network tiers are used in this system. The first tier includes wearing of different body sensor devices such as accelerometer, ECG, or SpO2 for sensing the physiological data. The second tier is based on sensing the environmental conditions like sensing of temperature, light, motion, dust, etc. The third and the last tier is the fundamental one which includes an Internet protocol-based network comprising of Stargate gateways that are known as AlarmGate. The concept behind this AlarmGate is that the body sensors broadcast the data with a single hop to the next nearest node and that node further communicates it using the technique of multihopping and hence reaching the AlarmGate [23]

The most promising replacement to RSA-based algorithms being developed is the elliptic curve cryptography (ECC). The small size of ECC keys [24] makes them much shorter and even takes up less space for the same security level. Many advances have been made for providing extreme level of security to the data. The use of it is suggested in the BlueCode architecture as well for good encryption pattern and technique of ECC. Moreover, as it has higher magnitude energy requirements, it is better to use it in applications where the use is not frequent and the security operations are required [25]

9.3 Proposed work

The proposed "aggregatable distributed database with blockchain technology is using advanced encryption standard (AES) and hash-based message authentication code (HMAC) algorithms for message encryption" framework. This framework focuses on achieving the patients' records securely and efficiently and hence accounts for integrity. The physician who needs to look over the status of a particular patient's health requests the health data by signing in with their own unique ID. The particular patient's specific private key also needs to be entered which is unique for every patient. The data that are retrieved by the doctor can further be used for identifying and checking for various parameters of a person's health. The data will only be retrieved when the ID provided by the doctor matches or is verified from the data access list (DAL) which forms the first layer of security. Once the physician has been verified according to the list and signature verification of the user has been done, the aggregator comes into play for verifying and authenticating any kind of illegal or malicious packets that might be injected into the data to corrupt it or for modifying the original data. Then it performs unsigncryption of the data available in the blockchain to provide its access to the user.

The data are basically made to enter the blockchain after passing through the aggregator and the application gateway for further processing and ensuring the authentication of the data. Basically, the data that are made to enter the blockchain contain several blocks and the health data of each patient are stored in a block. These blocks store all the information about the data, such as the date, time, and the unique sensor ID which is responsible for passing on the information. A single block can store up to a specified size of information in it after which a new block is created. The block storing the information stores a code that is unique in every sense and is called "hash" which thereby uniquely identifies the block. So, every computer that is participating in the blockchain network has a copy of the blockchain available thus making several copies (in millions) of the same blockchain on thousands of computers. This makes the manipulation of data in the blockchain difficult. This is because, for a data to be manipulated, the data will have to be modified in at least 51% of the computers before it can actually be modified. This helps in maintaining the privacy and integrity of the patient's health care data.

In this framework, we take on three data servers to which the deployed IoT sensors send all the information in the first stage. Any number of data servers can be used according to the type of data and security issues. The medical sensor networks are predeployed with unique sharing keys which are preshared with the three

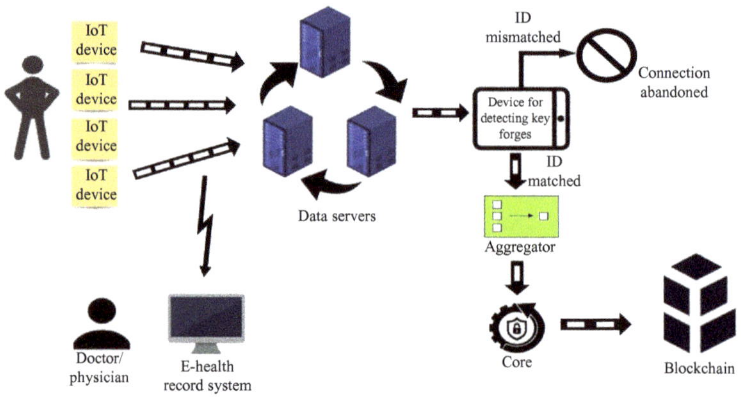

data servers in a way which made a secure connection of each medical sensor network to any one of the data servers. The shared key is unique for each. One more key that is secret had been predeployed for the generation of random numbers. When a medical sensor sends any numeric data of the patient, it splits the reading of the patient's data (z) into three integers (t, y, r) such that $t + y + r = z$ and sends to those three servers through three secure channels so that no data server could completely understand the patient's data and hence prevents from revealing any of the patient data, basically avoiding any kind of possible inside attacks.

Assuming that a medical sensor network sends a sequence of an attribute of the patient data z1, z2, ... each being less than 48 bits, to those selected data servers and random numbers are generated q1, w1, q2, w2, ... each of 60 bits using SHA-3 algorithm. Let $|ti|$ ($|yi|$) be the first 32 bits of qi (wi). Then it is the responsibility of the medical sensor to compute

$$ri = zi - ti - yi$$

where $i = 1, 2, 3, \ldots$

Suppose $Qi = \{$patient ID, attribute, data unit$\}$ and suppose the medical sensor sends $\{Qi, ti\}$ to the first server S1 through a secure channel. Similarly, $\{Qi, yi\}$ is sent to S2 and $\{Qi, ri\}$ is sent to S3 through secure channels where $i = 1, 2, 3, \ldots$ For the privacy of the data of the patients', it is necessary that the three data servers do not put their information together and at least one of the data servers does not leak the information.

Each of the data servers passes on the data to the application gateway which checks the standards of the data and passes it to the aggregator. The aggregator groups the data for efficient storage and passes it on to the blockchain for easy storage and the retrieval of the data. For maintaining network security, encryption is done and digital signatures are provided. The key used for encrypting the data is generated through HMAC algorithm and a digital signature is also provided. The aggregator uses its key for performing signcryption and forwards it to the core and finally to the blockchain. The encrypted data are verified for the digital signature produced and if it the signature matches, no modification in the encrypted data has taken place. Many cryptographic techniques are used for encrypting the data before storage. Modern cryptography includes public key algorithms like RSA and the symmetric key algorithms like DES and AES. The public key algorithm is based on the use of a public key and a private key from which public key is available to everyone but the holder of the private key is the only one who would be able to decrypt the data. On the other hand, symmetric key algorithms use a single key only which must be kept private if the data need to be kept private. Algorithms used here for encryption are advanced encryption standard (AES) and hash-based message authentication code (HMAC).

Algorithm for retrieval of patient information from the three data servers:

Now, when the physician or any authorized user requests for the data, a session key and the unique ID associated with the physician is verified by the DAL and if verified, further back-end authentication process proceeds to provide the required data requested by the end user. The data go to the aggregator for unsigncryption process after the signatures get matched and the encrypted contents of the data get decrypted. The three servers then find the three shares of the data t, y, r in accordance with the specified user, his identity, and the attributes required. This is followed by the running of the above algorithm by the three data servers to make the data reach the user in an efficient and completely secure way.

The proposed system uses Blockchain for maintaining the security of the data. The proposed system using blockchain is evaluated for the performance using various measures. The reliability is an important factor that contributes to the performance. The proposed system with blockchain compatibility provides better reliability than the system without blockchain facility.

9.4 Conclusion

The healthcare data are of utmost importance which need to be maintained safely. The health records of the patients are very confidential and need periodical updates. The data are to be stored in an efficient and secure manner such that it ensures the privacy of the data. Thus the WSNs play a vital role in the medical field in increasing the services for maintaining healthcare. The application of the emerging technology such as blockchain enables drastic improvement in security services thereby enabling efficient healthcare services for effective treatment.

References

[1] Ng, H. S., Sim, M. L. and Tan, C. M. "Security issues of wireless sensor networks in healthcare applications." *BT Technology Journal*. 2006; 24(2): 138–144.

[2] Shnayder, V., Chen, B. R., Lorincz, K., Fulford-Jones, T. R. and Welsh, M. "Sensor networks for medical care." Technical Report TR-08-05, Division of Engineering and Applied Sciences, Harvard University, 2005.

[3] Darwish, A. and Hassanien, A.E. "Wearable and implantable wireless sensor network solutions for healthcare monitoring." *Sensors*. 2011; 11(6): 5561–5595.

[4] Lee, J. D., Yoon, T. S., Chung, S. H. and Cha, H. S. "Service-oriented security framework for remote medical services in the Internet of Things environment." *Healthcare Informatics Research*. 2015; 21(4): 271–282.

[5] Dimitriou, T. and Ioannis, K. "Security issues in biomedical wireless sensor networks." *2008 First International Symposium on Applied Sciences on Biomedical and Communication Technologies* (pp. 1–5). Athens, Greece: IEEE. October 2008.

[6] Shi, E. and Perrig, A. "Designing secure sensor networks." *IEEE Wireless Communications*. 2004; 11(6): 38–43.

[7] Grover, J. and Sharma, S. "Security issues in wireless sensor network: A review." *2016 5th International Conference on Reliability, Infocom Technologies and Optimization (Trends and Future Directions)(ICRITO)* (pp. 397–404). Amity University Uttar Pradesh (AUUP), Noida, India: IEEE. September 2016.

[8] Al Ameen, M. and Kwak, K. S. "Social Issues in wireless sensor networks with healthcare perspective." International Arab Journal of Information Technology. 2011; 8(1): 52–58.

[9] Ko, J., Lu, C., Srivastava, M. B., Stankovic, J. A., Terzis, A. and Welsh, M. "Wireless sensor networks for healthcare." *Proceedings of the IEEE*. 2010; 98(11): 1947–1960.

[10] Deebak B. D., Al-Turjman F., Aloqaily M. and Alfandi O. "An authentic-based privacy preservation protocol for smart e-healthcare systems in IoT." IEEE Access. 2019; 7: 135632–135649.

[11] Roseline Mary J., and Buvana M. "A survey: Security issues and design challenges in healthcare monitoring system using wireless sensor network." International Journal of Innovative Research in Science, Engineering and Technology (An ISO 3297: 2007 Certified Organization). 2015; 4(11): 10758–10765.

[12] Nagasubramanian, G., Sakthivel, R. K., Patan, R., Gandomi, A. H., Sankayya, M. and Balusamy, B. "Securing e-health records using keyless signature infrastructure blockchain technology in the cloud." *Neural Computing and Applications*. 2020; 32: 639–647.

[13] Benharref, A. and Serhani, M. A. "Novel cloud and SOA-based framework for e-health monitoring using wireless biosensors." *IEEE Journal of Biomedical and Health Informatics*. 2013; 18(1): 46–55.

[14] Amin, R., Islam, S. H., Biswas, G. P., Khan, M. K. and Kumar, N. "A robust and anonymous patient monitoring system using wireless medical sensor networks." *Future Generation Computer Systems*. 2018; 80: 483–495.

[15] Kumar, P., Lee, S. G. and Lee, H. J. "E-SAP: Efficient-strong authentication protocol for healthcare applications using wireless medical sensor networks." *Sensors*. 2012; 12(2): 1625–1647.
[16] Yi, X., Bouguettay A. A., Georgakopoulos, D., Song, A. and Willemson, J. "Privacy protection for wireless medical sensor data." *IEEE Transactions on Dependable and Secure Computing*. 2015; 13(3): 369–380.
[17] Al-Turjman, F. "Intelligence and security in big 5G-oriented IoNT: An overview." *Future Generation Computer Systems*. 2020; 102(1): 357–368.
[18] Saleem, S., Ullah, S. and Yoo, H. S. "On the security issues in wireless body area networks." *International Journal of Digital Content Technology and its Applications*. 2009; 3(3): 178–184.
[19] Malan, D., Jones, T. F., Welsh, M., and Moulton, S. "CodeBlue: An ad-hoc sensor network infrastructure for emergency medical care." *Proceedings of the MobiSys 2004 Workshop on Applications of Mobile Embedded Systems (WAMES 2004)*, Boston, MA, USA. June 6–9, 2004.
[20] Shnayder, V., Chen, B. R., Lorincz, K., Fulford-Jones, T. R. and Welsh, M. Sensor networks for medical care. Technical Report TR-08-05, Division of Engineering and Applied Sciences, Harvard University, 2005. Available from: http://nrs.harvard.edu/urn-3:HUL.InstRepos:24829604.
[21] Kumar, P. and Lee, H.J. "Security issues in healthcare applications using wireless medical sensor networks: A survey." *Sensors*. 2012; 12(1): 55–91.
[22] Wood, A., Virone, G., Doan, T., *et al.* "ALARM-NET: Wireless sensor networks for assisted-living and residential monitoring." Technical Report CS-2006-01. Charlottesville, VA: Department of Computer Science, University of Virginia; 2006.
[23] Wood, A., Virone, G., Doan, T., *et al.* "ALARM-NET: Wireless sensor networks for assisted-living and residential monitoring." *University of Virginia Computer Science Department Computer Science Department Technical Report*. Vol. 2. 2006. p. 17.
[24] Malasri, K. and Wang, L. "Addressing security in medical sensor networks." *Proceedings of the 1st ACM SIGMOBILE international workshop on Systems and networking support for healthcare and assisted living environments* (pp. 7-12). ACM. June 2007.
[25] Liu, A. and Ning, P. "TinyECC: A configurable library for elliptic curve cryptography in wireless sensor networks." *Proceedings of the 7th International Conference on Information Processing in Sensor Networks*. St. Louis, MO, USA, April 22–24 2008. pp. 245–256.

Chapter 10

Acoustic glass for deaf people: a new device

Ahmad Mohammed[1], *Shaif Zahrah*[1], *Muaadh Al-Bahri*[1], *Basil Bartholomew Duwa*[1], *Fadi Al-Turjman*[2,3,4], *Ilker Ozsahin*[1,4] *and Dilber Uzun Ozsahin*[1,4]

Incomplete or complete deafness is viewed as the most widely recognized tangible disability. This deformity develops because of numerous internal ear variations from the norm. In any case, various methods for treating this variation from the norm have been created in order to help the hard of hearing. Analysts have searched for the most secure, solid, and moderate methods for treating and overseeing hearing misfortune. Some of these methods for treating the hard of hearing are the utilization of hearing gadgets which give an impermanent answer for the patient, utilization of print composing gadgets that empower hard of hearing individuals to send instant messages and is received as voice note, the utilization of video calls, and cochlear embed. These procedures could be pitiless and risky, living the patient with brief or lasting harm.

This investigation proposes sharpened glass (acoustic glass) that is worn like an ordinary eyeglass with helping highlights for the hard of hearing. This gadget is made out of various parts that are associated with it, that guide in distinguishing sounds and deciphering this sound to different segments and act as needs be for the client through a vibration.

The parts of this gadget are Arduino Nano which also serves as a handling unit, a potentiometer and mouthpiece, which fill in as the two data sources, RGB LEDs and the vibration motors, which are the yields, and a battery. Different segments comprise of the glasses with a casing and a phony focal point that holds the remainder of the parts, for example, the little receiver which identifies the vibrations brought about by the sound around and converts to electrical sign to be prepared. The handled sign is passed to the microcontroller which has been customized to comprehend the encompassing sound and interpret the prepared signs that are sent to the actuators. The actuators are of two kinds: the light emitting

[1]Department of Biomedical Engineering, Near East University, Nicosia / TRNC, Mersin-10, Turkey
[2]Department of Artificial Intelligence, Near East University, Nicosia / TRNC, Mersin-10, Turkey
[3]Research Center for AI and IoT, Near East University, Nicosia / TRNC, Mersin-10, Turkey
[4]DESAM Institute, Near East University, Nicosia / TRNC, Mersin-10, Turkey

diodes (LEDs) and the vibration motors. Another segment of the acoustic glass is the Potentiometer, which empowers the user to control the degree of sound. The microcontroller utilized is the Arduino, which has parts like the amplifier, the RGB LEDs, and the vibration motors appended.

The aim of this work is to utilize an acoustic glass as a gadget to support the hard of hearing individuals. This gadget is intended to give the client (quiet) signal when the sound is extremely uproarious, medium, and low. These signs are spoken to by hues known as the RGB LEDs. The acoustic gadget is sheltered, solid, and reasonable for patients.

10.1 Introduction

The world has gotten modern in innovation improvement and scaling down of other mechanical gadgets, which are utilized tremendously in medication [1,2]. The Internet of Medical Things (IoMT) is likewise acknowledged and utilized in medication and another para-drug. Up to this point, the regular or ordinary correspondence frameworks tackled restorative issues. Be that as it may, with advancement and development in innovation, correspondence frameworks don't offer enough flexibility to manage the large measure of grouped data [3,4]. These offer the chance to make better innovation that can productively improve the personal satisfaction.

The IoMT has made way for new opportunities in medicine to save and improve the quality of life. Its application extends to people suffering from deafness. IoMT is used in monitoring of patients and to direct them using mobile application or SMS [5,6].

Different investigations have uncovered the significance of the Internet of Nano-Things (IoNT) in medication. It improved nanotechnology and its application particularly in wellbeing ventures. These improved parts of nanotechnology incorporate nanosensors, nanointerfaces, etc. The nanosensors above all gather the electrocardiographic signs. Savvy spaces are arrangements where the Internet of Things (IoT)-empowering innovations have been utilized toward further advances in the way of life. It firmly coordinates with the current Cloud framework to affect a few fields in the scholarly community and additionally industry [7,8].

The human faculties are known to be indispensable for an appropriate body to work. These faculties are utilized for various observations, which incorporate the faculties for tasting, smelling, looking, contacting, and hearing. The body without one will in general be deficient [9]. In this work, we are worried about the issue including the feeling of hearing. The issue of having a feeble feeling of hearing called hearing hindrance, or even totally losing the capacity to hear which is called deafness. This can be gained by birth or procured after birth because of irresistible infections, mishap, or the in-take of improper medications during pregnancy [10].

In another research, vibration earphones were invented to help deaf people "hear" music as haptic vibration in the ears [11].

10.2 Literature review

Hearing debilitation: Around 360 million individuals comprehensively have been recorded to be in part or totally hard of hearing [12]. These people are said to be living in a low-and-center pay nations around the globe [13].

Kinds of hearing disability

1. **Conductive hearing loss (CHL)**
   This conference misfortune happens when there's a disturbance of the sound waves in the internal ear, cochlea, and the ear drum. This blockage can be brought about by disease, aggravation, or outer unsettling influence which causes puncturing of the ear drum [14].
2. **Sensorineural hearing loss (SNHL)**
   This sort of hearing misfortune is caused when the internal ear is influenced or harmed because of harmed hair cells. This sort of hearing misfortune significantly influences the old [15].
3. **Mixed hearing loss**
   This consultation misfortune happens when the conductive and the sensorineural hearing misfortune are joined. At the end of the day, the eardrum is harmed and tainted simultaneously [16].

10.3 Causes of hearing loss

Hearing impairment can be caused either as **congenital** or **acquired**.

1. **Congenital hear loss:** It is a result of hereditary and nonhereditary factors. The nonhereditary factors include the evolution of infectious diseases during pregnancy, lack of oxygen during birth, and inappropriate intake of drugs during pregnancy.
2. **The acquired cause:** It is mainly caused by diseases such as meningitis, accident on the head, noise and infection in the ear, aging, and heavy sound [17].

10.4 Diagnosis and treatment of hearing loss

10.4.1 Diagnosis of hearing loss

1. **Physical examination:** This type of diagnosis enables the physician to self-examine the patient to know the possible cause of the disability. This could be as a result of infection, inflammation, or the accumulation of earwax [18].
2. **Sound screening test:** The physician applies a whisper-test technique to screen the sound on the patient's ear at different sound levels. This will enable the physician to know the level of damage that occurred.
3. **Mobile-app hearing test:** This is the use of mobile application to screen the level of hearing loss and damage done. Various mobile-apps are available online for androids and iPhones.

4. **Audiometer screening test:** This involves an audiologist, who screens the patient using the audiometer. During these diagnoses, the patient puts on earphones and words are said to direct the patient on each of the ears [19].

10.4.2 The treatment of hearing loss

Related work

1. **Assistive hearing aid**
   - (i) **Hearing aid devices:** These devices are used in helping the deaf to improve in their speech and hearing abilities. These are devices designed to make surrounding sounds loud and audible to the user. However, hearing aid devices do not cure deafness but give a temporal solution.
   - (ii) **Teletypewriters:** This kind of device allows the user (deaf) to convert text messages into voice note for the receiver. In other words, it enables the receiver to listen to the message sent [20].
   - (iii) **Video relay services (VRS):** This is also referred to as video remote interpreting (VRI). It is a video telecommunication service that uses devices such as videophones or webcams to assist the deaf in interpreting messages into sign language through a video call by an interpreter [21].
2. **Earwax blockage removal:** The physician uses a tool to remove the blockage in the inner ear to reverse the cause of the hearing loss.
3. **Surgery:** This method is used in treating hearing loss to correct the abnormalities in the ear drum or the ossicles. The fluid is drained using small tubes inserted in the ear [22].
4. **The infrared sensor (IR) framework:** This gadget also is extremely valuable and significant for individuals with hearing issues and individuals who are hard of hearing. Like the FM gadget, this gadget needs a beneficiary to be conveyed with the individual or some of the time now and again it ought to be worn [23].
5. **Cochlear implant:** This is one of the known surgical remedy for hearing loss. This process is not like other hearing aid devices that amplify sound. This process involves the complete implant of the damaged ear loop and stimulates the hearing nerves directly. The device is implanted in the ear directly [24].

10.5 Methodology

The study was led by choosing the segments dependent on their highlights and proficiency. The parts were picked dependent on their connective capacities with each other.

The acoustic glass: the block diagram and components

This gadget comprises of various parts that are assembled to give a required output. These segments are as per the following: Arduino Nano which is a preparing unit, a potentiometer and receiver which fill in as the two information sources, RGB LEDs and two vibration engines are the yields, and a battery. Different parts comprise of the glasses with an edge and phony focal points that hold the remainder of the segments together. Another part that distinguishes sound vibration is the receiver. It changes over the sound into an electrical sign that is prepared. The handled sign is later passed into a microcontroller and dissects the sounds. The broke down sound is then used to give the criticism to the client. The microcontroller interprets the procured prepared sign to enactment signal which is later sent to the RGB LEDs. The microcontroller additionally controls the two vibration engines that vibrate at various sound levels. The potentiometer is a component controlled by the user. The surround sound, sound level, and the calibration of the potentiometer are all done by the user [25].

Components of the acoustic glass

Figure 10.1 consists of the architecture of the acoustic glass device used for the deaf. It involves the components used in the design of this device. The following are the outlined components of the device:

1. **Microphone:** This gadget picks the encompassing sound waves and sends the sound to be interpreted. As shown in Figure 10.2, this investigation utilizes a Maximum 9812 amplifier. The gadget is portrayed by its little size and simple to oversee. The Maximum 9812 receiver gadget has one info which makes it a mono-channel gadget. It has a coordinated intensifier that assists with enhancing the encompassing sounds that are picked. It channels and disposes of undesirable sound that is not required on the glasses. This mouthpiece is likewise described by three pins, one associated with the battery, ground level, and the last one associated with the Arduino Nano. The mouthpiece has a capacitor that is associated with an adaptable stomach on the article. The adaptable stomach vibrates when sound arrives at the receiver. The mouthpiece is fixed on the glasses on the left focal point in order to plainly pick signals.

Figure 10.1 Block diagram of the acoustic glass

Figure 10.2 MAX9812 microphones. Adapted from [26]

Figure 10.3 Arduino Nano. Adapted from [27]

2. **Arduino Nano:** Figure 10.3 speaks to the Arduino Nano as a microcontroller that comprises of a microchip, capacitors, resistors, voltage controllers, and other info and yield sticks that are computerized or simple. Arduino Nano changes in size, multifaceted nature, and capacities. Arduino is in various sorts, Arduino Uno (the known and productive), Arduino Mega (biggest), Arduino DUE, and Arduino Nano. This work utilizes Arduino Nano which has similar capacities with Arduino Uno. Arduino Nano utilizes one simple contribution to this work. It peruses sound sign picked by the receiver and handles to decide the shading (RGB LEDs) meant for the client. The identified quality of the sound decides the vibration.
3. **RGB LEDs:** Figure 10.4 is an image that speaks to the RGB LEDs. RGB is a shortened form for Red, Green and Blue while LEDs is a truncation for light emitting diodes. The hues all show the degree of sound passed. These hues

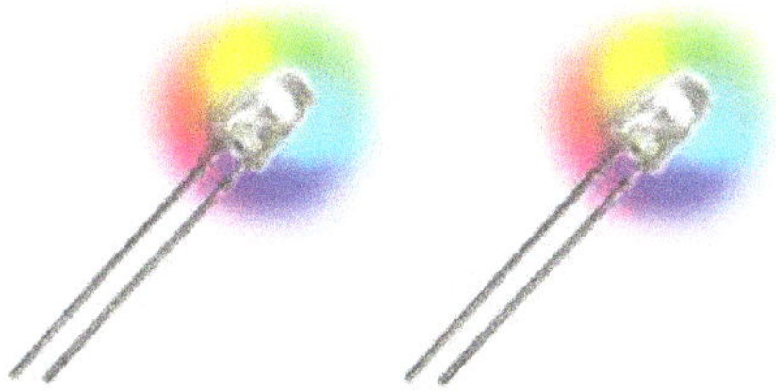

Figure 10.4 RGB LEDs. Adapted from [28]

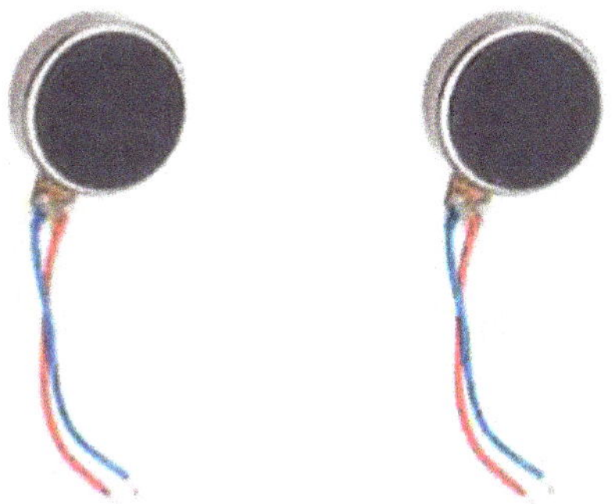

Figure 10.5 Vibration motors. Adapted from [29]

speak to the quality of the sound. For instance, when the sound is high, the LEDs show a red light, when it is low it discharges a green light, and when it is blue light it is in balance. These hues give diverse sound power.

4. **Vibration motor:** These are significant parts of the acoustic glass. They are little engines with qualified body shaft that create vibration to the client. They comprise of two terminals that are associated with different segments. This work utilizes two vibration engines that are fixed on either side of the glasses behind the client's ears as shown in Figure 10.5.
5. **Potentiometer:** Figure 10.6 is a picture of the potentiometer utilized. It is an instrument that is utilized in estimating the distinction in electric potential. In this work, the potentiometer utilized is described by a resistor. The light and vibration initiation is constrained by the client, which is aligned at an ideal level.
6. **Battery:** Figure 10.7 is the battery utilized as a wellspring of intensity supply. Force supply is required for any gadget. A gadget can't work without power. This work utilizes a 3.7-volt battery with 1,000 mAh, which is battery-powered and productive for this work.

Figure 10.6 Potentiometer. Adapted from [30]

Figure 10.7 Li-Po battery. Adapted from [31]

10.6 Discussions

A perceptible element of this gadget is its capacity to give elective signs to the hard of hearing. This remarkable component isn't found in different gadgets. Different investigations just enhance the sounds without alarming the client. Be that as it may, this gadget gives both sound intensification and the discovery of sounds to the client. Analysts have utilized different methods for helping the hard of hearing hear by supplanting the harmed organs with useful ones through medical procedure. This procedure is perilous for the patients. After connecting the various components, this device was effective. Subsequent to associating the circuit, the model gives better than expected to generally excellent outcomes: this incorporates the force utilization of every part alone and the model as a whole, and the reaction time of every part which is viewed as a significant theme.

Another way of solving this medical condition is the use of audiovisual subtitle in movies. Recently, investigations of dynamic visual investigation in hard of hearing grown-ups have additionally been directed utilizing less difficult visual improvements, for example, specks or lines, that activated plain consideration shifts [32]. Audiovisual interpretation researchers center around movie naming, subtitling, and voice-over as types of multimodal correspondence [33]. Other disadvantages of this method are that it makes sensory information accessible in diverse semiotic codes and offers the opportunity to comprehend information through different channels [34].

Acoustic glass has been used in solving a couple of research problems. Other research studies depend only on acoustic data. For instance, Alonso *et al.* [35] portray a framework ready to order anuran utilizing acoustic highlights. For deciding the start of a current competitor sound, it incorporates an initial step as an edge-based vitality indicator, which can be considered as an essential acoustic saliency identifier [35]. This made it easier for this work to be accomplished.

The most noticeable preferred position of this gadget is its capacity to be utilized effectively by anybody. The battery is battery-powered and truly reasonable. The highlights make it look extremely special when worn by the patient. It vibrates to provoke the client when there is peril.

This device exhibits almost 90% accuracy. Obviously, it has its flaws which are the high vibration for the user, which can also irritate.

The gadget was planned in an advanced manner. Be that as it may, later on certain highlights will be included based on the need. The highlights will be scaled down and the location of sound dependent on course utilizing GPS will be included.

10.7 Conclusion

This gadget is structured as an endeavor to build up a modernized gadget to support the hard of hearing. The gadget has unfathomable highlights that unmistakable it from other portable hearing assistant gadgets utilized for the hard of hearing.

Specialists have created gadgets that can enhance sounds without proffering answer for sound discovery for the hard of hearing. This gadget is created to recognize and intensify sounds at various levels. It likewise offers the client the chance to modify the potentiometer at various degrees of sound got [36].

Other means of treating partial or complete hard of hearing are cochlear implant (which is mainly done for those having complete of hearing), visual subtitle (important to those who can read), and using other available devices [37,38].

This gadget is exceptionally considered, relating it to different gadgets. It is likewise intended to give certainty when worn due to its extravagant and present-day highlights.

The point of this work is to use an acoustic glass as a device to help the almost deaf people. This contraption is proposed to give the patient signal when the sound is amazingly uproarious, medium, and low. These signs are addressed by tints known as the RGB LEDs. The acoustic contraption is protected, strong, and sensible for patients.

Recently, similar research on investigations of dynamic visual investigation in hard of hearing grown-ups has likewise been led utilizing less difficult visual boosts, for example, spots or lines, that activated plain consideration shifts [39]. This, however, cannot match the outcome of this research. Relating it to other devices used in treating the hard to hearing and the deaf, it is efficient, cost effective, and reliable.

References

[1] Al-Turjman F. "Intelligence and security in big 5G-oriented IoNT: An overview." *Future Generation Computer Systems.* 2020. 102: 357–368
[2] Al-Turjman F., Zahmatkesh H., and Mostarda L. "Quantifying uncertainty in Internet of Medical Things and big-data services using intelligence and deep learning." *IEEE Access*. 2019; 7(1): 115749–115759.
[3] Al-Turjman F., Nawaz M. H., and Ulusar U. D. "Intelligence in the Internet of Medical Things era: A systematic review of current and future trends." *Computer communications.* 2020; 150: 644–660.
[4] Al-Turjman F., and Alturjman S. "Context-sensitive access in Industrial Internet of Things (IIoT) healthcare applications." *IEEE Transactions on Industrial Informatics*. 2018; 14(6): 2736–2744.
[5] Jacob S., Menon V. G., Al-Turjman F., Vinoj P. G., and Mostarda L. "Artificial muscle intelligence system with deep learning for post-stroke assistance and rehabilitation." *IEEE Access*. 2019; 7: 133463–133473.
[6] Al-Turjman F. "A rational data delivery framework for disaster-inspired Internet of Nano-Things (IoNT) in practice." *Springer Cluster Computing*. 2019; 22(Suppl. 1): 1751–1763.
[7] Scheetz N. "Review of preparing deaf and hearing persons with language and learning challenges for CBT: A pre-therapy workbook." *Journal of Deaf Studies and Deaf Education*. 2019; 24(2): 188.
[8] Al-Turjman F. "A cognitive routing protocol for bio-inspired networking in the Internet of Nano-Things (IoNT)." *Springer Mobile Networks and Applications*. 2017. doi: 10.1007/s11036-017-0940-8.
[9] "Hearing loss may cause reassignment of auditory brain areas." *The ASHA Leader*. 2015; 20(9): 19–19. doi: 10.1044/leader.rib2.20092015.19
[10] McCracken. Teaching language to a boy born deaf: The Popham notebook and associated texts. Oxford, Clarendon Press: 2019.
[11] Hawkins S. "Phonological features, auditory object, and illusions." *Journal of Phonetics*. 2010; 38(1): 60–89.
[12] National Institute on Deafness and Other Communication Disorders (NIDCD). "Hearing aids." Retrieved from "Hearing Aids". Archived *from the original on 2011-11-13*. 2013. Available from https://www.nidcd.nih.gov/health/hearing-aids [Accessed December 2, 2011].
[13] Abu Samra Y., and Maghari A. "Enhancing FP-growth performance using multi-threading based on comparative study." *International Journal of Intelligent Computing Research*. 2015; 6(3): 613–620.

[14] "Hearing loss". *HealthCentral*. Available from https://www.healthcentral.com/category/hearing-loss [Accessed June 8, 2013].

[15] Beukes E., and Manchaiah V. "U.S. media portrayal of hearing loss and hearing aids." *The Hearing Journal*. 2019; 72(6): 36.

[16] Colucci D. "Cannabis and hearing care." *The Hearing Journal*. 2019; 72(8): 43.

[17] How Hearing Loss—and Aids—Affect Learning New Words. (2019). *The ASHA Leader*, 24(4).

[18] Carvalho G. M., Guimaraes A. C., Lucia F. D., *et al.* "Evaluation of the Digisonic® SP cochlear implant: Patient outcomes and fixation system with titanium screws." *Brazilian Journal of Otorhinolaryngology*. 2012; 78(6): 56–62.

[19] Papadakis C., Hajiioannou J., Kyrmizakis D., and Bizakis J. "Bilateral sudden sensorineural hearing loss caused by Charcot-Marie tooth disease." *The Journal Of Laryngology & Otology*. 2003; 117(5): 399–401. doi: 10.1258/002221503321626465

[20] Burnham D., Leigh G., Noble W., *et al.* "Parameters in television captioning for deaf and hard-of-hearing adults: Effects of caption rate versus text reduction on comprehension." *Journal of Deaf Studies and Deaf Education*. 2008; 13(3): 391–404. doi: 10.1093/deafed/enn003

[21] Briggs R. J. S, Tykocinski M., Xu J., *et al.* "Comparison of round window and cochleostomy approaches with a prototype hearing preservation electrode." *Audiology and Neurotology*. 2006; 11(Suppl.1): 42–48.

[22] Avci E., Nauwelaers T., Lenarz T., Hamacher V., and Kral A. "Variations in microanatomy of the human cochlea." *The Journal of Comparative Neurology*. 2014; 522(14): 3245–3261.

[23] Al-Turjman F., Altrjman C., Din S., and Paul A. "Energy monitoring in IoT-based ad hoc networks: An overview." *Computers and Electrical Engineering*. 2019; 76: 133–142.

[24] Bridges B. "Deaf-Heart in the interpreting field." *The Journal of Deaf Studies and Deaf Education*. 2019: 24(3): 317–317.

[25] Hearing aid buying guide. Consumer reports. February 2017. Archived from the original on 12 February 2017. Available from https://www.consumerreports.org/cro/hearing-aids/buying-guide/index.htm [Accessed February 13, 2019].

[26] Sessler G. M., and West J. E. "Self-biased condenser microphone with high capacitance." *Journal of the Acoustical Society of America*. 1962; 34(11): 1787–1788. doi:10.1121/1.1909130

[27] Arduino – Introduction. 2020. Available from https://www.arduino.cc/en/Guide/Introduction?setlang=en [Accessed January 22, 2020].

[28] US $33.99 |Free shipping 10pcs/lot Animal fashion Korea stationery cartoon wooden pencil pen holder with notes & clip 8.5*7*5.2cm-in Pen Holders from Education & Office Supplies on AliExpress. 2020. Available from https://www.aliexpress.com/item/1740787721.html [Accessed January 22, 2020].

[29] See https://shop.edwinrobotics.com/motors/595-vibrating-minimotor-disc.html [Accessed January 10, 2020].
[30] See https://www.switchelectronics.co.uk/5k-3362-single-turn-cermetpotentiometer-10-tolerance [Accessed January 12, 2020].
[31] Lithium Ion Battery - 1Ah - PRT-13813 - SparkFun Electronics. 2020. Available from https://www.sparkfun.com/products/13813 [Accessed January 22, 2020]
[32] Tardieu H., and Gyselinck V. "Working memory constraints in the integration and comprehension of information in a multimedia context." in H. van Oostendorp (ed.), *Cognition in a digital world*. London: Lawrence Erlbaum; 2003. pp. 3–24
[33] Carroll M., Gerzymisch-Arbogast H., and Nauert S. (eds.). "Audiovisual translation scenarios." *Proceedings of the Second MuTra Conference*, Copenhagen May 1–5, 2006, pp. 1–8 [Accessed April 1, 2016].
[34] Kress G., and Van Leeuwen T. *Reading images: The grammar of visual design*. New York: Routledge; 1996.
[35] Alonso J. B., Cabrera J., Shyamnani R., *et al*. "Automatic anuran identification using noise removal and audio activity detection." *Expert Systems with Applications*. 2017; 72: 83–92.
[36] Lindsay J. R., Kohut R. I., and Sciarra P. A. "Menière's disease: Pathology and manifestations." *Annals of Otology, Rhinology & Laryngology*. 1967; 76: 5–22.
[37] Anmyr L., Larsson K., and Olsson M. "Parents' stress and coping related to children's use of a cochlear implant: A qualitative study." *Journal of Social Work in Disability & Rehabilitation*. 2016; 15(2): 150–167.
[38] Blau V., Reithler J., van Atteveldt N., *et al*. "Deviant processing of letters and speech sounds as proximate cause of reading failure: A functional magnetic resonance imaging study of dyslexic children." *Brain*. 2010; 133: 868–879.
[39] Bottari D., Valsecchi M., and Pavani F. Prominent reflexive eye-movement orienting associated with deafness. *Cognitive Neuroscience*. 2012; 3(1): 8–13. 10.1080/17588928.2011.578209

Chapter 11

A framework for blind people using wireless medical sensors network

Mostafa Fakhouri[1], Ameer Jubran[1], Rashad Ghaleb[1], Timipawopri Adada[1], Fadi Al-Turjman[2,3,4], Ilker Ozsahin[1,4] and Dilber Uzun Ozsahin[1,4]

This audit explores a wireless framework for blind people in other to aid them with exploring with little/no help from others. With the assistance of this gadget, a client can move autonomously and ready to walk uninhibitedly practically like an ordinary individual. This helps the blind people who face challenges in identifying data about territory, particularly of stairs and gaps. In identifying territory, scientists have attempted various sensors such as ultrasonic, laser, camera and so forth, and distinctive calculation; in any case, agreeable result has not yet been accomplished and white stick is as yet the most utilized strolling support for the visually impaired individuals. This is on the grounds that practically all the examination results neglected to address the urgent prerequisites of the strolling emotionally supportive networks, similar to low vitality necessity, light weight, opening, and stair location abilities. Taking this issue into consideration in design, a belt, wearable around the abdomen is furnished with ultrasonic sensors. These sensors are associated with a microcontroller alongside a PC so we can get adequate information for breaking down territory on the walkway of the visually impaired outfitted with ultrasonic sensors. The accomplishments in the gadget are light, modest, and devour less vitality. Be that as it may, this gadget is restricted to standard pace of portability of the client and can't separate among vivify and lifeless impediments. In this manner, further research is prescribed to conquer these insufficiencies to improve versatility of visually impaired individuals.

[1]Department of Biomedical Engineering, Near East University, Nicosia / TRNC, Mersin-10, Turkey
[2]Department of Artificial Intelligence, Near East University, Nicosia / TRNC, Mersin-10, Turkey
[3]Research Center for AI and IoT, Near East University, Nicosia / TRNC, Mersin-10, Turkey
[4]DESAM Institute, Near East University, Nicosia / TRNC, Mersin-10, Turkey

11.1 Introduction

With the methodology of the maturing society, settling the observation of the long haul, interminable sicknesses has become a significant issue. Some intense ailments, for example, cardiovascular ailment, the everyday care of the old, and the wellbeing checking of the pregnant ladies, babies, and newborn children, additionally need the help of observing framework [1–3]. Existing restorative consideration frameworks, the majority of which utilize fixed therapeutic screens, complex types of gear, and wired information transmission frameworks, would bring about the patient's mental pressure and strain, and influence the rightness of gathered information and analysis of the illness [1,4]. Particularly in the ward care, an assortment of associations not just make the patients feel awkward, yet additionally cause the wards to seem disarranged, influencing the patients' disposition [5,6]. Subsequently, a viable therapeutic checking framework needs an ease, high unwavering quality remote transmission plot rather than the conventional wired methodology. The development of wireless medical sensor network system and its applications in the therapeutic field have infused new essentialness into the investigation of remote restorative observing framework [7,8].

Persistent observation of basic essential vitals of patients is a key procedure in emergency clinics. Today, this is typically performed by means of various cabled sensors, being appended to the patient and associated with bedside screens [7,9]. These arrangements have clear restrictions—the measure of estimations is restricted by accessible gadget plugs, quiet treatment is upset by cabling, and lastly understanding versatility is seriously confined, as the patient is tethered to gadgets at the bedside [10]. This opens up another application zone for sensor systems. Little remote sensors, joined to the patient body, measure vital sign information, and transmit them by means of the built-up sensor system to an outer perception unit [11]. Current sensor organize arrangements need to develop to be appropriate for the therapeutic area, which requires adaptable, yet safe set-up of the framework just as solid, safeguard activity [12]. The present sensor systems, basically utilized for natural checking, are either statically prearranged for a specific errand or construct unconstrained specially appointed systems. For therapeutic utilization, this framework conduct must be changed to a solid and characterized framework set-up, working naturally yet by and by being under unequivocal control of a clinician [13,14]. For this, two issues must be settled: first, an appropriate correspondence convention is required for the wide assortment of medicinal sensors, and second, an advantageous and safeguard set-up component must be created [15]. This chapter presents the engineering for a dispersed therapeutic body sensor system, and afterward for the most part centers around the framework set-up process [15]. The framework arrangement incorporates programmed tolerant recognizable proof and hence permits a protected gadget affiliation. Clinicians, who will be the clients of the sensor organize, require an adaptable and simple to-go through the methodology to set patient sensors [16]. Be that as it may, they additionally should be consistently in control of the framework, which bars totally programmed revelation and setup systems. Giving the clinician an instinctive way to control affiliation and disassociation of remote sensors to the system [17], the main aim is to present a useful engineering set-up and the itemized convention of the set-up strategy, in which a wireless medical sensor framework can be integrated with to aid blind people in navigation and environmental awareness [18].

11.2 Related works

11.2.1 White cane

This is a stick used by the blind or visually impaired. It helps the blind person to check for any obstacles in the close environment. This cane is used as an extension of the blind person's arm to feel the environments. It has no electrical component [19]. Other canes are used in the medical field but only in blind cases, the color white is used. It also enables people around the patient, whether in the street, public transport, or in markets, to identify the blind person and they can help by clearing the path (Figure 11.1) [20].

11.2.2 Ultrasonic-based blind assisting system

The type of sensor used is ultrasonic distance sensor, without the use of Servomotor or global positioning system (GPS), for alerting the user that there are obstacles

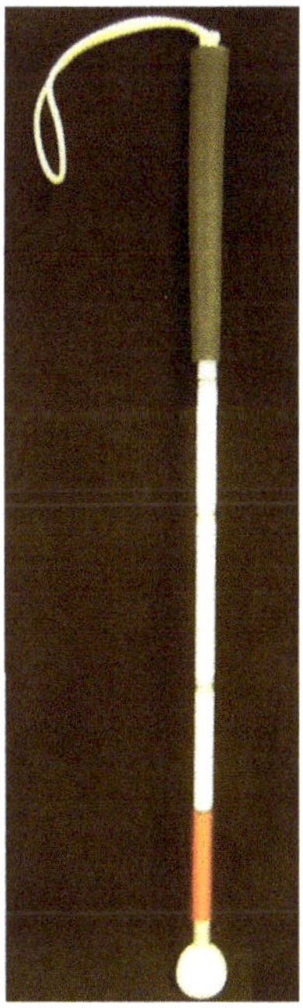

Figure 11.1 White cane. Adapted from [53]

ahead of him/her. Buzzers, vibration motors, or both were used here and when the user starts operating on this device, the transmitter in the ultrasonic sensor sends or emits the signals which hit the obstacles, and the signals are refracted/reflected back [21]. The reflected/refracted signals are detected by the receiver in the ultrasonic sensor, and these detected signals will be converted to electrical signals by the microcontroller (ARDUINO or any processing device or system), and the converted signals go to the alarming system which consists of buzzers, vibration motors, or both of them [22].

This system gives the user the advantage of the ultrasonic sensor's wide range of detection which is between 20 cm and 4 m, but this will only show the obstacles without providing any information about the locations of the obstacles which may not be useful for the user if the obstacle is located outside the peripheral detecting range of the ultrasonic sensor [23].

11.2.3 Infrared-based blind assisting system

The type of sensor used in this system (Figure 11.2) is infrared distance sensor, without the use of Servomotor or GPS, for alerting the user that there are obstacles ahead of him/her. Buzzers, vibration motors, or both were used here and when the user starts operating on this device, the first IR-LED that acts like a transmitter sends or emits the signals that will hit the obstacles, and after hitting the obstacles, the signals are refracted/reflected back [24]. The reflected/refracted signals are detected by the second IR-LED that acts as a receiver, and these detected signals will be converted to electrical signals by the microcontroller (Arduino or any processing device or system), and the converted signals go to the alarming system which consists of buzzers, vibration motors, or both of them [25].

By using infrared sensor, the range of detection is smaller than the detection range in the Ultrasonic sensor (15–80 cm). It is also more expensive than ultrasonic sensor, and less efficient in the circuit because it malfunctions faster and more often than ultrasonic sensor (Figure 11.3) [25].

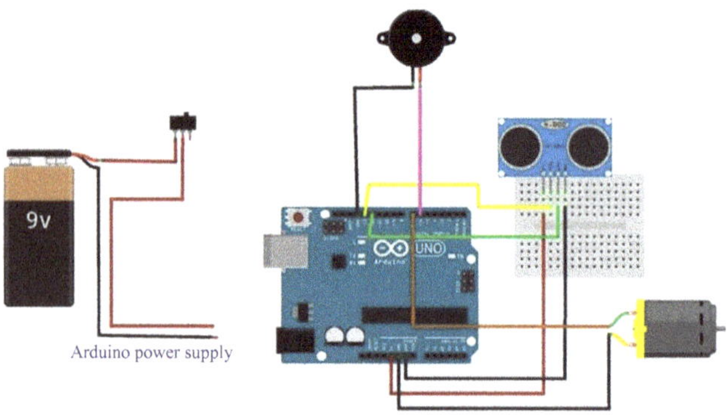

Figure 11.2 Ultrasonic-based system. Adapted from [23]

11.2.4 Sensor-based blind assisting system with global positioning system

The type of sensor (Figure 11.4) used is ultrasonic distance sensor or infrared distance sensor, without a Servomotor, for alerting the user that there are obstacles ahead of him/her. Buzzers, vibration motors, or both are used here, and when the user starts operating on this device, the transmitter in the sensor sends or emits the signals that will hit the obstacles, and after hitting the obstacles, the signals are refracted/reflected back [28]. The reflected/refracted signals are detected by the receiver in the sensor, and these detected signals will be converted to electrical signals by the microcontroller (Arduino or any processing device or system), and the converted signals go to the alarming system which consists of buzzers, vibration motors, or both of them [28].

Since GPS will be used, it will need to be connected with other components via Wi-Fi wireless or Internet from a sim card, and this will be more difficult in coding and the Wi-Fi component is more expensive than other components [29]. The main advantage of this is that it provides information about the user's location so that his family or caretakers know his place or location, and it can also have a memory card that has voice instructions to guide the user by giving him/her the directions along the route so that he/she can go to any place they want without anyone's help (Figure 11.4).

Since GPS will be connected via Wi-Fi, then it will need continuous satellite or network coverage so that the Wi-Fi keeps working flawlessly, [30] but it also needs a certain type of power to operate the Wi-Fi chip, so a battery should be used, but this battery may die unexpectedly, and if that happened it will be a flaw instead of being a good feature.

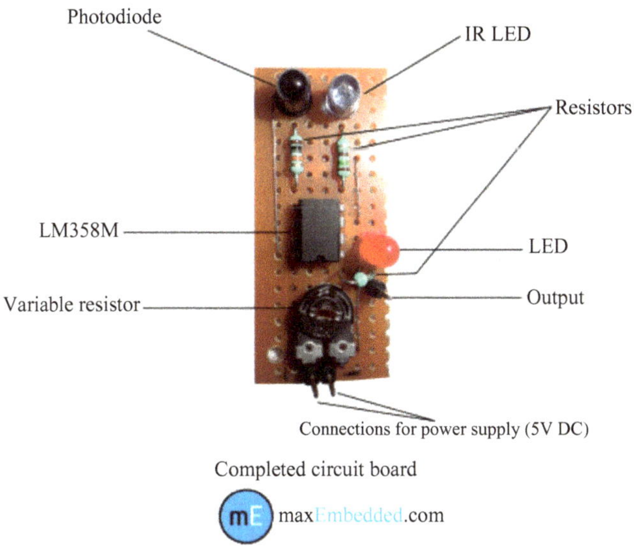

Figure 11.3 Infrared-based system. Adapted from [26]

Figure 11.4 Sensor-based system with global positioning system. Adapted from [27]

Using GPS doesn't mean that it will show or inform the user with the location of the obstacle, and it means that it will provide information about the location of the user and if he needs any help or not.

Unlike the previously mentioned techniques, in this audit we propose the use of an ultrasonic distance sensor instead of an infrared distance sensor because the ultrasonic distance sensor is more compatible and more efficient with the servomotor, which is a special motor that turns 180° to give a more accurate and precise detection of the obstacles around the user [31]. The buzzer and vibration motors will be controlled by switches that the user presses depending on the type of alarm that the user wants. Instead of using two ultrasonic sensors that are placed on different sides of the walking cane, we will use the servomotor to replace them to make the circuit smaller and lighter in weight. Where the GPS is used for locating the user by family or healthcare giver, it can be combined/replaced with the servomotor and vibration motors that will locate the position of the obstacles and alarm the user, because it is more efficient to locate the obstacles than locating the user of the cane since it will give the user more freedom in movement without asking for help. Although replacing the GPS with the servomotor and vibration motors would defeat the whole wireless purpose [32], by using two different methods or ways to alarm the user (buzzer and vibration motors), we can guarantee that the user will be alarmed at the same moment as the obstacle is detected, which will help the patient respond better and faster to the obstacles ahead of him/her.

11.3 The method

11.3.1 The circuit

Arduino is the masterpiece in our work, it will receive and analyze all the data that comes from the input components and send them to the output components in the circuit [33]. It has two lines of pins, six analog pins (Inputs), 15 (including one ground pin) digital pins (Outputs), and six power pins (including ground pin and

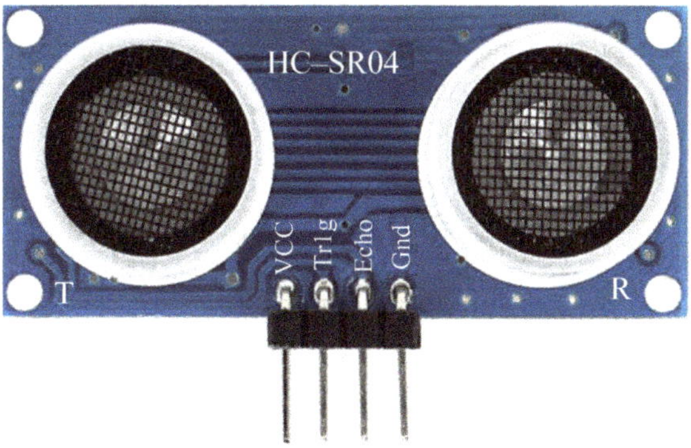

Figure 11.5 An ultrasonic distance sensor. Adapted from [34]

5 V pin), and it also has a special slot for USB cable to connect the Arduino with computers or tablets to upload the codes from the computer to the Arduino so the circuit can be activated.

A mini breadboard is used to minimize the size of circuit to make the cane lighter for the patient. On the breadboard, all components will be connected. After connecting the circuit, it will be placed inside the small box or container that will protect it from damages caused due to the movement of the user, and this container should also be waterproof to prevent liquids from leaking into the circuits.

Two ultrasonic sensors can be used for this work. They should be placed on two different sides so they can detect obstacles from different directions. But using a servomotor to make the circuit smaller in size in comparison with using two ultrasonic sensors, a 180° field of detection which will be wider than the field of detection provided by two ultrasonic sensors is provided. The servomotor will increase the reliability of the device and decrease the time needed for detecting the obstacles since it has fast acceleration and deceleration.

An ultrasonic distance sensor (Figure 11.5) is connected because it is easier to implement in the circuit with the servomotor, and it is also easier in coding and has a wider detection range (from 2 cm to 4 m) in comparison with infrared distance sensor (from 15 cm to 1.5 m depending on the input voltage).

Ultrasonic distance sensor (HC-SR04) has two main parts: transmitter and receiver. The transmitter will send ultrasonic waves or signals that will hit the objects or obstacles [35]. When the transmitted signals hit the objects, they will be refracted/reflected, the reflected/refracted signals will be detected by the receiver on the sensor, and the duration of time between transmitting the signal and receiving the reflected signal is used to determine the distance between the sensor and the obstacle or object (Figure 11.6).

11.3.2 Connecting the circuit

First, the Arduino Uno is connected with the breadboard by using cables or wires; we connect the negative part of the breadboard with the ground pin in the Arduino, we also connect the positive part of the breadboard with the 5 V pin in the Arduino, and this will make our connections later easier, faster, and less crowded by using less cables. Then we attach the ultrasonic sensor on top of the servomotor so that we can connect them on the breadboard together to take less space. On the ultrasonic sensor, we have four pins (VCC, TRIG, ECHO, and GND), and VCC is connected with the positive part of the breadboard, TRIG and ECHO are connected to a certain pin on the digital side of Arduino, and GND is connected with the negative part of the breadboard. After that we connect the buzzer with both the breadboard and Arduino by connecting the positive part of the buzzer to a pin on the digital side of Arduino and the negative part will be connected to the negative part of the breadboard (ground). Then we connect two vibration motors with the Arduino and breadboard by connecting the positive part to a pin on the digital side of the Arduino and the negative part will be connected to the negative part of the breadboard (ground). Each of the vibration motors will be attached to a special piece that is will be on top of the cane. Finally, we will use switches to control our circuit and organize it, and one of the switches will be used to activate the circuit (ON and OFF), while the other switch will control the buzzer (ON and OFF) without affecting the vibration alarm. This switch can be used to turn off the buzzer in crowded places to avoid disturbing other people while vibration will keep working continuously. The circuit will be powered by a 9 V battery, which will provide the voltage needed by the components to work properly so that we don't have any malfunction that may affect the user or people around them. The battery will be connected with both Arduino and breadboard and it will consider the power source or power supply (Figure 11.7).

The ultrasonic sensor which consist of four pins is connected as follows:

trigPin to Pin 6 in the digital part which is an output,

echPine to Pin 5 in the analog part which is an input VCC pin to the 5 V on the Arduino Uno,

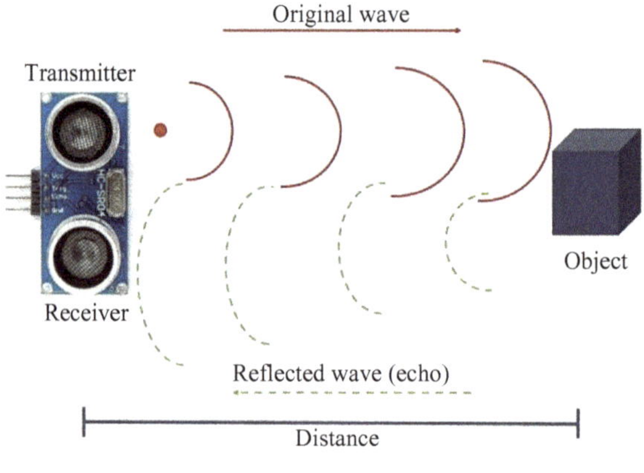

Figure 11.6 Working principle of the ultrasonic sensor. Adapted from [36]

GND pin to the ground on the Arduino Uno.

Also

Connect the buzzer to Pin 13.

The vibration motor to pin (10, 11, and 12) which all of them are output.

Connect the servo motor to pin 3 on the Arduino Uno.

The block diagram illustrated in Figure 11.8 shows how the circuit is being connected.

11.3.3 Long cane (white cane)

This is a cane used by blind people in their movements from one place to the other, and it will be the base since all other materials and components will be attached to it (Figure 11.9) [37].

11.3.4 Distance sensor

Ultrasonic sensor, as shown in Figure 11.10, generates and detects ultrasound. Infrared sensor is also an option but will cost more and has a smaller detection range, while ultrasonic sensor has a wider detection range and costs less than the infrared sensor (Figure 11.10) [19].

11.3.5 Buzzer

It will be used to give out a sound alarm when any obstacle is detected in the range of the ultrasonic distance sensor or ultrasonic distance sensor, and this will be our first type or method in alerting the blind about obstacles so that they can avoid them (Figure 11.11) [40].

11.3.6 Switch

It is used to turn off the buzzer in the crowded places to avoid any inconvenience caused by the buzzer or alarm, or it can be used to activate either sound alarm or vibration alarm or both (Figure 11.12).

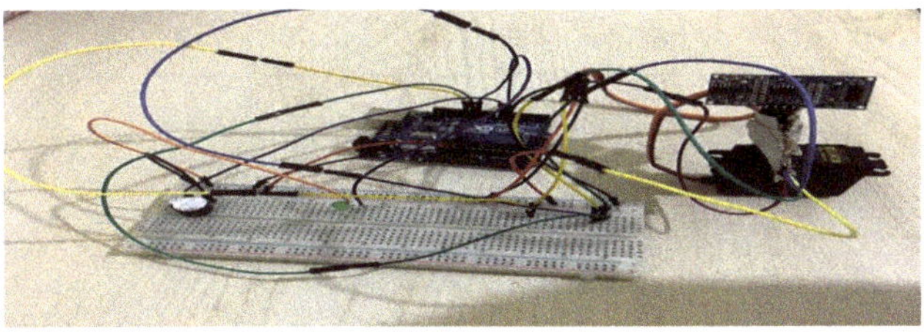

Figure 11.7 Sample of connection of circuit

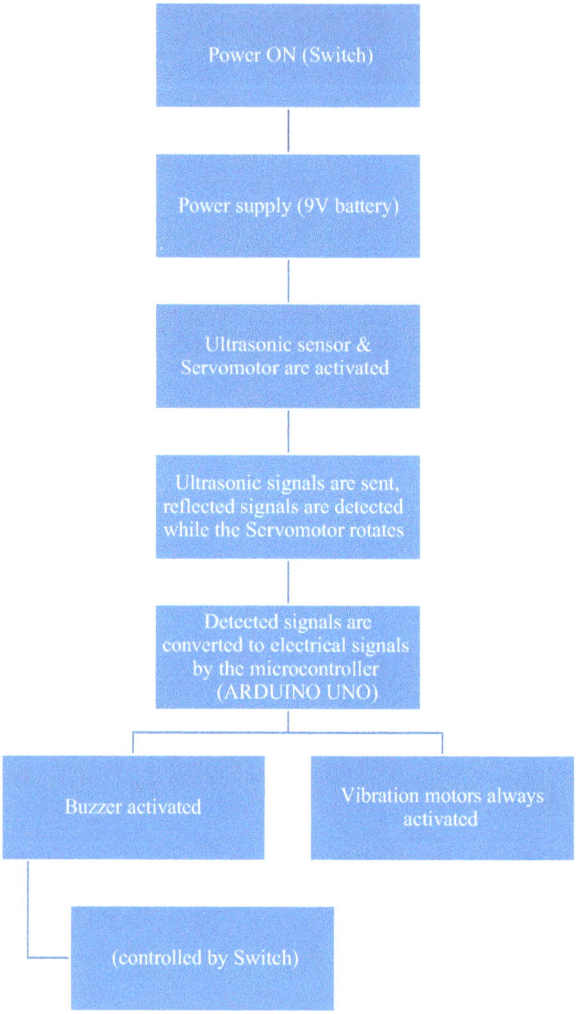

Figure 11.8 Block diagram of the circuit

Figure 11.9 White cane. Adapted from [38]

Figure 11.10 Ultrasonic sensor. Adapted from [39]

Figure 11.11 Buzzer. Adapted from [41]

Figure 11.12 Switch. Adapted from [42]

11.3.7 Vibration motor

A small electronic motor that works when it receives an electrical current. This generates a force that is interpreted as vibrations. We will use two vibration motors, and each motor will be attached to one hand; thus it will be our vibration alarm (Figure 11.13) [43].

11.3.8 Arduino Uno

Arduino is a microcontroller that receives all the data from the circuit (sensors, capacitors, switches, etc.) and analyzes them. After analyzing them, other components like (vibrators, motors, buzzers, LEDs, etc.) will work based on the code that is written earlier in the Arduino, so the Arduino will receive the data, analyze it, and translate it to orders that are done by other components in the circuit (Figure 11.14) [33].

11.3.9 Breadboard

A board where all the capacitors, resistors, cables, buzzers, and vibrators will be connected on it and the board will be connected with the Arduino by using the analog or digital pins that are on the Arduino (Figure 11.15) [46].

11.3.10 Belt or bracelets

It will be worn by the patient, and the vibrators will be attached to the bracelet on one hand (right or left) and will vibrate according to the data from the sensor that is attached to the servomotor.

11.3.11 Servomotor

A motor that moves in a certain way, horizontally or vertically depending on the use. It will be attached to the ultrasonic sensor and it will move horizontally, and by moving horizontally, it will act as a radar to determine the location of the obstacles so that blind people can avoid them (Figure 11.16) [48].

11.3.12 Resistors, cables, capacitors, and battery

These parts will complete our circuit, the battery will be our energy source. The motor starts moving with the sensor in order to detect the obstacles. The battery

Figure 11.13 Vibration motor. Adapted from [44]

Figure 11.14 Arduino Uno. Adapted from [45]

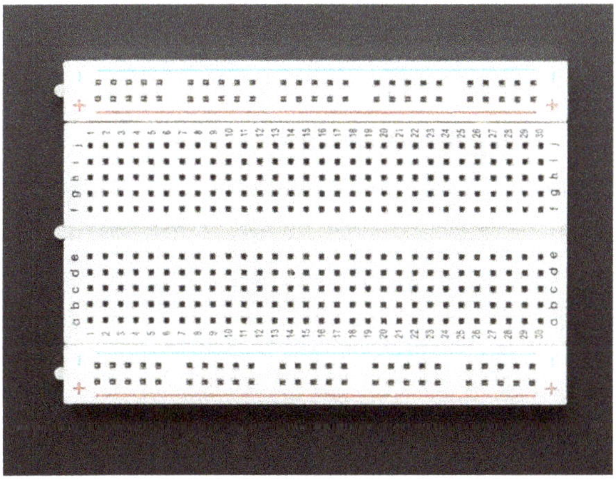

Figure 11.15 Breadboard. Adapted from [47]

used in this study is cost effective and it can be replaceable by rechargeable battery if needed (Figure 11.17–11.19) (6) (17) [50].

11.4 Results and discussion

After finishing the connections of our circuit and writing the codes for the microcontroller (Arduino Uno), the entire work is checked, and this included checking the status of the components separately and then checking the status of the

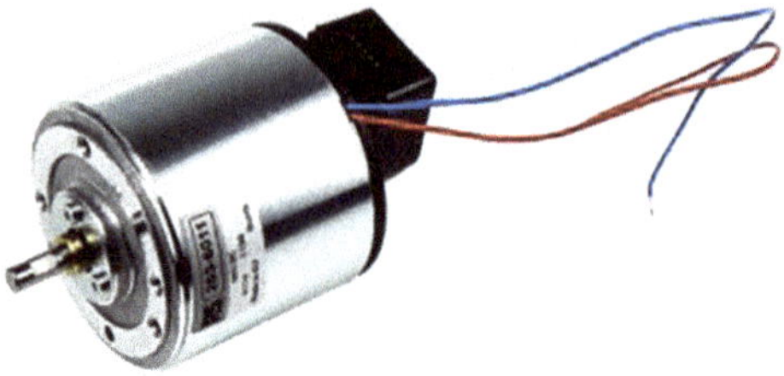

Figure 11.16 Servomotor. Adapted from [49]

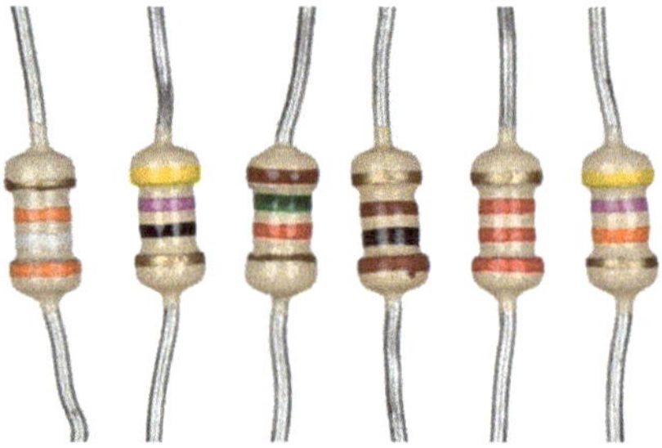

Figure 11.17 Resistors. Adapted from [51]

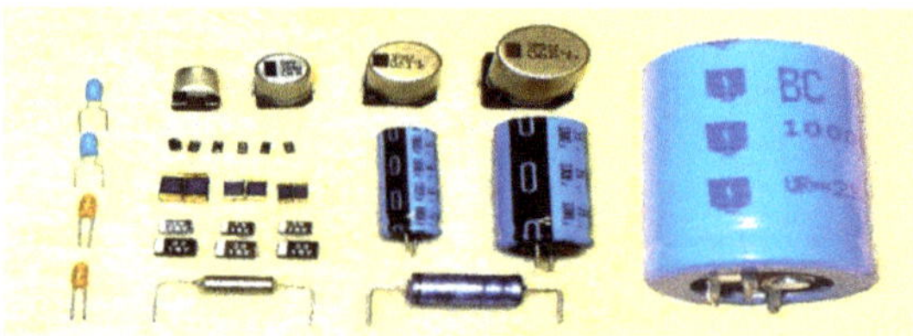

Figure 11.18 Capacitors. Adapted from [52]

Figure 11.19 9V battery. Adapted from [53]

Figure 11.20 The prototype

components together during the connection of the circuit. All components used are compatible with each other, and this makes the prototype more efficient, takes less time for operating, and will give better results because of the compatibility of the components.

When the operating switch is turned on, power will be supplied to the Servomotor, and the Servomotor will rotate at an angle of 180°, and as it rotates, the ultrasonic distance sensor which is located on top of the servomotor moves according to the rotation of the servomotor. While the servomotor rotates, the ultrasonic distance sensor sends ultrasonic signals that will hit the obstacles or objects within a range that is chosen earlier in the coding part which is 50 cm.

After obstacles or objects were detected, the alarming system which consists of the buzzer and the vibration motors is activated. A sound alarm from the buzzer will be noticed and the vibrations from the vibration motors and these alarms work simultaneously as we said earlier in the earlier chapter.

The buzzer can be turned off by a special switch and making the prototype depends only on the vibration motors to check if they work properly. This switch gives the user the freedom of choice in which type of alarm he/she wants depending on their preferences (Figure 11.12).

The prototype (Figure 11.20) has a fast response time. It also has a very small gap in the period between detecting the objects or obstacles and activating the alarming systems (buzzer and vibration motors), and the components consume little amount of power that is supplied by the 9 V battery that is in the prototype (Figure 11.20).

In summary, the working principle of our prototype is as follows:

First, the main switch is turned on, which will activate the circuit, and when the circuit is activated the battery will provide the power needed by the components to work properly. One of these components is the servomotor that holds the ultrasonic sensor that sends signals and detects them when they are reflected. These signals are converted to electrical signals by the microcontroller which is Arduino Uno, and the converted signals are then sent to output components which are three vibration motors that will vibrate and a buzzer that will generate sounds

Table 11.1 The prototype result table based on power usage and response time

Components	Main feature	Power consumption	Time response
Prototype	Servomotor	Low	Fast
Battery	Power source	None	Fast
Servomotor	Rotation for sensor	Low	Fast
Ultrasonic sensor	8–50 cm	Low	Fast
Arduino Uno	Microcontroller	Medium	Relatively fast
Vibration motors	Alarm system	Low	Fast
Buzzer	Alarm system	Low	Fast

simultaneously with the vibration motors. We have three vibration motors to determine the direction of the detected obstacles so that the user can avoid them.

After connecting the circuit, the prototype gives between above average to very good results, and this includes the power consumption of each component alone and the prototype as a whole. The response time of each component is considered an important topic in our prototype. The detection range of our prototype may differ between 8 cm and 4 m but we chose it to be between 8 cm and 50 cm because it is more convenient for our idea.

Table 11.1 explains the prototype in general does not consume lots of energy and these results are obtained by manually connecting each of the components separately to a power source in a functioning electric circuit and measuring their power consumption and response time. This is done to make sure the system energy supply (battery) can easily supply the required voltage to power the system. As we can see from Table 11.1, the power consumption of servomotor, the ultrasonic sensor, and vibration motor/buzzer which functions as an alarm system is low. The Arduino Uno board functioning as the microprocessor is medium having the highest power consumption rate which is to be expected since it does all the processing. The power supply does not consume any power obviously, because it supplies the power hence the result is none in the table. Every component mounted on the prototype has almost instant response time except for the Arduino Uno which should be expected since processing takes place there. This audit serves as a guide to building the system perfectly and the choice to add GPS to the system is open and is very compatible with the porotype too. This would include a sim card to bring about a Wi-Fi service. The device as said earlier is installed in a belt bringing convenience to its user, and because it is a wearable device in can be fashion into any accessory chosen by the user to be connected to the cane. This result is very practical because the device was tried and tested walking a short distance by some students from Near East University in Northern Cyprus.

11.5 Conclusion

Humans interact and respond to conditions and circumstances around them based on what they see, hear, or smell from their environment. This is why it is important to maintain our senses and protect them from any harm or damage. Any damage we take will either eliminate our sensory organs or cause them to malfunction.

Visually impaired people form a large portion of our society. Science and engineering branches are concerned for helping and improving their lives by designing many techniques that can provide them with their daily needs with minimal/no help from other people, which increases their independence and freedom. This design is not a fix but would help to take a step forward in making their life easier and can be a platform for a major breakthrough.

The prototype differs from earlier prototypes that use infrared sensors alone, ultrasonic sensors alone, and infrared or ultrasonic sensors with GPS implemented with them. Each prototype has its advantages, disadvantages, materials, and working principle, as it is explained earlier in detail. But the combination of them would improve on the device.

The battery in this prototype can be considered as the main disadvantage because it can die unexpectedly anywhere and anytime if it wasn't checked periodically by the user, his family, or caretaker. One of the ways this can be solved with is replacing the battery with solar cells, in any type, normal or bendable, but this will affect the complexity of the device, the whole cost of the device will also increase and this is the main reason they weren't used in the prototype. Voice navigation system can be used in the future to help the visually impaired in a better way and to give them more choices.

References

[1] Teng X. F., Zhang Y. T., Poon C. C., and Bonato P. "Wearable medical systems for p-health." IEEE Reviews in Biomedical Engineering. 2008; 1: 62–74.

[2] Al-Turjman F., Zahmatkesh H., and Mostarda L. "Quantifying uncertainty in Internet of Medical Things and big-data services using intelligence and deep learning." *IEEE Access*. 2019; 7(1): 115749–115759.

[3] Al-Turjman F., Ulusar U., and Nawaz M. "Intelligence in the Internet of Medical Things era: A systematic review of current and future trends." *Elsevier Computer Communications Journal*. 2020; 150(15): 644–660.

[4] Al-Turjman F., and Alturjman S. "Context-sensitive access in Industrial Internet of Things (IIoT) healthcare applications." *IEEE Transactions on Industrial Informatics*. 2018; 14(6): 2736–2744.

[5] Mukhopadhyay S. C. "Wearable sensors for human activity monitoring: A review." IEEE Sensors Journal. 2014; 15(3): 1321–1330.

[6] Al-Turjman F. "Intelligence and security in big 5G-oriented IoNT: An overview." *Elsevier Future Generation Computer Systems*. 2020; 102(1): 357–368.

[7] Buckley J. L., McCarthy K. G., Gaetano D., Loizou L., O'Flynn B., and O'Mathuna C. "Design of a compact fully-autonomous 433 MHz tunable antenna for wearable wireless sensor applications." Microwaves Antennas & Propagation IET. 2017; 11(4): 548–556.

[8] Al-Turjman F. "A rational data delivery framework for disaster-inspired Internet of Nano-Things (IoNT) in practice." *Springer Cluster Computing*. 2019; 22(Suppl. 1): 1751–1763.

[9] Yang D., Cheng Y., Zhu J., *et al.* "A novel adaptive spectrum noise cancellation approach for enhancing heartbeat rate monitoring in a wearable device." Access IEEE. 2018; 6: 8364–8375.

[10] Yang N., Wang Z., Gravina R., and Fortino G. "A survey of open body sensor networks: Applications and challenges." *2017 14th IEEE Annual Consumer Communications & Networking Conference (CCNC)*. Las Vegas, NV. 2017; 65–70. DOI: 10.1109/CCNC.2017.7983083.

[11] Bonato P. "Wearable sensors and systems." IEEE Engineering in Medicine and Biology Magazine. 2010; 29(3): 25–36.

[12] Wu M., and Xie Q. "The design of wireless medical monitoring network based on ZigBee." *Network computing and information security*. Berlin, Heidelberg: Springer; 2012.

[13] Yang G., Xie L., Mäntysalo M., *et al.* "A health-IoT platform based on the integration of intelligent packaging, unobtrusive bio-sensor, and intelligent medicine box." IEEE Transactions on Industrial Informatics. 2014; 10(4): 2180–2191.

[14] Al-Turjman F. "A cognitive routing protocol for bio-inspired networking in the Internet of Nano-Things (IoNT)." *Springer Mobile Networks and Applications*, 2017. DOI: 10.1007/s11036-017-0940-8.

[15] Pawar A., and Ghumbre S. "A survey on IoT applications, security challenges and counter measures. *2016 International Conference on Computing, Analytics and Security Trends (CAST)*. 2016; 294–299. DOI: 10.1109/CAST.2016.7914983.

[16] Zhai Y., Liu Y., Zhou T., and Shen P. "Identification of key factors in health service adoption based on Internet of Things and empirical test." *2017 29th Chinese Control and Decision Conference (CCDC)*. 2017. pp. 7257–7262.

[17] Zhou B., Hu C., Wang H. B., Guo R., and Meng M. Q.-H. *A wireless sensor network for pervasive medical supervision.* IEEE: Shenzhen, China; 2007.

[18] Bharadwaj S. A., Yarravarapu D., Charan Kumar Reddy S., Prudhvi T., Sandeep K. S. P., and Siva Dheeraj Reddy O. "Enhancing healthcare using m-care box (monitoring non-compliance of medication)." *2017 International Conference on Innovative Mechanisms for Industry Applications (ICIMIA)*. Bangalore, 2017; 167–171. DOI: 10.1109/ICIMIA.2017.7975594.

[19] Faria J., Lopes S., Fernandes H., Martins P., and Barroso J. "Electronic white cane for blind people navigation assistance." *2010 World Automation Congress*. Kobe, IEEE. 2010; 1–7.

[20] [Online] Available from https://maxembedded.files.wordpress.com/2013/07/circuit-discription-new.png?resize=470%2C415 [Accessed December 13, 2019].

[21] Bunnan S., Singh G. P., and Tondare S. P. "Ultrasonic blind walking stick for the visually impaired." IJRET. 2016; 5: 350–352.

[22] Batarseh D. T., Burcham T. N., and McFadyen G. M. "An ultrasonic ranging system for the blind." *Proceedings of the 1997 16 Southern Biomedical Engineering Conference* (pp. 411–413). IEEE. April 1997.

[23] "How to make a smart cane for the visually impaired with Arduino." Marker pro. [Online] Suhail X, August 23, 2016. Available from https://maker.pro/arduino/projects/arduino-smart-cane-for-the-blind [Accessed December 13, 2019].

[24] Bhardwaj P., and Singh J. "Design and development of secure navigation system for visually impaired people." International Journal of Computer Science & Information Technology. 2013; 5(4): 159.

[25] Ertan S., Lee C., Willets A., Tan H., and Pentland A. A wearable haptic navigation guidance system. In Digest of Papers. Second International Symposium on Wearable Computers (Cat. No. 98EX215). Pittsburgh, PA, USA. 1998; 164–165. DOI: 10.1109/ISWC.1998.729547.

[26] [Online] Available from http://maxembedded.com/2013/08/how-to-build-an-ir-sensor/ [Accessed December 14, 2019].

[27] [Online] Available from https://i.ytimg.com/vi/vJGg6AuJ3fo/hqdefault.jpg [Accessed December 13, 2019].

[28] Velázquez R., Pissaloux E., Rodrigo P., Carrasco M., Giannoccaro N., and Lay-Ekuakille A. "An outdoor navigation system for blind pedestrians using GPS and tactile-foot feedback." Applied Sciences. 2018; 8(4): 578.

[29] Li X., Zhang X., Ren X., Fritsche M., Wickert J., and Schuh H. "Precise positioning with current multi-constellation global navigation satellite systems: GPS, GLONASS, Galileo and BeiDou." Scientific Reports. 2015; 5: 8328.

[30] Cao Y. F., Cheung S. W., and Yuk T. I. "A multiband slot antenna for GPS/WiMAX/WLAN systems." IEEE Transactions on Antennas and Propagation. 2015; 63(3): 952–958.

[31] Przybyla R., Flynn A., Jain V., *et al.* "A micromechanical ultrasonic distance sensor with >1 meter range." *2011 16th International Solid-State Sensors, Actuators and Microsystems Conference*. Beijing, IEEE. 2011; 2070–2073. DOI: 10.1109/TRANSDUCERS.2011.5969226.

[32] Latha N. A., Murthy B. R., and Kumar K. B. "Distance sensing with ultrasonic sensor and Arduino." International Journal of Advance Research, Ideas and Innovations in Technology. 2016; 2(5): 1–5.

[33] Radha P. S. D., Rao D. S., Rajitha S., Sharmila Y., Sailaja T., and Murali K. "Implementation of wireless AD-HOC network using Arduino controller." International Journal of Management, IT and Engineering. 2012; 2(5): 80–95.

[34] [Online] Available from https://image3.geekbuying.com/ggo_pic/2015-1217/20151217153112sb1d7g2.jpg [Accessed December 14, 2019].

[35] [Online] Available from https://i0.wp.com/randomnerdtutorials.com/wp-content/uploads/2013/11/How-ultrasonic-sensor-works.jpg?w=700&ssl=1 [Accessed December 14, 2019].

[36] Mahmud M. H., Saha R., and Islam S. "Smart walking stick-an electronic approach to assist visually disabled persons." International Journal of Scientific & Engineering Research. 2013; 4(10): 111–114.

[37] [Online] Available from https://ae01.alicdn.com/kf/HTB1F6r9bmYTBKNjSZKbq6xJ8pXax/Aluminum-Mobility-Folding-Cane-White-Cane-for-the-Blind-Folds-Down-4-Sections-123-cm-48.jpg [Accessed December 14, 2019].

[38] Static generation robots. [Online] Available from https://static.generation-robots.com/2653-large_default/hc-sr04-sonar-sensor.jpg [Accessed December 14, 2019].

[39] Adriansyah A., and Dani A. W. "Design of small smart home system based on Arduino." *2014 Electrical Power, Electronics, Communications, Control and Informatics Seminar (EECCIS)*. (pp. 121–125). IEEE. August 2014.

[40] Solarbotics. [Online] Available from https://solarbotics.com/?s=buzzer [Accessed December 14, 2019].

[41] [Online] Available from https://cdn.sparkfun.com//assets/parts/9/0/00097-03-L.jpg [Accessed December 13, 2019].

[42] Do Kweon S., Park I.O., Son Y. H., Choi J., and Oh H. Y. Samsung Electro Mechanics Co Ltd. 2008. Linear vibration motor using resonance frequency. U.S. Patent 7,358,633.

[43] Seeedstudio Mini Vibrating Motor. Robotshop. [Online] Available from https://www.robotshop.com/en/seeedstudio-mini-vibrating-motor.html [Accessed December 13, 2019].

[44] [Online] Available from https://cdnlearn.adafruit.com/assets/assets/000/035/419/medium800thumb/components_halfbb.jpg?1472966532 [Accessed December 13, 2019].

[45] Otto J. C., Jones H. A., and Harris Corp. 1988. Multilayer circuit prototyping board. U.S. Patent 4,772,864.

[46] Available from https://cdnlearn.adafruit.com/assets/assets/000/035/419/medium800thumb/components_halfbb.jpg?1472966532 [Accessed December 13, 2019].

[47] Murphy J., McMullin F., Egan M., Brosnan M., and Kollmorgen Technologies Corp. 1987. Servomotor control systems. U.S. Patent 4,661,756.

[48] [Online] Available from https://media.rs-online.com/t_large/F2636011-01.jpg [Accessed December 14, 2019].

[49] Freeborn T. J., Maundy B., and Elwakil A. S. "Fractional-order models of supercapacitors, batteries and fuel cells: A survey." Materials for Renewable and Sustainable Energy. 2015; 4(3): 9.

[50] Electrolytic capacitor. Wikipedia. [Online] 2015. Available from https://en.wikipedia.org/wiki/Electrolytic_capacitor [Accessed December 13, 2019].

[51] [Online] Available from https://upload.wikimedia.org/wikipedia/commons/thumb/7/75/Electronic-Axial-Lead-Resistors-Array.jpg/1200px-Electronic-Axial-Lead-Resistors-Array.jpg [Accessed December 13, 2019].

[52] Duracell plus Power Battery 9V. Tool station. [Online] Available from https://www.toolstation.com/duracell-plus-power-battery/p91814 [Accessed December 13, 2019].

[53] Hossain E., Md Raisuddin K., Muhida R., and Ali A. Analysis and implementation for a walking support system for visually impaired people. IJIMR. 2011; 1: 46–62. DOI: 10.4018/ijimr.2011070104.

Chapter 12

Medical sensor capabilities in smart cloud networks: state-of-the-art approaches

B.D. Deebak[1], Fadi Al-Turjman[2] and Patruni Muralidhara Rao[1]

For the sake of medical-based wireless sensor networks (WSNs), several challenges have been addressed. However, the major shortfall is lacking data management. Since medical sensor networks generate and collect a huge amount of highly sensitive data, it always introduces several challenges with existing architectures that are still unresolved. The challenges include scalability, availability, and security to improve system efficiency. Besides, the application systems based on WSNs provide authentic information about patients' health. Therefore, this authentic information must be readily accessible to ease the system responses and to enhance the medical-rescue process especially during the patient's emergency situation by healthcare providers.

As a result, rescue management has become a challenging issue for medical wireless sensor networks. In this chapter, the modern architecture of wireless medical sensor networks is studied to review the key challenges, which primarily focuses on data collection, and analysis. The goal of the architecture is not only to overcome the aforestated challenges but also to facilitate the easy sharing of information between medical experts and emergency situations. Moreover, an efficient and flexible security mechanism is studied to assure integrity, confidentiality, and fine-grained access control to outsourced medical data. A framework based upon publication/subscription model is preferred to ease the convergence of Smart-Cloud, which uses a strategic framework model to monitor the activities of the patient's like physiological data. It uses a dedicated server database to authorize the patients' authenticity before the data being collected and stored in the Cloud. However, it has limited storage capabilities and service availabilities, whereby a single-point-of-failure is highly experienced. In the future, this conceptual idea is extensively studied to highlight the integration benefits of the Cloud-IoT.

[1]School of Computer Science and Engineering, Vellore Institute of Technology, Vellore, Tamil Nadu, India
[2]Artificial Intelligence Department, Research Center for AI and IoT, Near East University, Nicosia, Mersin 10, Turkey

12.1 Introduction

Of late, digitalization of the health care system has brought comprehensive transformation to the healthcare industry. It can store the medical data, which are easy to access by the various beneficiaries such as private, public, and government sectors. The contributors of the healthcare organization may demand more transparency to protect the sensitive data in the secure cloud environment. Moreover, the medical data can be more sensible, searchable, and producible to meet the set objectives of the medical institute. Mobile devices including smartphones, wearables, suitable medical applications, and the development of various wireless monitoring systems have come up with healthcare services everywhere [1]. Primarily, modern healthcare system endeavors to offer potent services that optimize the functionality of the monitoring systems to save the human lives. This can be considered as a ubiquitous healthcare system that can easily monitor health status, nurture medical services, and assure substantial health irrespective of the locality of the patient [2]. Transmission Control Protocol/Internet Protocol (TCP/IP) is widely used to adopt the set of networking protocols, which enable Internet access to connect the computer systems. Initially, the Internet was connected between computers only, but later on, mobile devices are also connected to the Internet. Today the Internet connection is available for many devices. The network connection transformation has been changing from Ethernet to wireless sensor networks (WSN). The present state of the Internet can be able to communicate between large numbers of Internet of Things (IoT) devices. It can likely connect 20.5 billion IoT devices by 2020 upon 3 trillion US dollars spent specifically on hardware [3].

Sensor networks (SN) have been emerging to assure the monitoring of physical world using self-organized networks of limited-powered wireless sensors and actuators to sense and process the information via wireless communication to store in a central location [4]. Since the sensor networks come up with limited battery and processing power, there is a need for optimized network architecture to minimize the resources for the applications. The requirements of sensor networks make their architecture to be challenging from the requirements of traditional architecture [5]. WSN spatially distributes the group of physical objects that are self-configured to organize the collected data at a dedicated location. There are various types of WSNs that have been introduced and wireless medical sensor networks (WMSN) is one among them.

The advancements in WMSN can rapidly be opening enormous opportunities to the modern healthcare system. Subsequently, increased integration of extensive medical resources and wireless medical sensor technologies address the major challenges to meet the requirements of the modern healthcare system. Since the traditional health care systems concentrate on hospitals and clinics, the new way of healthcare moves to the patient's residence, which can combine the modern wireless medical sensor technologies with traditional human caring to [6]. Currently, various kinds of sensor networks are available especially for medical applications including wearable, implanted, and environment underlying devices. These devices indigenous in the real world to track the temporal conditions for the patient to monitor regular health conditions using cameras, sensors, and other related tools. In the health monitoring system [11], security and privacy are the major concerns, because the user's data of these health care systems are highly sensitive and important [17]. Various applications of WSN have been focusing on patient health monitoring systems that are in demand. Recently, research in the medical field elastically growing and also improved in sensor device applications are being rapidly developing everywhere. Various projects developed in early 2015 and some of them are in the developing stage. Most of the recent projects have focused on specific health devices namely wearable devices, chronic condition support mobile applications, remote patient systems, etc. Most of these projects are funded by both public and private agencies.

Figure 12.1 shows the fast growth of WSN and worldwide revenue increasing rapidly for medical and health-care domains. It is predicted that the growth rate of sensor networks and Chipset sales will be highly demanded in the future with a tremendous growth rate [7]. Today most of the human activities are directly connected to Internet-based applications for instance humans directly involved in Internet-based medical applications. These impacts undoubtedly influence the lives

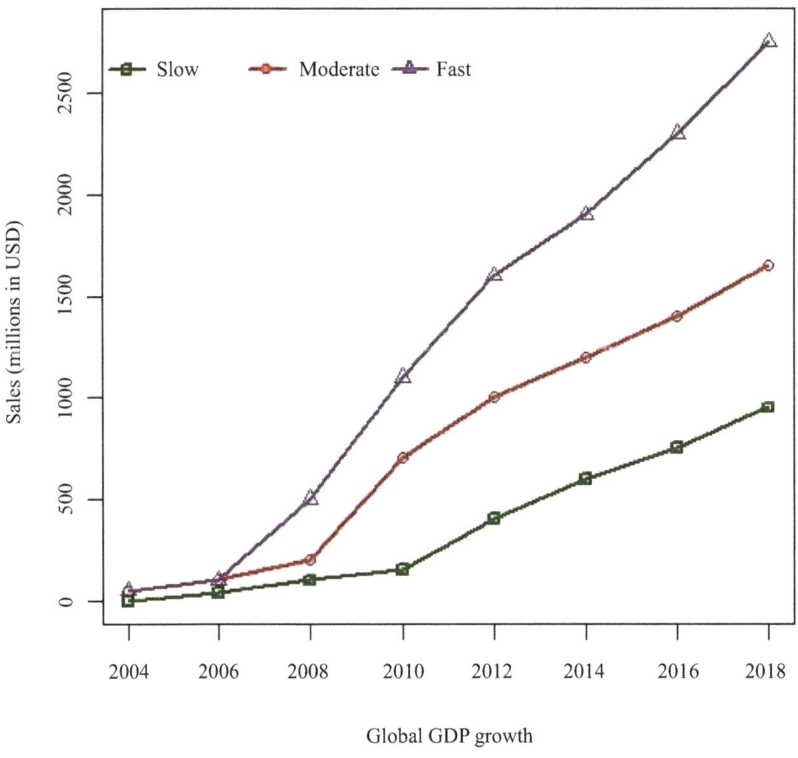

Figure 12.1 Chipset sales in the market growth

of individuals. In WSN, various health-care applications depend on several sensor technologies that can show security vulnerabilities such as eavesdropping, man-in-the-middle, user anonymity, denial of service, smart-card lost, password, etc. [8]. Smart cloud integrated IoT-enabled sensor network addresses several security challenges to authenticate service functionalities of the end users. Today, various concerns of health pitfalls especially for the instinctive sensor devices are directly related to social implications. In order to provide security to the health care systems, various mechanisms have been introduced. We broadly discuss the security mechanisms based on authentication, authorization, and integrity at the highest layer of the network stack. But the traditional network stack cannot be sufficient network architecture to the IoT environment. In order to improve the security efficiencies of the healthcare systems, a novel security framework is highly demanded.

12.2 Background

The motivation of this chapter is to furnish the current progress in the domain of WMSN and smart cloud environments. Broadly, we focus on state-of-the-art approaches in the on-going research. Elderly and patients with chronic diseases such as heart disease, diabetes, asthma, high blood pressure, and hypertension will gain the advantage of closely monitoring health-care systems to take control of them. WMSN has been a widely used network mechanism to connect these health-care applications. In this network, a large amount of biomedical data to be collected using various sensors from patients for a long period of time [7]. The largely collected bio-medical data can be efficiently stored and managed in a security-enabled cloud platform for further processing.

Since 1940, the evolutions of computing systems have been exponentially increasing in terms of computing power and storage to portability. The rapid change is seen in the use of a computer from the beginning to the present. There are various classes of personal computing systems including workstations, desktops, laptops, and PDAs have brought to meet the marketplace challenges. According to the prediction of "Bell Law" [12] based on the volume, it is expected that future computing devices will be even smaller than tokens. According to current technological trends, the application prototypes of various systems with many sensory devices can be accomplished in tiny mm-scale volumes. These tiny systems would function mainly to sense the objects and simple data processing, encapsulating intelligence, and sensing capabilities in such small devices. These can create an opportunity to work with IoT especially to meet the future requirements of the healthcare system in integrating "intelligence" to "things." Millions of small computing devices such as sensors are expected to be largely used in the IoT era. Current research says the mm-scale systems can be established with fewer resources such as processing power, low-power electronic design, and portable packaging. These small devices will drive the future needs of new IoT applications including wearable devices, implanted diagnosis sensors, and various monitoring systems (Table 12.1).

Table 12.1 List of abbreviations

Notation	Description
WSNs	Wireless sensor networks
IoT	Internet of Things
WMSN	Wireless medical sensor networks
OSI	Open system interconnect
RFD	Reduced-function device
FFD	Full-function node device
I/O	Input/output
LAN	Local area network
WAN	Wide area network
LPWAN	Low power wide area network
BLE	Bluetooth low energy
IPV6	Internet protocol V6
PLD	Programmable logic devices
IC	Integrated circuits
COPD	Chronic obstructive pulmonary disease
NFC	Near field communication
RFID	Radio-frequency identification
MQTT	Message queuing telemetry transport
CoAP	Constrained application protocol
AMQP	Advanced message queuing protocol
Rest	Representational state transfer
XMPP	Extensible messaging and presence protocol
LoRaWAN	Long range wide area network
OS	Operating system
SOS	Sensor operating system
PD	Parkinson's disease
BLESS	Beaconless
FSM	Finite state machine
IPC	Inter-process communication
EDF	Earliest deadline first
RM	Rate-monotonic
SSI	Single system image
ADC	Analog-to-digital converter
WAP	Wireless access point
DoS	Denial of service
SoC	System on chip
SN	Sensor networks
QoS	Quality of service
GPSR	Greedy perimeter stateless routing
INF	Intermediate node forwarding
GEAR	Geographic and energy aware routing
CTR	Critical transmitting range

Early technological advancements of various applications namely air traffic control, national weather stations, electric power grids, etc. deployed over traditional standard configuration. Moreover, the use of standard communication protocols, specialized computing systems, and security mechanisms are highly expensive. Researchers consider WSNs can be an emerging domain for large networked systems with low-power wireless tiny devices for high-resolution sensing environment [26]. Various features, functionalities, purposes, and capabilities of many sensors can be a major advantage of WSN. The less expensive WSN are now being trending for novel application developments in the field of healthcare and physical security. WSN can be a multidisciplinary domain that includes signal processing, radio, and networking, artificial intelligence, systems architecture for infrastructure administration, power management, platform technologies, resource optimization, and many more. In the current IT development RAM, a high-power processor, digital signal processing is combined with recent advancements to provide a new trend of low cost, high-performance sensors, and actuators that are able to achieve high-resolution spatial and terrestrial data.

12.3 Monitoring system architecture

Today, according to recent research reviews, remote health monitoring systems are evolving very quickly. Subsequently, we have reviewed various monitoring mechanisms for different applications such as climate monitoring, smart city monitoring, smart power monitoring systems, and majorly agriculture monitoring system. But as per research studies, remote health-care monitoring systems are completely different from the above-mentioned systems. The remote health-care monitoring system should comprise of sensor nodes to collect the medical data from several medical applications such as implanted and other applications, gateway is to permit the dataflow between the sensor nodes and WSN, wireless access point (WAP) is to make the communication between the user to the network specially deployed at users premises, Cloud environment is to provide services for largely scalable resources which could leverage the needs such as infrastructure and platform to develop and deploy applications, medical servers are to provide the real-time concurrent resource utilization to various remote applications, Database is used to store and analyze the data and to display it, and finally Monitoring devices are used to monitor the changes that are being made. The proposed generic architecture possesses these required components is shown in Figure 12.2.

The sensor node helps to acquire the data from various physical locations and processes before it transmits to the next node. To maintain a reliable and accurate monitoring system, we need reliable data from various sensor nodes. However, there are various challenges that we need to be specific to network management and data management. Henceforth, we require a good design of WSN to make through data analysis and remote monitoring.

12.3.1 Design issues and security challenges

According to the statistics in [4], the 1990s band population of 370 million will multiply in 2025 to approximately 760 million; the fast-growing population and emerging healthcare systems have caused several related challenges to the healthcare personnel, manufacturers of healthcare products and governments. Therefore,

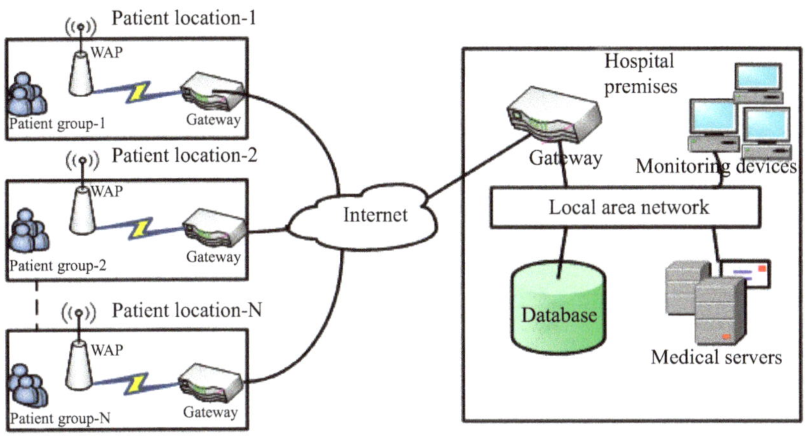

Figure 12.2 Generic system architecture for the healthcare system

effected stakeholders use e-Health systems to expand healthcare services, disease tracking, remote monitoring, and other related medical applications [5].

12.3.1.1 Design issues

In a broad sense of the remote health monitoring system, we have some design issues of WMSN related to real world includes:

1. **Fault-tolerant communication** because of the sensor deployments in an uncontrolled environment.
2. **Scalability** may be another challenging issue in a sense a large number of sensor nodes deployed in a sensory environment may be in the order of dozens, hundreds, or thousands.
3. **Coverage challenges** can address underlying issues in WSNs that reflect the QoS that can be accommodated by specified SNs.
4. **Transmission media** is another important issue since it is connected in a multihop fashion and communicating sensor nodes are likely linked with wireless connection medium. As many standard issues related to wireless communication such as fading and high error rate may influence the WMSN [25].
    (i) **Data aggregation** is another important issue as the sensor nodes are situated close to every other node. There is a high possibility of the same data being generated by the next to each other. Hence the data required to be aggregated and the redundant data should be ignored as response and data transmission rates are highly cost-effective. Hence the necessary data only need to transmitted and received.
5. **An autonomous operation** can be a common feature and challenge as well because WSNs are obliged to operate and organize autonomously as they are deployed in a location in which human intervention may not be possible.

6. **Lifetime** can be a common challenge in WSN, which should work for a very significant time duration with low power utilization. They should last a minimum of a half year to 1 year. Every sensor node in WSN might be controlled utilizing only a 3 V battery and this ought to be adequate for the lifetime of the sensor node. The implementation of protocols for WSN ought to be with the end goal that the nodes consume as less energy as could be allowed. This will help in making the WSN last more.

12.3.1.2 Topological challenges

Today WSNs are extremely exacted in distributed systems and also deployed in a variety of geographical locations. Topologically related challenges have gained more concern about optimal network topological issues. Figure 12.3 depicts various challenges of WSN related to topology [28].

1. **Geographic routing** can be dependent on geographical location information that can predominantly work for WSN, based upon the idea which a source node can communicate with a destination node using geographical location rather than traditional network address.
2. **Sensor coverage topology** shows in what way the sensors can monitor the regions. The communication and connectivity challenges have gained a lot of attentiveness in recent research works. This challenge can be legislated like a decision-making problem of which the objective is to cover at least the required number of sensors for every point in the service area of the WSN.
3. **Sensor holes** can be a routing hole comprise of a location in the WSN in which the nodes are not contactable or available nodes are not active in the routing based on impetus. So to avoid this attack to the packet delivery, various geographic routing protocols have been introduced namely GPSR, INF, GEAR, Compass Routing, etc.

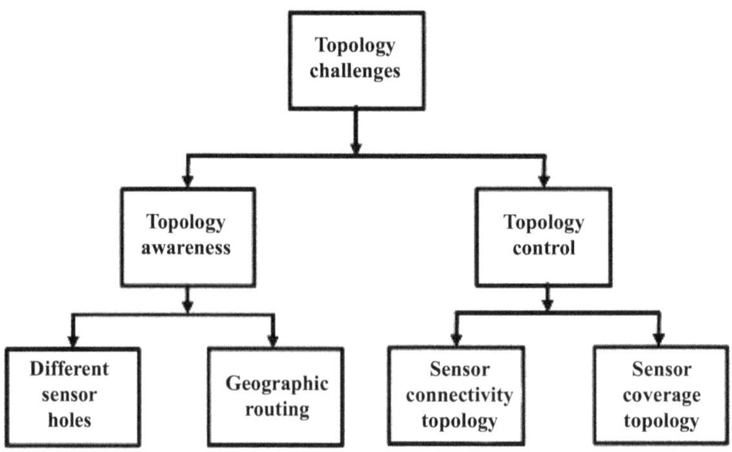

Figure 12.3 Topology challenges in WSNs

4. **Sensor connectivity topology** mainly comprises of power control and power management schemes. The power control scheme aimed to be ample network energy efficient. In order to achieve this, it distinguishes to examine the transmission range between nodes of homogeneous and heterogeneous scenarios that have the same range or not. For homogeneous networks, CTR issues have been examined in a theoretical and practical manner. Researchers proposed various solutions to determine the minimum common transmission required to assure network connectivity. However, heterogeneous networks becoming highly influential as the sensor nodes have different transmission ranges to communicate WSN.

Alongside, various other design issues are becoming crucial for sensor nodes with respect to WSN such as deployment, localization, database-centric and querying, programming models for SN, architecture, synchronization, calibration, network layer, data aggregation, data dissemination, hardware, and OS for WSN.

12.3.1.3 Security challenges

Today WMSNs feature a diverse set of security challenges including physical level, communication level, code level, and sink-node and routing protocol level challenges.

1. **Physical attacks** caused us to destroy the sensor nodes physically or to slink from the location. The adversary may even do more damage such that they can examine internal code and modify, extract secret keys, and inject vulnerable information. Hence this challenge may lead to going future communications in the attacker's hands.
2. **Logical attacks** are known to be communication attacks. These may cause to damage the nodes remotely without physical intervention of the attacker.
3. **Indisposed programming behavior** leads to susceptible software that may cause to be a code level security challenge that a malicious user can easily attack. The OS should be appropriately designed, implemented, and tested to defend software bugs. Reference [27] states the majority of vulnerabilities can be related to the application level.
4. **Sink-node** accumulates the data from other surrounding nodes and reports to the system via wireless network access. If this is compromised the entire system will lead to a failed state because the entire deployment cost completely depends on the sink-node.

A robust OS is responsible for securing all its operations and the susceptive OS may give advantage to the attacker to access it. In the same sense, the routing mechanism does so. In WMSNs, routing protocols commands whole communications whereas weak routing mechanism is susceptible to attacks. Table 12.2 depicts major security problems, requirements, potential security solutions, and a few standard protocols [5,23].

Several vulnerabilities have been identified based on WSNs especially working with medical sensor networks such as Sybil attack, new node injection attack, jamming attack, DoS attack, collision attack, content analysis attack, black-hole attack, and hello flood attacks [9].

12.3.2 Sensor node design

The sensor node can be a node in SN that is able to do a proportion of processing, gather the sensory information, and communicate with other nodes. A WSN can jointly form with dozens of sensor nodes and it can be a good model of system on chip (SoC) that has constrained resources such as bandwidth, memory, computing power, and battery power. Figure 12.3 describes the generic view of sensor node components. Generally, every sensor node comprises sensors, memory units, microcontrollers, radiofrequency transceiver, and power supply units. These sensor nodes can be utilized in a wide variety of applications majorly in remote monitoring applications. Using multihop communication, these sensor nodes can have the ability to communicate with other nodes (Figure 12.4).

1. **Power supply unit**

   In the wireless medical sensor node, each component is powered by the power supply unit. In spite of its limited capacity, the power supply unit needs efficient operations for the task performed by each component. WSN's power is spent away by sensors transceivers, in the process of transmitting or receiving sensor signals.

Table 12.2 Security problems and potential solutions in WMSN

Security problems	Requirements	Potential solutions	Protocols
Unauthorized access (authentication and authorization)	Secure key establishment	Cryptographic schemes and random key distribution	EAP, PAP, Kerberos, TACACS
DoS	Availability	Intrusion detection system and data redundancy	OSPF, BGP
Compromised node	Resistance to node compromise	Node tamper detection	LEAP
Message disclosure	Confidentiality and privacy	Link encryption and access control	SNMP
Message modification	Integrity and authentication	Secure hash key and digital signatures	SHA, MD5
Intrusion and vicious activities	Secure group management Intrusion detection system	Secure group communication intrusion detection system	PIDS, AB-IDS
Routing challenges	Secure routing	Secure routing protocols	SEAD, DSDV, SEF

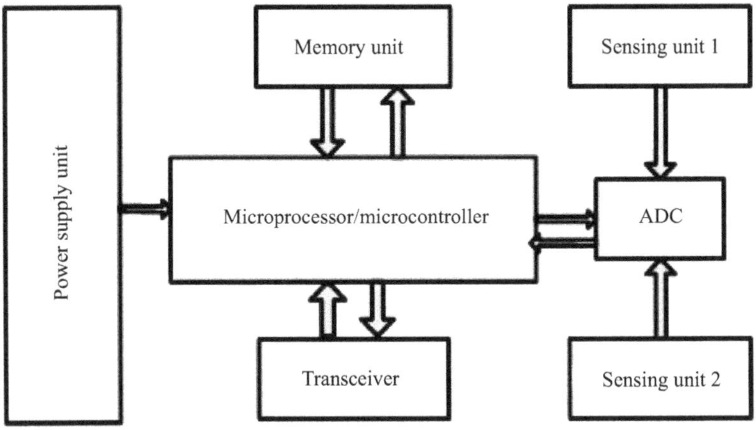

Figure 12.4 Sensor node design

2. **Transceiver unit**
   This unit plays a dual role as both the transmitter and also the receiver. It is responsible for communication between two sensor nodes. It acts as a bridge among wireless sensor networks and other networks [20]. It is composed of the four states of operations namely transmitting, receiving, sleep, and idle.
3. **Memory unit**
   Sensor nodes have a memory that incorporates RAM, internal flash, EEPROM, and external flash.
4. **Sensor**
   The sensor can be defined as hardware devices that collect information from the physical world. According to information gathered, distinct sensors are used. A sensor holds on two nodes, they are (1) active node in which the sensor node is awake and receive information and (2) power save mode in which node sleeps and frequently wakes and checks pending messages.
5. **Analog to digital converter**
   Generally, sensors receive information from the physical world and real-time objects or things; these data are comprised of analog signals. ADC helps in converting analog signals to digital signals and then transfers them to the processing unit.

12.3.3 Security requirements

The proposal of an unacknowledged authentication result for the WSN is to solve three kernel problems as mentioned below:

1. Obtain identity authentication;
2. Assure the security of identity authentication; and
3. Accomplish anonymous identity authentication efficiently.

In order to justify the above-mentioned problem, the proposed solutions should meet the following requirements.

Security: It can be a fundamental requirement, and it directs to the ability to resist your system from various vulnerabilities. With respect to WSN, security must be provided majorly in three concerns namely data level, node level, and network-level security. In this context, data-level security describes to assure safety from unauthorized access and protects from deviation and fraud. The node-level security describes maintaining proper security standards to avoid intruders to access a node and obtain cryptographic keys. Network-level security describes in a broad sense that network security consists of a set of policies assigned by the network administrator to protect unauthorized access to the system and data as well.

Anonymity: It can ensure that the sensor node must be untracked underneath of attacks performed by the adversary. It comprises of two popular methods: symmetric/asymmetric encryption algorithm and a pseudonym.

Efficiency: It can be observed by several indicators such as computation cost, storage space, interaction times, and energy management.

Authentication: Authentication is an important security measure of WSN to provide protection for healthcare-monitoring infrastructures from various potential vulnerabilities. It is mandatory especially for WSN because it uses a shared communication mechanism. But still, authentication cannot rectify the problem of the endangered node. An endangered node can have secret key information of authorized nodes so that it can easily authenticate itself. Therefore, we may need standard intrusion detection techniques to identify these endangered nodes in the system.

Key establishment and management: The applications of WSN communications can be passively monitored; moreover, the nodes are likely to capture by the attackers for unauthorized use. To overcome this kind of scenario, cryptographically secured communications are highly demanded. Key establishment technique can be utilized to provide the cryptographic technique like symmetric key and asymmetric key cryptographic techniques. In order to establish communication between two sensor nodes, we require a private and authenticated connection between them. Therefore, the creation of a secret shared key is very significant to provide a secured distributed environment.

Availability: Availability is to provide guarantee the users to access reliable information by the authorized ones. Commonly, DoS attacks can affect the failure of availability and resultantly allow node capture attacks. Due to the loss of availability, the remote health monitoring system may have serious consequences. For instance, developing a health-care application loss of availability may have a failure to observe the critical conditions of the patient and monitory mechanism may result in several consequences like financial loss, reputation, etc.

Confidentiality: Confidentiality refers to protect information from divulge to the unauthorized presence and should be realized by encryption. Basically, data confidentiality can be compromised by privacy attacks.

Privacy: Privacy enables you to have access rights to control your information and provides how your personal information can be protected. Generally, the major threats of WSN might allow malicious identities to deploy secret spying networks to spy unfamiliar victims.

Secure routing: The main objective is to assure that every intermediate node should not remove the existing node or cannot add any additional nodes to the route. In the real-time scenario, the need for secure routing mechanisms is highly demanded in a way to ensure confidentiality, authenticity, integrity, and availability. However, secure routing can be an extremely challenging job in WSN because of its recurrent changes in the topology. Since the majority of WSN routing protocols are simple, it is highly disposed to various attacks. Moreover, the maximum number of attacks against WSN includes selective forward, spoofing and acknowledgment spoofing, Sybil, sinkhole, and wormhole attacks. In order to protect the system against these attacks, a novel routing protocol can be highly demanded especially for WSN [29].

12.3.4 Hardware components

Basically, hardware components comprise of physical-components and computer-based platforms. Sensors, actuators, network connections (LAN, WAN), access points, routers, gateway, operating systems (OS), etc. are categorized under physical hardware components. The computer-aided platforms may amble applications on Linux, Windows, Macintosh, or any other OS. However, all these components predominantly connect over the LAN network of its own requirements. These platforms allow users to use high-level programming languages to develop software components, which simplifies the integration of enterprise applications.

Moreover, the wireless devices can be more constrained about processor and memory capabilities to make them fit with real-time processing. The microcontroller can be the best suitable unit which can frequently be programmed in C Language.

12.3.4.1 Gateway

The gateway is an interface specially designed to allow the data flow between applications and the nodes on the WSN. The information collected from various wireless sensor nodes can be manipulated by the wireless access point, that is, thorough a gateway and broadcast to the application. Gateway is completely different from routers and switches in that they may use more than one protocol to communicate and operate at any layer of Open System Interconnect (OSI) network stack. The applications may work on local machines or remotely connected machines, in the vice versa, the gateway relay the data to the WSN. In general, all gateways can be used to accomplish protocol conversion and authorize with other wireless network remote protocols.

12.3.4.2 Leaf node

This is also called as end-point and can be advised as a reduced-function device (RFD). This can be exclusively deliberate to furnish physical interaction among the sensor and actuator that is connected to WSN. Generally, these nodes are equipped with many I/O connections for communicating with sensors and actuators.

12.3.4.3 Relay node

This can be a full-function node device (FFD) and also named as routers and which can be used to widen broad area network exploration. This FFD can be a server or workstation or an embedded device to provide core functionalities of sensors and/or actuators. These can also be connected to provide the same input/output (I/O) operations of a leaf node.

12.3.4.4 Sensor or actuator

These are basically electronic devices used to interact with the physical system to monitor and control the data. For instance, sensors monitoring crop growth in the agriculture field, medical sensors to monitor continuous health conditions, etc. These distributed applications can enable WSN-based applications to provide some of the basic functionalities such as signal conditions and data acquisition of various sensors, data storage, processing capabilities, data analysis, actuation, communication-based scheduling and execution of networking tasks, etc.

12.3.4.5 Network topologies

1. **Star topologies:** A star topology can be a single-hop system in which every sensor node communicates with a gateway. These gateways provide easy connections for various tiny networks. In the case of star configuration, if a node fails it does not influence other nodes unless it is a central node. However, in furthermore nodes can be easily added without interruption of its own network. An example of a star topology is shown in Figure 12.5.
2. **Mesh topology:** It connects from each node to every other node and has the capacity to make use of less power exhaustion against other topologies within the range. However, mesh topology can be exceedingly error resistant because from each separate node it has compound connections to other nodes and multiple paths to the gateway. An example of a mesh topology is shown in Figure 12.5.
3. **Ring topology:** In this network configuration, the connections create a circular path. Every network device can be connected to two other devices in a circular manner. It transmits data packets from one device to another till they reach the destination. Most of these topologies permit data packets to unidirectional in the ring network. It can be used in either LAN and/or WAN depending upon the network configuration that connects to the computers together is shown in Figure 12.5.

12.3.5 Operating systems design specifications

The variations of sensor nodes, mentioned in [13,14], with their specifications including the capacity of sensor, bandwidth, size, and frequency of a microprocessor, and OS that should provide control abstraction to the application software and control over hardware components. Since the standard OS is the system

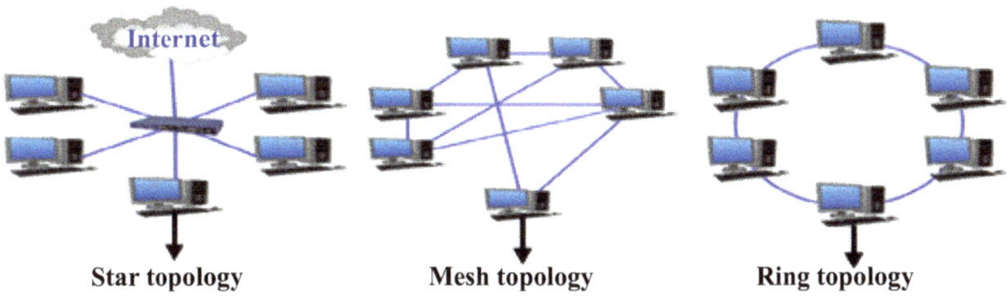

Figure 12.5 Network topologies

software, it is advised for traditional computers like PCs and workstations with lots of peripheral devices and computing resources. However, the purpose of a standard OS consequently serves to manage CPU time, file systems, devices, and manage processes and memory. Generally speaking, WSN has limited-resources, variable topology, and various data-centric applications. Thus, WSN cannot be suitable for standard OS because of its implementation in a modular layered format with Kernels and Libraries. While considering limited characteristics of WSN, we require a novel OS for the development of their applications and run on it. However, there are various design issues for the development of these new OS for WSN such as process management and schedule, energy efficiency, memory management, kernel model, application programming interface (API), no external disk, and code upgrade and reprogramming. During the design of WSN OSs, there is a need for considering all the above-mentioned distinctive issues to fulfill their necessary resources, network management architecture, and data-centric application restraints [15]. Bearing with the limited resources of sensor nodes, the sensor operating system (SOS) is obliged to incorporate various functions such as providing real-time support, efficient resource management, generic API, support reliable and efficient code distribution, and power management and must be concise in small size.

1. **Magnet OS:** It is a distributed OS especially developed for the application adoption and energy prudence. Magnet provides a special mechanism for applications to handle the underlying resources. Generally, the load of creating adoption mechanisms is on the application itself but this approach is not energy-efficient. However, Magnet OS has mainly four goals such as adoption of the fundamental resources and its changes in a static manner, efficient with respect to energy preservation, put up general abstraction for the application, and it is scalable for prominent networks. The implementation of Magnet OS is based on single system image (SSI) or individual unified JVM that includes static and dynamic components. The use of static components is to rewrite the application in byte code level whereas dynamic components are utilized to object creation, invocation, migration, and application monitoring. The beauty

of SSI abstraction is to provide extensive freedom in object placement and it simplifies application development.

2. **Tiny OS:** It is a tiny microthreaded OS that can directly permit application software to access hardware. It helps an event-based model to defend the high level of synchronous application in a miniature memory, achieved higher throughput because it is associated with an event and can rapidly create multiple tasks without pooling. Also, it has a tiny scheduler that schedules the operations of fixed components [16]. Every component comprises four major parts namely fixed-size frame, event handler, command handler, and a group of tasks. Tiny OS has few advantages of its evolution including very efficient because it needs little code and small-scale data, events can propagate easily and quickly, context switching is high, and finally efficient modularity.
3. **Mate OS:** It is one of the components of Tiny OS and exclusively works on top of it. This can be a byte-interpreter intended to make Tiny OS available to nonexpert programmers to provide fast and efficient programming of the whole WSN system. Mate has an execution environment that helps UC-Berkeley Mote since this Mote does not have proper hardware protection. The implementation of mate is little varied from other OS that this code can be made-up of capsules, every single capsule has 24 instructions and each instruction has 1-byte length. Each and every capsule holds type and version details that make code injection easy. Capsules can be categorized in four ways such as message sender, message receiver, timer, and subroutine. Mate can be able to implement new routing protocols and predominantly implements Beaconless (BLESS) ad hoc routing protocol [21]. Subsequently, the mate can be able to program the whole network.
4. **Sen OS:** It is an FSM-based OS that can well distinguish concurrency and reconfiguration, and also can be extended to network management. It has majorly three components: first, it has a kernel that holds a state sequencer and an event. Here state sequencer waits for an input from the event queue (a FIFO queue). Second, it has a state transition that keeps track of information on state transition and call-back functions. Third, it has a call-back library of call functions.
5. **Contiki OS:** It is GUI-based OS that can support multitasking and built-in TCP/IP stack. It is specially developed for networked and resource-constrained application devices. This can mainly be used for real-time IoT applications such as street lighting, sound monitoring system for smart cities, smart alarms, and radiation monitoring.
6. **Emeralds OS:** It is a micro-kernel-based OS developed [20] in C++ exclusively for embedded and real-time distributed applications running on constrained resource platforms. It uses semaphores and conditional variables for priority synchronization and also expands full semaphore semantics to reduce the quantity of context switching. IPC can be recognized based on mailboxes, shared memory, message passing, and inter-task communication and specially optimized for intra-node. But Emeralds does not use mailbox and network issues. Here global variables can be utilized to exchange selective information

between tasks. Moreover, the implementation of device drivers can be done at the user level and interrupt handling can be done at the kernel level. However, it supports multithreaded processing, complete memory protection and they can be scheduled using two renowned schedulers namely EDF and RM schedulers.

12.4 Standard technologies in WMSN

Generally, the core functionality of WMSN is to collect a large amount of medical information from various implanted and other sensors and to give support for specific healthcare applications depending upon the task of WSN deployment [24]. It comprises of centralized sink node and millions of sensor nodes to sense the physical circumstances. Widely available on-demand sensors nodes can be classified in the following ways.

1. The **Berkeley motes** [13] can be generic sensing platforms to perform generic sensing tasks.
2. The **spec node** [14] developed at the University of California, Berkeley. This specialized sensor node designed using a single chip with low-power and low-cost operation.
3. The **Stargate** [14] can be a gateway platform utilized as a sink and can directly connect to the Internet with low-level sensor nodes.
4. The **iMote** [14] is a high-bandwidth sensing platform that can be used to handle the flow of sensed data with high-bandwidth.

12.4.1 Communication protocols

First, IoT communication protocols operating at physical and data link layers are one of the key components of the IoT communication system. According to the current scenario, a couple of protocols and standards have been designed especially for IoT-based medical applications. These standard protocols can be grouped into two categories, that is, IoT data protocols and IoT network protocols.

12.4.1.1 IoT data protocols

These IoT-based WMSN protocols enable physical devices to exchange information from one to another. MQTT, CoAP, AMQP, Rest, XMPP, and Stomp are some of the numerous legacy protocols to support data transfer.

12.4.1.2 IoT network protocols

IoT network protocols work at lower level connecting IoT devices to each other and with IoT cloud platform [22]. There are various IoT network protocols that can be worked in short- and long-range medium. Bluetooth, ZigBee, Sigfox, LoRaWAN, Z-Wave, 6LowPAN, Thread, Wi-Fi, Cellular, NFC, Neul, RFID, LTE cat 0,1&3, ANT&ANT+, DigiMesh, MiWi, EnOcean, and Dash7 are the renowned IoT networking protocols [8,9]. In general, Internet protocol version 4 and 6 supports nearly all IoT implementations at network layer of the OSI model [10]. As far as wireless sensor networks-based IoT is concerned, several wireless communication protocols have been accustomed to connect the smart-devices over Internet protocol version 6 (IPv6) such as Bluetooth Low Energy (BLE), 6loWPAN, ZigBee, Z-Wave, etc. These are short-range standard protocols, whereas Sigfox and Cellular are the wide-range standard protocols operating based on low power wide area network (LPWAN) protocols [1]. Generally, each IoT protocol has its own advantages, features, and specifications based on their networking standards are discussed in Table 12.3.

12.4.2 Programmable logic devices (PLDs)

1. **PLD**

   PLD can be defined as an IC that comprises a large number of gates and flip-flops which can be configured by the user to operate different functions. The internal logic gates and/or connections of PLD can be configured by a programming process.
2. **PROM**

   One time programmable chip that once programmed which cannot be erased or altered is called programmable ROM. In PROM, AND-plane has all min-terms generated and connections of all AND-plane outputs to OR-plane gate inputs. In the OR-Plane, the application of high-voltage transistor refers to the minterms that are not required to burn certain outputs in particular (Figure 12.6).
3. **PAL**

   This device has programmable AND array and fixed connections for the OR array. AND plane constitutes of product terms, out of which three terms are inputs to OR-gate in OR-plane. Programming fuses of the AND-plane helps in the implementation of expressions.
4. **PLA**

   In PLA, programmable connections are accessed through both AND & OR gates. Henceforth, it is a most flexible PLD. "k" number of AND gates can be used, where $k < 2n$; n is the number of inputs. By programming AND gates, product terms to the input variables are generated. Inside PLA, the fabrication of AND & OR gates is performed with the help of links or fuses among them. By opening suitable links and ignoring the specific connections, the desired Boolean functions are implemented in the form of the sum of products.
5. **CPLD**

   CPLD is a combination of a bunch of PLD blocks where inputs and outputs are connected together by the global interconnection matrix. The programmability of CPLD can be categorized into two levels. Each block of PLD can be programmed and then Programming of interconnections between PLDs is performed.

12.4.3 Microcontroller unit

A microcontroller can be a low-cost oriented, little, and independent PC on-a-chip that can be utilized as an embedded system. A couple of microcontrollers

Table 12.3 Standard IoT network protocol specifications

Protocol	Key specifications	Standard
Bluetooth	Version: BLE 4.0, 4.2 Frequency: 2.4 GHz (ISM) Range: 50–150 m (Smart/BLE) Data rates: 1 Mbps	IEEE 802.15.1: WPAN/ BLUETOOTH
ZigBee	ZigBee 3.0 Frequency: 2.4 GHz Range: 10–100 m Data Rates: 250 kbps	IEEE 802.15.4
6LowPAN	Frequency: Bluetooth Smart (2.4 GHz)	RFC6282
LoRaWAN	Frequency: Various Range: 2–5 km (urban environment) 15 km (suburban environment) Data rates: 0.3–50 kbps	IEEE 802.15.4
Thread	Frequency: 2.4 GHz (ISM)	IEEE 802.15.4 6LowPAN
Z-wave	Frequency: 900 MHz (ISM) Range: 30 m Data rates: 9.6/40/100 kbit/s	IEEE 802.11, IEEE 802.15, IEEE 802.16
Sigfox	Frequency: 900 MHz Range: 30–50 km (rural environments), 3–10 km (urban environments) Data rates: 10–1,000 bps	IEEE 802.15.4
Cellular	Frequencies: 900/1800/1900/2100 MHz Range: 35 km max for GSM; 200 km max for HSPA Data rates (typical download): 35–170 kps (GPRS), 120–384 kbps (EDGE)	GSM/GPRS/EDGE (2G), UMTS/HSPA (3G), LTE (4G)
NFC	ISO/IEC 18000-3 Frequency: 13.56 MHz (ISM) Range: 10 cm Data rates: 100–420 kbps	ISO/IEC 18000-3
MiWi	Frequency: 2.4 GHz Range: 20–50 m	IEEE 802.15.4
Neul	Frequency: 900 MHz (ISM), 458 MHz (UK), 470–790 MHz (White Space) Range: 10 km Data rates: Few bps up to 100 kbps	IEEE 802.11af
RFID	Frequency: 120–150 kHz (LF) Range: 10 cm	ISO/IEC2024 & ISO/IEC JTC 1/SC 31

may use 4-bit expressions and work at clock rate frequencies, which normally include 8- or 16-bit microprocessors, small RAM, Programmable ROM, flash memory, parallel and serial I/Os, analog-to-digital and digital-to-analog converters, timers, and signal generators. In the WSN signal chain, a microcontroller

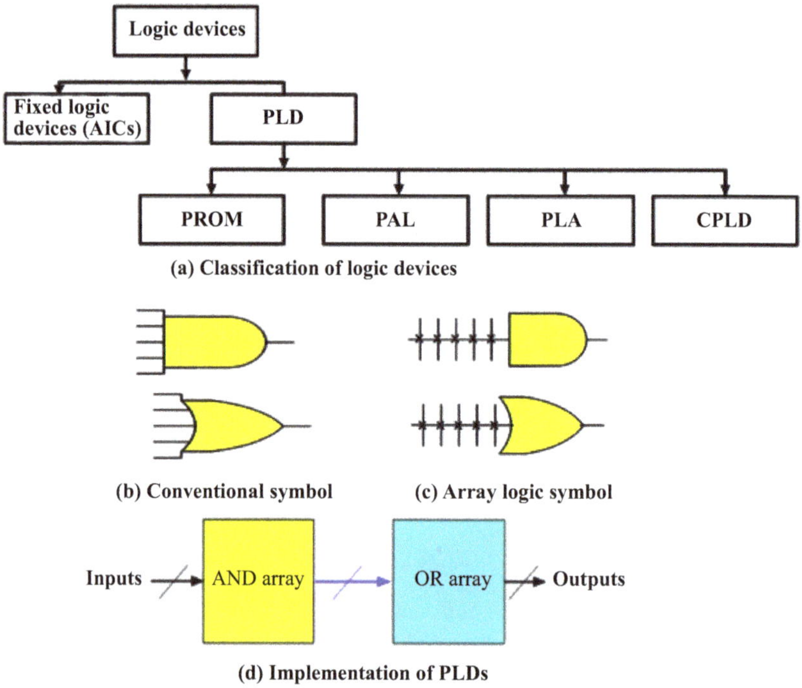

Figure 12.6 Classification and implementation of programmable logic device (PLD)

is a central component that communicates with all of the sensors to read and process the data, which sensors are measuring. The main components of the microcontroller are an 8- or 16-bit microprocessor, a small amount of RAM, programmable ROM and/or flash memory, parallel and/or serial I/O, timers and signal generators, and Analog-to-Digital (A/D) and D/A converters. In general, they are used for the purpose of executing a specific code that controls one or more tasks in the device operation. Usage of microcontrollers often occurs in devices like car engines, consumer electronics, measurement equipment, etc. Microcontrollers must satisfy low power requirements, and like many devices, their monitoring is operated by batteries.

12.5 Applications of WMSN

12.5.1 Patient monitoring

For desperate unhealthy patients, frequent measurement of patient specifications like heart rate, respiratory rate, blood pressure, blood oxygen saturation, and other distinct metrics are necessary. With the rapid growth of mobile technologies, the patient monitoring system came into existence, which supports the health care industry. Mobile and sensor devices will be helpful to monitor the health condition of patients by medical experts using wireless communication. Health information of patients will be sent to medical experts by using sensor devices through the gateway node of WMSN. Generally, sensor nodes are powered by tiny batteries. Hudson defined the patient monitoring system as "frequent observation of the patient, his or her physiological function, and the function of life support system equipment for the purpose of guiding management decisions, including when to make therapeutic interventions." These systems help in giving alert to medical experts about life-threatening events and even provide physiologic input data to monitor life support devices.

12.5.2 Heart attack monitoring system

Changes in ECG patterns are helpful in the diagnosis of cardiovascular diseases such as angina, heart failure, heart attack, etc. Real-time monitoring of cardiovascular patients is performed with the assistance of wireless sensor technology [18]. For potential and fast delivery of health alerts, wireless sensor networks are combined with other technologies like cellular networks, wireless LAN, and broadband networks. ECG signals are analyzed by real-time data analysis, visualization, and warning system. This system provides services like remote monitoring and delivery of warning to doctor, relative, and hospital.

12.5.3 Handling COPD and PD patients

The development of wearable ambulatory systems that monitor physiopathological parameters is a rapid technology growing day by day. Cigarette smoking is the primary cause of diseases like chronic obstructive pulmonary disease (COPD) and Parkinson's disease (PD). In COPD, airways are narrowed inside the lungs. Most Common symptom of COPD is associated with the abnormal inflammatory response of the lung to toxic particles or gases.

PD is a central nervous system disorder that affects motor skills and speech of the patient. Muscle rigidity, tremor, and slowing of physical movements are characteristics of PD. Ambulatory monitoring contains wearable physiological sensors that are integrated into a telemedical system. This system helps COPD patients and PD patients since it has continuous monitoring in an ambulatory setting, early detection of abnormal conditions, etc. Heart rate monitoring, respiratory rate, and oxygen saturation are very important to analyze the health condition of COPD and PD patients. WSNs have potential use in building ambulatory systems for the above systems. Sensor nodes equipped with flex sensors are placed on the right hand for heart rate and acceleration measurement and on armpits for temperature.

The collection of physiological data and movement data from sensors is performed by acceleration measurement [19]. If any abnormal activity occurs, an alert is sent to emergency departments and doctors. Table 12.4 illustrates the physiological conditions of COPD and PD patients that cause alerts.

Table 12.4 Alert detection parameter for COPD and PD patients

Alert detection parameter	COPD metrics	PD metrics
Heart rate	$\lvert\Delta HR/5/min\rvert > 15$ bpm	$\lvert\Delta HR/5min\rvert > 15$ bpm
Body temperature	BT > 99°F	BT > 99°F
Movement	Active	Active
Others	Brady cardiac – HR < 55 bpm	B.P-Systolic or diastolic changes > ±11%

12.6 Conclusion

In this chapter, we have addressed the important aspects of remote healthcare systems. In order to ensure secure communication and data transfer, various design issues and security challenges have been discussed in detail. Moreover, we have presented a generic architecture for remote healthcare system that ensures better usability for the patient's community. Various design guidelines for WMSN construction and utilization have been considered for the outsourced medical data that claim to achieve integrity, confidentiality, and fine-grained access control. With the advancement of WMSN, the optimized network architecture can rapidly be opening enormous opportunities to the modern healthcare system. Subsequently, increased integration of extensive medical resources and wireless medical sensor technologies addresses the major challenges to meet the requirements of modern healthcare systems.

References

[1] Pramanik, P. K. D., Upadhyaya, B. K., Pal, S., and Pal, T. "Internet of Things, smart sensors, and pervasive systems: Enabling connected and pervasive healthcare." *Healthcare data analytics and management* (pp. 1–58). Academic Press; Elsevier; Amsterdam; Netherlands; 2019.

[2] Hollnagel, E., and Braithwaite, J. *Resilient health care*. London: CRC Press; 2019.

[3] Nalini, N., and Suvithavani, P. "A study on data analytics: Internet of Things & health-care." International Journal of Computer. 2017; 6(3): 20–27.

[4] Deebak, B. D., and Al-Turjman, F. "A hybrid secure routing and monitoring mechanism in IoT-based wireless sensor networks." Ad Hoc Networks. 2019; 102022.

[5] Al-Turjman, F., Ever, Y. K., Ever, E., Nguyen, H. X., and David, D. B. "Seamless key agreement framework for mobile-sink in IoT based cloud-centric secured public safety sensor networks." IEEE Access. 2017; 5: 24617–24631.

[6] Zhang, Y., Sun, L., Song, H., and Cao, X. "Ubiquitous WSN for healthcare: Recent advances and future prospects." IEEE Internet of Things Journal. 2014; 1(4): 311–318.

[7] Cardullo, P., and Kitchin, R. "Being a 'citizen' in the smart city: Up and down the scaffold of smart citizen participation in Dublin, Ireland." GeoJournal. 2019; 84(1): 1–13.
[8] Al-Turjman, F., and Mostarda, L. "A hash-based RFID authentication mechanism for context-aware management in IoT-based multimedia systems." Sensors. 2019; 19(18): 3821.
[9] Gartner Inc. "Hype cycle for the Internet of Things 2017." Technical report, July 2017.
[10] El-hajj, M., Fadlallah, A., Chamoun, M., and Serhrouchni, A. "A survey of Internet of Things (IoT) Authentication schemes." Sensors. 2019; 19(5): 1141.
[11] Milenković, A., Otto, C., and Jovanov, E. "Wireless sensor networks for personal health monitoring: Issues and an implementation." Computer Communications. 2006; 29(13–14): 2521–2533.
[12] Al-Turjman, F. "Intelligence and security in big 5G-oriented IoNT: An overview." Future Generation Computer Systems. 2020; 102(1): 357–368.
[13] Li, M., and Yang, B. "A survey on topology issues in wireless sensor network." *ICWN*, Las Vegas, NV, USA. June 2006. p. 503.
[14] Deebak, B. D., Ever, E., and Al-Turjman, F. "Analyzing enhanced real-time uplink scheduling algorithm in 3GPP LTE-advanced networks using multimedia systems." Transactions on Emerging Telecommunications Technologies. 2018; 29(10): e3443.
[15] Al-Janabi, S., Al-Shourbaji, I., Shojafar, M., and Shamshirband, S. "Survey of main challenges (security and privacy) in wireless body area networks for healthcare applications." Egyptian Informatics Journal. 2017; 18(2): 113–122.
[16] Deebak, B. D. "A secure-ware system for web server: ensuring platform interoperability, security, privacy, usability, and functionality." National Academy Science Letters. 2017; 40(3): 157–160.
[17] Al Ameen, M., Liu, J., and Kwak, K. "Security and privacy issues in wireless sensor networks for healthcare applications." Journal of Medical Systems. 2012; 36(1): 93–101.
[18] Bell, C. G., Chen, R., and Rege, S. "Effect of technology on near term computer structures." Computer. 1972; 5(2): 29–38.
[19] Hill J., Szewczyk R., Woo A., Hollar S., Culler D., and Pister K. "System architecture directions for networked sensors." ACM SIGOPS Operating Systems Review. 2000; 34(5): 93–104.
[20] Hill, J., Horton, M., Kling, R., Krishnamurthy, L., and Krishnamurthy, L. "The platforms enabling wireless sensor networks." Communications of the ACM. 2004; 47(6); 41–46.
[21] Barr, R., Bicket, J. C., Dantas, D. S., et al. "On the need for system-level support for ad hoc and sensor networks." ACM SIGOPS Operating Systems Review. 2002; 36(2): 1–5.

[22] Abiodun, A. S., Anisi, M. H., and Khan, M. K. "Cloud-based wireless body area networks: Managing data for better health care." IEEE Consumer Electronics Magazine. 2019; 8(3): 55–59.

[23] Chin, J., Callaghan, V., and Allouch, S. B. "The Internet-of-Things: Reflections on the past, present and future from a user-centered and smart environment perspective." Journal of Ambient Intelligence and Smart Environments. 2019; 11(1): 45–69.

[24] Sohraby, K., Minoli, D., and Znati, T. *Wireless sensor networks: technology, protocols, and applications*. Hoboken, New Jersey: John Wiley & Sons; 2007.

[25] Coleri, S., Ergen, M., and Koo, T. J. "Lifetime analysis of a sensor network with hybrid automata modelling." *Proceedings of the 1st ACM International Workshop on Wireless Sensor Networks and Applications* (pp. 98–104). ACM. September 2002.

[26] Levis, P., and Culler, D. "Maté: A tiny virtual machine for sensor networks." ACM Sigplan Notices. 2002; 37(10): 85–95.

[27] Sinha, A., and Chandrakasan, A. P. "Operating system and algorithmic techniques for energy scalable wireless sensor networks." *International Conference on Mobile Data Management* (pp. 199–209). Berlin, Heidelberg: Springer; 2001.

[28] Zuberi K. M., Pillai P., and Shin K. G. "EMERALDS: A small-memory real-time microkernel." *Proceedings of ACM Symposium on Operating Systems Principles (SOSP'99)*, Kiawah Island, SC. 1999; 277–291.

[29] David, D. B. "Mutual authentication scheme for multimedia medical information systems." Multimedia Tools and Applications. 2017; 76(8): 10741–10759.

Chapter 13

Severity level classification and detection of breast cancer using computer-aided mammography techniques

Punitha Stephan[1], Fadi Al-Turjman[2] and Thompson Stephan[3]

Breast cancer is the main cause of increase in the cancer death rate among women globally. Early diagnosis of the breast tumors at premature stages prevents the increase in the mortality rate all around the world. Mammography is currently the most efficient and effective way of diagnosing breast cancers to prevent the patients from unwanted therapies and biopsies. The main objective of the proposed work is to help the radiologist to identify the severity level by appropriate grading of the breast cancers using computer-aided mammography (CAM) techniques. The accurate detection of the breast tumors in the proposed work is done using a modified region growing (MRG) followed by a semantic segmentation algorithm. The proposed system uses two-stage classification systems in which the first stage uses an optimized genetic fuzzy classifier (OGFC) to classify the mammograms as normal and abnormal taking statistical features and the wavelet features as input. Further, the abnormal images are classified to identify the stages of the breast cancer as stage I, stage II, stage III, and stage IV using the second stage classification system which consists of a hybrid neural network optimized using genetic algorithms and trained using the grading dataset taking the shape and size features of the malignant tissue as input. The performance of the proposed system will be analyzed using the true-positive (TP), true-negative (TN), false-positive (FP), and false-negative (FN) values, and accuracy. The proposed system is evaluated using various digital mammograms.

13.1 Introduction

In the recent times, all developed and developing industrialized countries concentrate mainly in the cancer health care application particularly in breast cancer

[1]Department of Computer Science and Engineering, Karunya Institute of Technology & Sciences, Coimbatore, India
[2]Artificial Intelligence Department, Research Center for AI and IoT, Near East University, Nicosia, Mersin 10, Turkey
[3]Department of Computer Science & Engineering, Amity School of Engineering and Technology, Amity University Uttar Pradesh, Noida, India

which occurs among women above the age of 30 leading to increase in death rate around the world. The dense regions of the breast are the victims of the breast cancer where the tumors occur. A tumor which is round with smooth appearance and well circumscribed boundary is usually benign. A malignant mass is one which is speculated with roughness and has a blurry boundary [1–3]. This disease became the most common cancer among women. If the patients who are suspected are detected with correct diagnosis at an earlier stage, appropriate treatment options will save the life of the patients. Recently, in sentinel lymph node biopsy, the intra-operative diagnosis of the disease plays a major role in diagnosing the severity [4].

In the case of developed countries, breast cancer became the second cause of death in ladies. In developing countries, the same has become the third leading cause. In developing countries like India, for every two women with breast cancer, one woman dies of it. In 2015, the new cases of breast cancer are 155,000 and out of it, 76,000 women in India died because of the aggressiveness of the disease. At the global level, a higher incidence of breast cancer is observed among women. This may expect to increase in the year 2016 and hence the urge for early detection and diagnosis is necessary. Approximately 3 lakh fresh cases of breast cancer are registered in India almost every year. The prevalence of the disease in urban and rural women is almost equal. Almost 115,000 new cases were registered in hospitals in India per year and in 2016 this number may increase to 250,000. The factors which least to breast cancer are increased age, women with no breastfeeding, hormonal imbalance, menopause at a later age, smoking, and regular alcohol intake. The treatment options for breast cancers are surgery of the affected organ, chemotherapy, and radiation [5]. After lung cancer, breast cancer disease becomes the second reason for death of women all over the world [6].

The chance of the second stage cancer called invasive cancer is 12.5% [7]. New cases of about 182,000 breast cancers are still diagnosed at different stages in which 46,000 ladies die every year in developed countries like the United States [8]. In developed countries, one in every four die because of the breast cancer [9]. The severity of the breast cancer is usually calculated based on the size of the tumor growth and based on this the breast cancers are divided into four stages. The stage I cancer accounts for the tumor till 2 cm size and the lymph nodes are not affected. The stage II cancer has tumor growth from 2 to 5 cm and also the lymph nodes are affected in similar side where the tumor occurred. The stage III cancer called locally advanced contains tumors of more than 2 inches in diameter and are spread to lymph nodes of underarm and also the nearby tissues and lymph nodes are affected. The stage IV called metastatic contains the tumor growth beyond the breast and has affected the underarm and other parts of the body which are close or distant to the breast [10].

There are no prevalent preventive measures available for the breast cancer. Early detection and appropriate diagnosis is the only way to treat the patients and save them from complicated treatments and death. In the clinical practice in recent times due to low cost and many people can easily access the digital mammography becomes the efficient and effective way for diagnosing the breast cancers. Digital mammography is more accurate in detecting the cancers which occur in fatty and dense breasts and still the accuracy is decreased for young women with less fatty breast and the women who have already undergone surgery for various reasons. It also provides more accuracy in finding the tiny calcium deposits called micro-calcifications [11]. Since large numbers of women are affected, the necessity for automatic systems increases since only a few radiologists cannot easily diagnose accurately large number of mammograms and there is a chance for misdiagnosis due to visual fatigue [12]. Radiologists consider computer-aided diagnosis as a secondary option in diagnosing the breast cancer [13]. It accurately finds the exact locations of the tumor for easy diagnosis giving a positive impact for treatments [14,15].

Since a lot of abnormalities happen in the breast area, mammography method is not the only solution to identify all of them but still it is preferred widely in India and other countries because its complexity is very less and also the cost is low. Eighty to ninety percent of the breast abnormalities are properly identified by the radiologist through automated systems attached with the mammography machines. In the case of Digital Mammography in every case, 100% result cannot be expected and occurs with some errors. Vector quantization used lots of segmented areas and information to detect the breast cancer. Thus, it was time consuming as well as complex.

The remaining part of the chapter is organized as follows. Section 13.2 gives the related works for the proposed methodology. Section 13.3 contains the problem definition of the existing techniques. Section 13.4 gives the detailed explanation of the proposed methodology in detail. Section 13.5 gives the evaluation metrics for the proposed system. Section 13.6 gives the discussion of the chapter. Sections 13.7 and 13.8 give the future enhancements and the conclusion of the chapter.

13.2 Related works

Breast cancer detection in the current time is a challenging issue in women. Breast cancer can be curable if detected at the starting stage. Lots of researchers tried to give different techniques for the given work. Here, some related works done by some researchers are described. Sulochana *et al.* [16] proposed a technique which was based on some experimental results for recognizing the abnormalities such as masses and calcifications. The clinical attributes were extracted and analyzed and used in an Artificial Neural Network optimized using a soft computing technique for classification of benign and malignant abnormalities. Visual detection method was used for the implementation and this technique can be further extended to be used with a data acquisition software and hardware interface systems.

Singh *et al.* [17] proposed an approach using image processing techniques called K-means and fuzzy C-means clustering for clear recognition of clusters and masses. By mixing these techniques, breast cancer area is detected in raw mammogram images. This system also provided the classification of the total cancer affected area and this also can be extended with proper data acquisition and hardware interface systems. O'Halloran *et al.* [18] suggested an artifact removal algorithm and the beam former for multistatic data. The identical early-stage artifacts are grouped and removed. The conventional MIST algorithm is adapted for the multistatic data. This system is well suited for tumors with less than 5 mm in size with the heterogeneous breast tissue. The optimal S/C ratio is obtained by using the signals from the three antennas located at the opposite side of the transmitting antenna and this optimal S/C ratio is achieved is good when compared to conventional systems.

Karabataka and Cevdet Ince [19] proposed association rules and neural network for finding out the breast cancer at an earlier stage. The proposed system consists of the optimal feature selection method which acts as the main target for pattern recognition and taxonomy and it reduced the feature vector set from the original vector set. The reduced feature set uses four features using AR. The system was given to Wisconsin breast cancer and tested with three cross validation method and the classification rate is 95.6% for four inputs and 97.4% for eight inputs. This model can be well suited as a prodiagnostic system in finding the breast cancer in patients. A CAD system is proposed based on particle swarm optimized neural networks (SONN) by Dheeba and Tamil Selvi [20] for the classification of microcalcifications. The feature extraction is based on the law feature extraction for texture energy points in the regions of interest. The clusters of the microcalcifications are classified using a PSO optimized neural network. The system is tested with MIAS database with multiple view mammograms and good sensitivity rate of 91% and specificity level of 86.1% are achieved when used for clinical images.

Rouhi *et al.* [21] proposed two methods for segmentation of the mammograms. In the first method, the segmentation is done using region growing in which the thresholds are adjusted using a trained artificial neural network and in the second method, the segmentation is done using cellular neural network optimized using genetic algorithms. Intensity, shape, and texture features are extracted in the feature extraction part which is again classified using an artificial neural network and it is tested with MIAS and DDSM databases and the sensitivity is 96.87% and specificity is 95.94% and the overall accuracy is 96.47%. Chen *et al.* [22] presented a system in which the topologies of the microcalcifications are considered for the classification of MCC. The topology of the clusters of microcalcifications is extracted and classified using K-nearest neighbor classifier to classify as benign or malignant. The system is tested for both MIAS and DDSM databases and analyzed under the receiver operating curve (ROC) which yielded an area of 0.96. This system helps the radiologists to find the microcalcification clusters position and location giving a better clinical understanding.

Servulo de Oliveira *et al.* [23] proposed a system based on taxonomic measures and support vector machine classifier. The images are pre-processed using filters and the taxonomic diversity index (Δ) and the taxonomic distinctness (Δn) used in ecology for phylogenetic trees are used to extract the texture features from the Regions of interest (ROI). The masses are classified as mass and nonmass using SVM classifier and gained an accuracy of 98.88% when tested for digital database for screening mammography (DDSM) images. A novel method is proposed by Kim *et al.* [24] where the stellate features played a major role for discrimination.

The images are pre-processed and the stellate features are further extracted from three subregions called the core, inner, and the outer part which rely on the statistical characteristics for individual subregions. The classification is done through AdaBoost learning for classifying the masses as normal benign and malignant masses. The method is tested for DDSM consisting of 140 mammograms and it is proved that the stellate feature region-based method outperforms the other region-based techniques.

Singh and Gupta [25] presented a segmentation technique based on averaging and thresholding in which a max mean and least variance technique is used to segment the tumour portions. An averaging filter and thresholding is used to find the malignant region in the first stage. A rectangular area is selected and chosen where the min-max technique is implemented in the output image. The segmentation stage is implemented through the detection of region boundary and the patches of the tumor are also detected using morphological closing operation along with the gradient technique which gains an execution time of 4.20 s.

13.3 Problem definition

In the existing works, identifying the benign and malignant tumors was not carried out with proper data acquisition and interface systems. Many techniques detect the size and location of breast tumors. The success rate of the segmentation algorithms increases when a computer-aided diagnosis system is attached to it. To design an effective automatic diagnostic system for breast cancer diseases is the main challenge for us. From the existing techniques studied so far, not many researchers have approached the staging of cancers in patients based on severity levels (size of the tumor). Sensitivity of many existing CAM techniques is high whereas the specificity levels decrease the overall accuracy. When microcalcifications are considered, the detection is more challenging and the large areas of the breast are eliminated and the correlation coefficient raises whereas the information of small areas still exists in the picture. In the case of the breast mass, there is a rise in correlation coefficient after the elimination of small areas and the information regarding the large areas exists in the picture and these are classified at a later stage. This difference can not only be used to detect the masses and microcalcifications. Most of the existing methodologies used to detect breast cancer are expensive and complex. They are based on the intensity changes which will not give better results and they are time consuming too.

13.4 Proposed methodology

The primary intention of this research is to identify the severity level of the abnormal image using computer-aided mammography (CAM). The proposed research makes use of the image processing techniques: preprocessing, feature extraction, classification, malignant segmentation, and grading as shown in Figure 13.1. Next, statistical and wavelet features are extracted from the mammogram image. The statistical

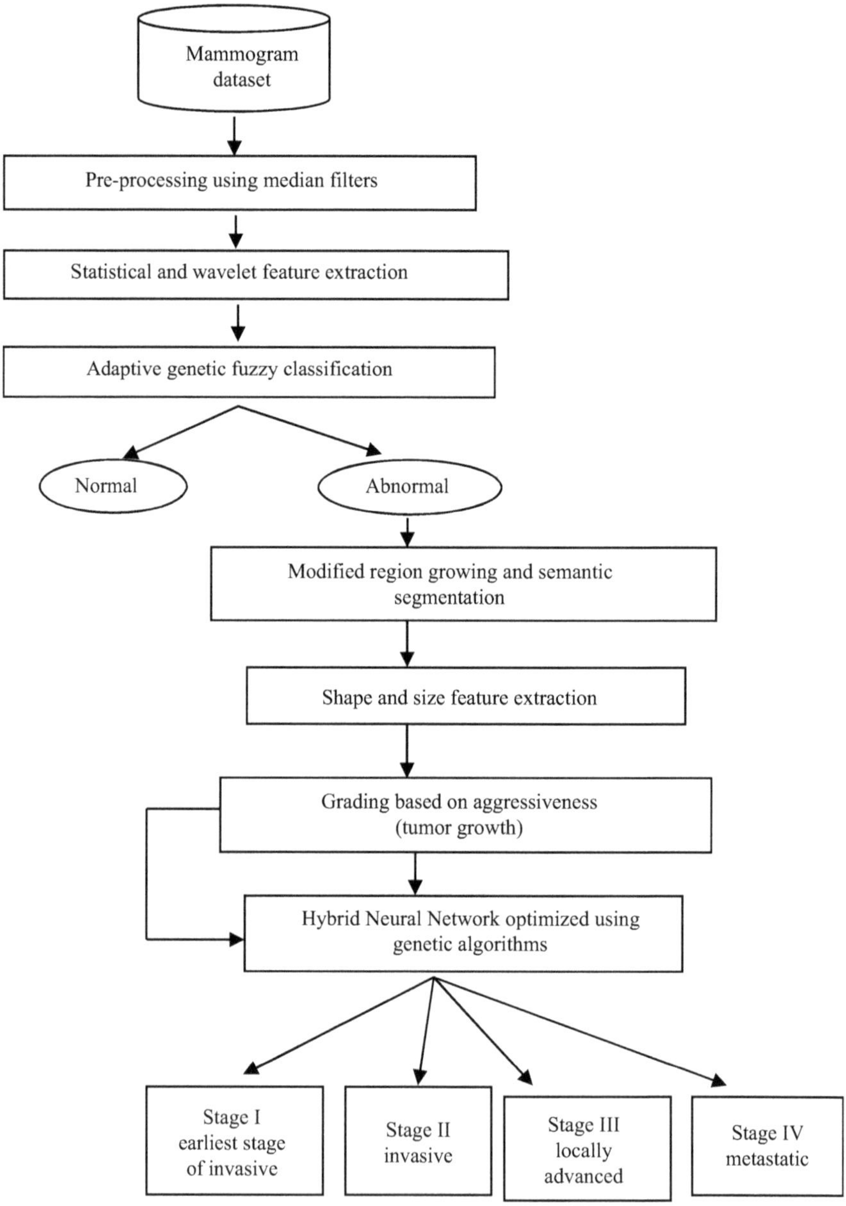

Figure 13.1 The proposed architecture of severity level detection and classification of breast cancers

features that are extracted in our proposed method are area, mean, skewness, texture feature using LGXP, and wavelet using discrete wavelet transform (DWT). Once these features are extracted, they will be given as an input for OGFC for the classification to identify whether the image is normal or abnormal. MRG algorithm followed by semantic segmentation, when the malignant tissue is segmented from the abnormal images. The malignant tissue is segmented efficiently into three dissimilar segments after the semantic segmentation.

Different dissimilar grades were offered, based on the aggressiveness of the malignant tissue. Based on the size and shape of the tissues, the stages of breast cancer like stage I, stage II, stage III, and stage IV are classified. The classifier utilized for this purpose is the hybrid neural network where optimization technique is employed to select the weight factor. The optimization algorithm employed here is genetic algorithm. With the aid of the proposed approach, breast cancer can be easily and effectively detected at an early stage. The performance of the proposed system will be analyzed by TP, TN, FP, FN, and accuracy and it will be compared with existing techniques. The proposed approach will be implemented in MATLAB® and planned to be evaluated using various digital mammography images.

13.4.1 Preprocessing

The preprocessing of the breast mammographic images is done using the 3×3 median filters where a 3×3 window is used and the output values are evaluated using the median of the neighboring pixels. These filters are used to remove the Gaussian, salt, and pepper noise in the breast region. The labels and artifacts are removed using the thresholding techniques.

13.4.2 Segmentation using modified region growing

Detection of tumors of varying sizes plays an important role in increasing the diagnosis rate of breast cancer of various stages. Region Growing is a classical iterative pixel classification approach which uses a set of initial seeds and checks the neighboring pixels and iteratively aggregates the pixels based on some constraints. The traditional region growing uses only the intensity constraints as a threshold for aggregation. The proposed work uses both the intensity and also the orientation constraints as shown in Figure 13.2 through which the problems of over-segmentation are eliminated. The proposed work also greatly distinguishes the combinations of shades in the images. The proposed work also segments the malignant images into three dissimilar segments using semantic segmentation to make the segmentation more accurate. The malignant tumors are further graded based on the aggressiveness.

13.4.3 Feature extraction

The classification accuracy mainly depends on the features extracted and selected which can be used for correctly identify and verify the breast tumors. The proposed work uses the statistical feature and wavelet feature extraction in the first stage

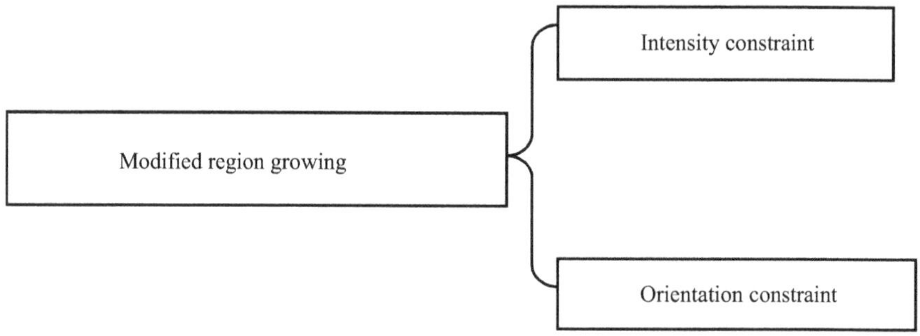

Figure 13.2 The modified region growing

classification. The statistical features such as area which is the measure of the size of the tumor, mean which is a measure neighboring pixel intensity, skewness which is a measure of symmetry, and texture feature which signifies the texture quality are extracted. The texture features are extracted using Local Gabor XOR pattern LGXP and wavelet features using Discrete Wavelet transform (DWT) in the first stage of classification and the second stage classification uses the shape and size features of the segmented abnormal image for appropriate grading and classification of the abnormal images.

13.4.4 Two-stage classification

The proposed methodology uses a two-stage classification system in which the first stage classifies the mammograms as normal and abnormal using an OGFC and the second stage targets only the abnormal images and groups it according to the stage of the breast cancer involved in the mammograms. This section describes the classification used in the proposed work in detail.

13.4.4.1 Optimized genetic fuzzy classification

The first stage classification of the proposed work consists of fuzzy classifier which is optimized using genetic computations such as generation of the initial chromosomes, fitness function computations, mutation operations which are adaptive, crossover, and the optimal selection. The classification is primarily divided into two steps where the first one is the generation of the fuzzy rules based on genetic algorithms and the next step is the classification of the mammograms into normal and abnormal. The genetic algorithm uses the training dataset as the input and generates the optimum rule set which is stored in the rule base of the fuzzy classifier. Using the required membership functions and the fuzzy rules, the fuzzy classifier classifies the images in the training phase and in the testing phase, the fuzzified input and rules are matched and the fuzzy score is calculated. Thus, the first stage fuzzy classifier of the proposed work consists of the generation of the fuzzy rule set using genetic computations, the calculation of the membership function used by the classifier, and the classification of the mammograms collected by the data acquisition. The architecture of the OGFC is shown in Figure 13.3.

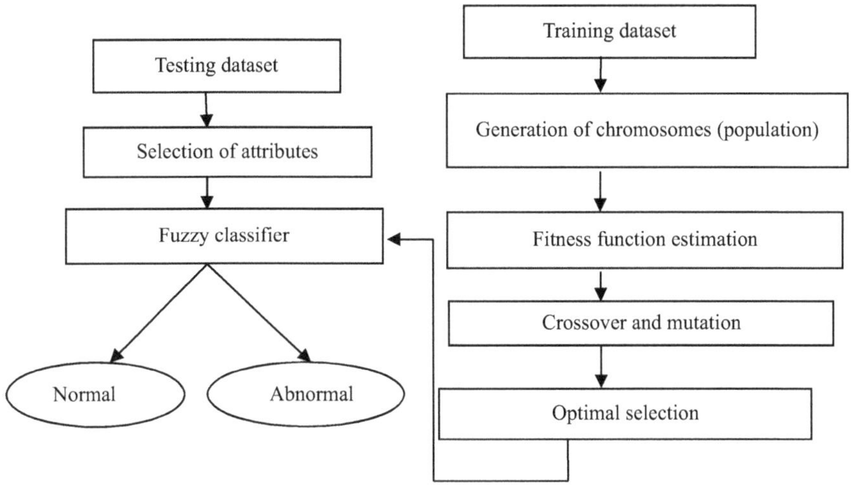

Figure 13.3 Block diagram of optimized genetic fuzzy classification

13.4.4.2 Genetically optimized hybrid neural classification

The proposed work uses a feed-forward neural network whose structure and weights are optimized using genetic algorithm for the accurate classification of the mammograms for finding out the exact stage of the breast tumor. The basic process involved in proposed genetically optimized hybrid neural network is as follows:

The optimized selection of the individuals is used for the adjustment of the input weights of the neural network with the aim of achieving the local minima. The classifier is trained using the grading system through which the stages are identified correctly for the input mammograms. The training algorithm used is the Levenberg–Marquardt back propagation algorithm where the number of input neurons, hidden layers, hidden neurons, and the output neurons are the parameters used. The learning rate, momentum, and the termination criteria are also included in the input parameters. The genetically optimized neural classifier uses the shape and size features and then classifies as stage I, II, III, IV cancers.

13.5 Evaluation metrics

The proposed work for diagnosing the breast cancers is evaluated as the means of sensitivity, specificity, and accuracy which are calculated as follows with the help of true-positive, true-negative, false-positive, and false-negative values.

13.5.1 Sensitivity or true-positive rate

It tells the capability to measure the people with cancer as cancerous giving a positive result. The sensitivity can be calculated using (13.1):

$$\text{Sensitivity} = \frac{\text{True positive}}{\text{True positive} + \text{false negative}} \tag{13.1}$$

13.5.2 Specificity or false-positive rate

It tells the capability to measure the people without cancer as noncancerous giving a negative result. The specificity can be calculated using (13.2):

$$\text{Specificity} = \frac{\text{True negative}}{\text{True negative} + \text{false positive}} \tag{13.2}$$

13.5.3 Accuracy

It tells the proposition of perfectly identified cases to the total number of input cases. Accuracy can be calculated by (13.3):

$$\text{Accuracy} = \frac{\text{True positive} + \text{true negative}}{\text{True positive} + \text{true negative} + \text{false positive} + \text{false negative}} \tag{13.3}$$

13.5.4 Positive predictive value or precision

It gives the measure of cancerous cases perfectly identified as positive by the classifier. The precision can be calculated using (13.4):

$$\text{Precision} = \frac{\text{True positive}}{\text{True positive} + \text{false positive}} \tag{13.4}$$

13.5.5 Negative predictive value or recall

It tells the measure of noncancerous cases perfectly identified as negative by the classifier. The negative predictive value can be calculated using (13.5):

$$\text{Recall} = \frac{\text{True negative}}{\text{True negative} + \text{false negative}} \tag{13.5}$$

13.5.6 False-negative rate or miss rate

It gives noncancerous cases wrongly identified as positive. The false-negative rate can be calculated using (13.6):

$$\text{Miss rate} = \frac{\text{false negative}}{\text{false negative} + \text{true positive}} \tag{13.6}$$

13.6 Discussions

Various research works have been carried out for health care applications that include classification and detection techniques, and various intelligent methods for Internet of Medical Things [26], Internet of Medial Things using deep learning

Table 13.1 Comparison with existing breast cancer diagnosis schemes

Authors	Algorithm	Purpose	Difference in our proposed work
Dheeba *et al.* [20]	ANN optimized with PSO	Optimizing the input weight and structure	Neural network optimized using genetic algorithms
Jaber *et al.* [29]	Improved region growing algorithm	The classical region growing is improved by considering the intensity and orientation constraints	The modified region growing is carried out using both intensity and orientation along with the semantic segmentation for segmenting the malignant tissues
Dennis *et al.* [30]	Adaptive genetic fuzzy classifier for medical imaging	Fuzzy classifier optimized using genetic algorithms and tested with lots of medical data such as liver, heart, etc.	Fuzzy classifier optimized using genetic algorithm using a modified cross over, mutation, and fitness functions to be used in the identification of the mammograms as normal and abnormal
Chen *et al.* [22]	Topological modeling and classification	KNN classification along with the topological feature extraction	Fuzzy and neural classification along with the statistical and wavelet feature extraction
Xu *et al.* [31]	Stacked sparse autoencoder	Nuclei detection technique for breast pathology images	Detection of breast tumors for breast mammograms

[27], and healthcare applications for Internet of Things in industries [28]. For comparison, we have chosen some research works which are listed in Table 13.1.

13.7 Future enhancements

The future enhancements of the proposed work can include the optimal selection of the feature selection in extracting the best features for a better accuracy. The optimization of the proposed two-stage classification can be carried out using other soft computing techniques. The MRG algorithm can be further improved by extending the constraints in addition to the intensity and orientation for aggregation of the pixels. The detection can be extended by the use of pseudo coloring techniques through which the segmentation parts can be represented using various color models for better visualization. The performance analysis can be evaluated more efficiently by the use of free response operating characteristic (FROC) curves.

13.8 Conclusions

Thus, the proposed work is needed for the diagnosis of breast cancer in the early stages to prevent the patients from unnecessary radiations and biopsies using CAM techniques. It accurately detects the location of the breast tumor in mammograms using image segmentation algorithms. It effectively grades the breast tumors based on the aggressiveness (growth of the tumor). It efficiently identifies the stages of breast cancer in patients through fuzzy and neural classification systems and optimizes the results of the classification system using soft computing techniques. The proposed work can be believed as a good combination of the classifier and detection systems for medical diagnosis.

References

[1] Homer M. J. *Mammographic interpretation: A practical Approach*. 2nd edn. Boston, MA: McGraw Hill; 1997.

[2] Reston V. A. *Illustrated breast imaging reporting and data system (BI-RADSTM)*. 3rd edn. American College of Radiology: 1998.

[3] Ireaneus Y., Rejani A., and Selvi S. T. "Early detection of breast cancer using SVM classifier technique." International Journal on Computer Science and Engineering. 2009; 1(3): 127–130.

[4] Saheb Basha S., and Satya Prasad K. "Automatic detection of breast cancer mass in mammograms using morphological operators and Fuzzy C–means clustering." Journal of Theoretical and Applied Information Technology. 2009; 704–709.

[5] "Breast cancer cases in India to double by 2015: experts." DNA India. 2011. Available from http://www.dnaindia.com/health/report-breast-cancer-cases-in-india-todouble-by-2015-experts-1600847 [Accessed February 15, 2020].

[6] Deserno T. M. "Fundamentals of biomedical image processing." in *Biomedical Image Processing. Biological and Medical Physics, Biomedical Engineering*. Springer, Berlin, Heidelberg; 2010.

[7] Breast Cancer Statistics. 2009. http://www.breastcancer.org.

[8] Cheng H. D., Cai X., Chen X., Hu L., and Lou X. "Computer-aided detection and classification of microcalcifications in mammograms: A survey." Pattern Recognition. 2003; 36(12): 2967–2991.

[9] Garcia M., Jemal A., Ward E. M., *et al.* Global cancer facts & figures. Atlanta, GA: American Cancer Society, 2007. Available from https://www.cancer.org/content/dam/cancer-org/research/cancer-facts-and-statistics/global-cancer-facts-and-figures/global-cancer-facts-and-figures-4th-edition.pdf

[10] http://www.cancercenter.com/breast-cancer/stages/

[11] Bottema M. J., Lee G. N., and Lu S. *Automatic image feature extraction for diagnosis and prognosis of breast cancer*. Series in Machine Perception and Artificial Intelligence. Vol 39. Singapore: World Scientific Publishing Co. Pt. Ltd.; 2000.

[12] Guzmán-Cabrera R., Guzmán-Sepúlveda J. R., Torres-Cisneros M., *et al.* "Digital image processing technique for breast cancer detection." 2012.

[13] Astley S. M., and Radiol B. J. "Computer based detection and prompting of mammographic abnormalities." US National Library of Medicines. 2004; 77: S194–S200.

[14] Burhenne L. J. W. "Potential contribution of computer aided detection to the sensitivity of screening mammography." Radiology. 2000; 215: 554–562.

[15] Freer T. W., and Ulissey M. J. "Screening mammography with computer aided detection: Prospective study of 2860 patients in a community breast cancer." Radiology. 2001; 220: 781–786.

[16] Wadhwani S., Wadhwani A. K., and Saraswat M. "Classification of breast cancer detection using artificial neural networks." Current Research in Engineering, Science and Technology (CREST) Journals. 2013.

[17] Singh N., Mohapatra A. G., Rath B. N., and Kanungo G. K. "GUI based automatic breast cancer mass and calcification detection in mammogram images using K-means and Fuzzy C-means methods." International Journal of Machine Learning and Computing. 2012; 2(1): 7–12.

[18] O'Halloran M., Jones E., and Glavin M. "Quasi-multistatic MIST beam-forming for the early detection of breast cancer." IEEE Transactions On Biomedical Engineering. 2010; 57(4): 830–840.

[19] Karabataka M., and Cevdet Ince M. "An expert system for detection of breast cancer based on association rules and neural network." 2009.

[20] Dheeba J., and Tamil Selvi S. "A swarm optimized neural network system for classification of microcalcification in mammograms." Journal of Medical Systems. 2012; 36: 3051–3061.

[21] Rouhi R., Jafari M., Kasaei S., and Keshavarzian P. "Benign and malignant breast tumors classification based on region growing and CNN segmentation." Expert Systems with Applications. 2015; 42: 990–1002.

[22] Chen Z., Strange H., Oliver A., *et al.* "Topological modeling and classification of mammographic microcalcification clusters." IEEE Transactions on Biomedical Engineering. 2015; 62(4): 1203–1214.

[23] Sérvulo de Oliveira F. S., Carvalho Filho A. O., Silva A. C., sode Paiva A. C., and Gattass M. "Classification of breast regions as mass and non-mass based on digital mammograms using taxonomic indexes and SVM." Computers in Biology and Medicine. 2015; 57: 42–53.

[24] Kim D. H., Choi J. Y., and Man Ro Y. "Region based stellate features combined with variable selection using AdaBoost learning in mammographic computer-aided detection." Computers in Biology and Medicine. 2015; 63: 238–250.

[25] Singh A. K., and Gupta B. "A novel approach for breast cancer detection and segmentation in a mammogram." Procedia Computer Science. 2015; 54: 676–682.

[26] Al-Turjman F., Ulusar U., and Nawaz M. "Intelligence in the Internet of Medical Things era: A systematic review of current and future trends." *Elsevier Computer Communications Journal.* 2020; 150(15): 644–660.

[27] Al-Turjman F., Zahmatkesh H., and Mostarda L. "Quantifying uncertainty in Internet of Medical Things and big-data services using intelligence and deep learning." *IEEE Access*. 2019; 7(1): 115749–115759.

[28] Al-Turjman F., and Alturjman S. "Context-sensitive access in Industrial Internet of Things (IIoT) healthcare applications." *IEEE Transactions on Industrial Informatics*. 2018; 14(6): 2736–2744.

[29] Jaber I. M., Jeberson W., and Bajaj H.K. "Improved region growing based breast cancer image segmentation." International Journal of Computer Applications. 2016; 135(8): 1–4.

[30] Dennis B., and Muthukrishnan S. "AGFS: Adaptive Genetic fuzzy system for medical data classification." Applied Soft Computing. 2014; 25: 242–252.

[31] Xu J., Xiang L., Liu Q., *et al*. "Stacked sparse auto encoder for nuclei detection on breast cancer histopathology images." IEEE Transactions on Medical Imaging. 2016; 35: 119–130.

Chapter 14
Biosensors in healthcare: an overview

R. Indrakumari[1], *T. Poongodi*[1], *B. Balamurugan*[1] *and Fadi Al-Turjman*[2]

14.1 Introduction

Biosensor is an analytical device with biologically derived sensitive recognition element incorporated with a transducer [1]. The biosensor consists of three parts: (1) a biological identification elements that distinguish the molecules with the help of certain chemical substances, (2) a transducer that converts the bio-recognition elements into electronic signals, and (3) a signal processor that converts the signal into usable form [2–4]. Monitoring biochemical progression is the most vital procedure in the biological applications. The demand for bio-sensors is high because of its characteristics such as cost-effectiveness and high sensitivity. Bio-sensors combine the bio-molecules with the help of modern optoelectronics and microelectronics [5].

The advancement in the micro- and nanotechnologies, improvement in the biological components, and novel methodologies in the integration of the bioreceptors and the transducers has made exponential progress in the biosensor technology [6] (Figure 14.1). The result of this makes biosensors an interdisciplinary concept that combines the state-of-the-art attainment in biology, physics, material science, engineering, information technology, and chemistry [7]. Earlier biosensors are not compact and used for the applications such as food analysis, medicine, environmental, biotechnology, agriculture, healthcare, production monitoring, security, and defense [8]. The recent applications of the biosensors are especially used for industry and environmental analysis to observe the chemical and microbial components of the water [9], recognition of various toxics substance such as dinoflagellate toxins, plant toxins, bacteria, etc. [10], and examining the contamination of pesticides in the food substance. Now the use of biosensors is extended to healthcare industries due to their precision in analysis to conclude the

[1]School of Computing Science and Engineering, Galgotias University, Greater Noida, Uttar Pradesh, India
[2]Artificial Intelligence Engineering Department, Research Center for AI and IoT, Near East University, Nicosia, Mersin 10, Turkey

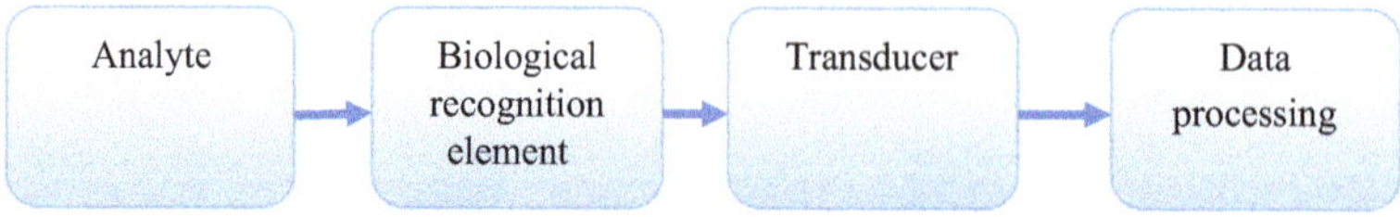

Figure 14.1 Biosensor components

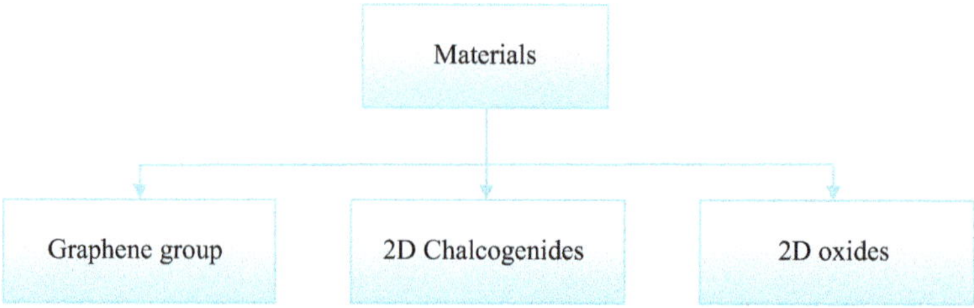

Figure 14.2 Biosensors material

urea, glucose, lactate, blood parameters [11], triglyceride monitoring [12], infectious diseases, cholesterol [13], mutational analysis [14], cancer diagnostics [15], skin allergy [16], and infectious diseases. The accuracy of the biosensors depends on the transducer material and its properties as it should possess high biological affinity and immobilization of the biosensitive layer. Latest development in the nanotechnology-oriented research gives several novel composition of material such as quantum dots (QDs), graphene, carbon nanotubes (CNTs), nanostructures, and several 2D materials which possess the maximum surface-to-volume ratio, making it more suitable for sensors application (Figure 14.2).

This chapter reviews the history of biosensors, its principles, materials used, and its application in glucose monitoring and respiratory airflow monitoring.

14.2 Monitoring principles: transducers

Transducer meant to convert the biological recognition event into a digital signal and process further to produce the result. Transducer adopts several principles such as piezoelectric, electrochemical, optical, or thermal concepts [17]. Electrochemical technology plays a dominant role in devising the transducer because of its reproducibility, sensitivity, low cost, and minimal maintenance. The subdivision of this sensor is conductometric or amperometric and potentiometric type [18–20]. The popular commercially available glucose biosensor is the enzymatic amperometric type. The working principles depend on the capabilities of two electrodes and the current flow in it (Figure 14.3).

Glucose measurement is done with the help of the enzymes such as glucose-1-dehydrogenase (GDH), hexokinase, and glucose oxidase (GOx) [21,22]. Glucose oxidase (GOx) is the most standard enzyme for biosensor as it has characteristics such as maximum selectivity for glucose, cheap, easy availability, and ionic strength.

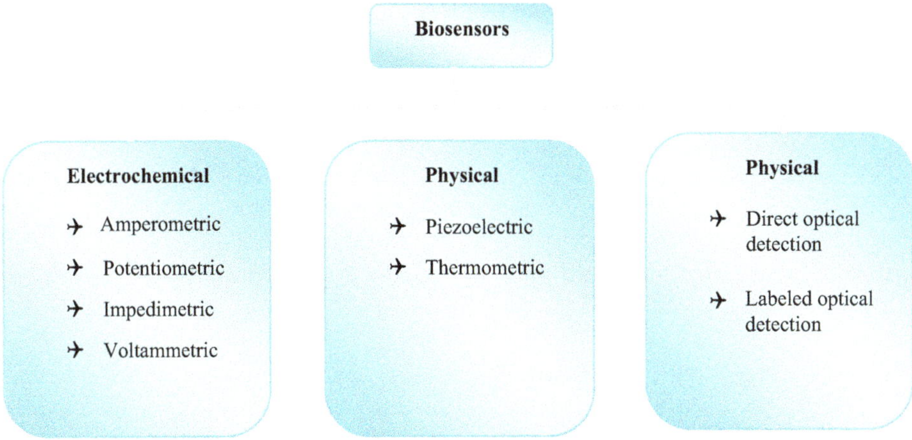

Figure 14.3 Biosensors: based on the transducer type

14.3 Diabetes and the need for glucose monitoring

Diabetes is a global problem resulting from deficiency in the insulin secretion causing failure and dysfunction of organs such as heart, nerves, blood vessels, kidney, and eyes. Diabetics involves many pathogenic processes that destruct the pancreatic beta-cells causing abnormalities in fat, protein, and carbohydrates. The symptoms include polydipsia, blurred vision, polyuria, and weight loss. Complications of diabetics include nephropathy leading to renal failure, retinopathy with potential loss of vision, peripheral neuropathy with risk of foot ulcers, autonomic neuropathy, Charcot joints, sexual dysfunction, genitourinary, and cardiovascular symptoms.

Diabetes falls under three etiopathogenetic categories namely type 1 diabetes caused by the deficiency of insulin hormone secretion. Human body requires insulin to utilize sugar for energy. Nearly 10 percentage of population have type 1 diabetes. Type 2 is the most prevalent class caused by the inadequate compensatory insulin secret response and the resistance to insulin action, approximately 90% of population is suffering from type 2 diabetics. Gestational diabetes occurs during pregnancy and affects up to 4% of people during pregnancies causing complications to fetus. With proper control, this disease can be controlled or managed and even prevented.

14.4 Biosensor for monitoring glucose

Leland C. Clark Jr. introduced the biosensor concept during the development of the oxygen electrode [23]. This work formed the base for the invention of several sensors in future when the extension of this work was addressed in the New York

Academy of Sciences symposium. In this symposium, Leland C. Clark Jr. explained the concept to make electrochemical sensors such as conductometric, polarographic, and pH sensors wisely embossing enzyme transducers as membrane-enclosed sandwiches.

The basic concepts of biosensors are to convert a biologically induced recognition incident into an exploitable signal. These biochemical signals are converted into electronic signals using a transducer (Figure 14.4).

In 1975, the Yellow Springs Instrument Company extended Clark's idea in the development of glucose analyzer; it was the first biosensor-based laboratory analyzer.

14.5 Historical perspectives of glucose biosensors

The first glucose meter was introduced in the year 1971 based on dextrostix and reflectometer. Dextrostix-based blood glucose reading strip is the first blood glucose strip available in the market since 1965 [24].

14.5.1 First generation of glucose biosensor

In 1962, Clark and Lyons proposed the concept of biosensor to display the blood glucose level [2]. This biosensor was comprised of an oxygen electrode, an inner oxygen semipermeable membrane, a thin layer of GOx, and an outer dialysis membrane. Here the glucose concentration is measured by considering the oxygen concentration. Later Updike and Hicks reduced the complications in the electrochemical glucose assay and stabilized GOx [25]. The first commercially available Clark's technology-based biosensor for glucose measurement was introduced by the Yellow Springs Instrument Company analyzer in 1975, and used the amperometric detection of hydrogen peroxide technique. It uses platinum electrode which makes it unaffordable for individual users, thus available only in clinical laboratories. Natural oxygen substrate and hydrogen peroxide play a vital role in the first-generation glucose biosensors as the calculation of peroxide formation is simple when considered miniature devices [26]. The limitation of the first generation biosensor was the amperometric calculation of hydrogen peroxide needs a maximum processing capacity for high selectivity.

14.5.2 Second generation of glucose biosensors

The limitations of the first-generation glucose biosensors were overcome by incorporating mediated glucose biosensors. Redox mediators, a non-physiological electron acceptor is used here, instead of oxygen as in first generation glucose biosensors [27]. Many variations of electron mediators such as tetracyanoquinodimethane (TCNQ), ferrocene, tetrathialfulvalene (TTF), methyl viologen, ferricyanide, thionine, quinines, and methylene blue were incorporated to enhance the performance of the sensor [28]. Among these, ferrocene is considered as the best mediator as it does not react with oxygen, stability in oxidized and reduced form holds reversible electron transfer kinetics property, independent of pH and easy reaction with other enzymes [28].

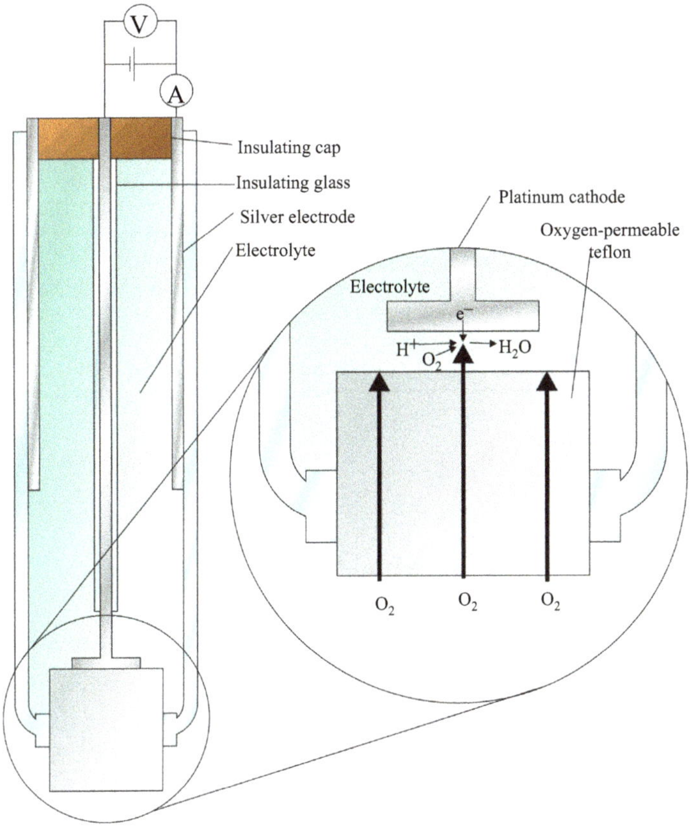

Figure 14.4 Clark oxygen electrode

During 1980s, the second-generation glucose biosensors based on mediator was available for commercial use with modified electrodes and customized membranes for improving the performance of the sensors [29]. In 1987, a pen-sized electrochemical-based blood glucose monitor for self-monitoring of diabetic patients was launched by Medisense Inc which used ferrocene and GDH-PQQ [30]. The success of this device opens up a new path in the revolution of glucose biosensors.

14.5.3 Third generation of glucose biosensors

The third-generation glucose biosensors transfer between the electrode and enzyme without the intervention of the mediator. Organic conducting materials are used here instead of mediators thus facilitating implantable glucose monitoring devices. Many researches have been done for direct electron transfer approaches such as GOx/polypyrrole system [31], tree-like crystal structure, and oxidized boron-doped diamond electrodes. The following Table 14.1 shows the history of glucose biosensors.

Table 14.1 History of glucose biosensors

Year	Biosensor Event
1962	Clark and Lyons proposed the first description of biosensor
1967	Updike and Hicks introduced the world's first practical enzyme electrode
1973	Hydrogen peroxide-based glucose enzyme electrode was introduced
1975	YSI analyzer was introduced
1976	World's first artificial pancreas
1982	Subcutaneous implantable first needle-type enzyme electrode was introduced
1984	Cass introduced ferrocene mediated amperometric glucose biosensor
1987	Blood glucose biosensor by MediSense ExacTech
1999	Commercial glucose sensor
2000	Wearable noninvasive glucose monitor

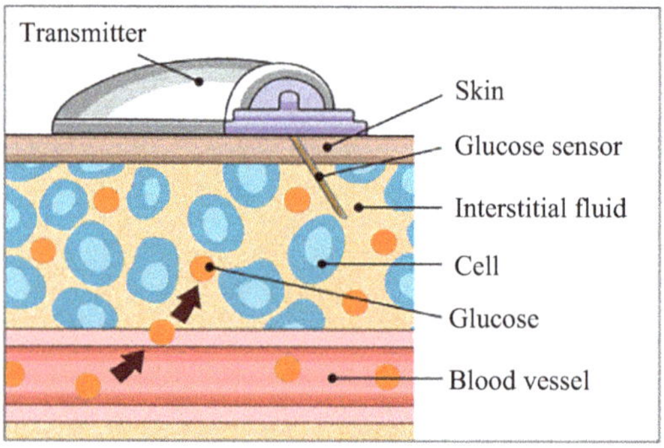

Figure 14.5 Cross-section illustration of a CGM contacting with skin

14.5.4 Continuous glucose monitoring systems

Continuous glucose monitoring system (CGMS) was proposed in 1974 which provided enhanced diabetic control [32]. There are two variations in the Continuous glucose monitoring system—a continuous blood glucose monitor and a continuous subcutaneous glucose monitor. In this method, the surface of electrode is contaminated by coagulation and proteins, and hence most of the Continuous glucose monitoring system does not calculate the glucose directly. This urges the invention of the subcutaneously implantable needle-type electrodes which reflect the glucose level in the blood (Figure 14.5) [33].

MiniMed, a US-based company introduced needle-type glucose biosensor with the limitation that it does not give real-time data. The microdialysis technique allows the continuous subcutaneous glucose monitoring system to reflect the blood glucose level without allowing direct contact between the transducer and the interstitial fluid [34]. Later, the German-based SCGM and Italy-based GlucoDay provided accuracy and precision with minimal drift when compared to needle-type sensors (Figure 14.6) [35].

14.5.5 Noninvasive glucose monitoring system

The invasive continuous glucose monitoring system draws blood from human body to measure the glucose level [36]. This method is disliked by many people as it pricks the skin to measure the glucose [37]. The invention of a noninvasive glucose monitoring technology provides an alternative comfortable solution to people with diabetes. The most common type noninvasive glucose monitoring system is the transdermal or the optical approaches [38]. The optical glucose sensors are based on the light properties in the anterior chamber of the eye using methodologies such as photoacoustics [39], Raman spectroscopy [40], infrared absorption spectroscopy [41], and optical coherence tomography [42]. Cygnus, a US-based manufacturer, developed the world's first transdermal glucose sensor, the GlucoWatch Biographer.

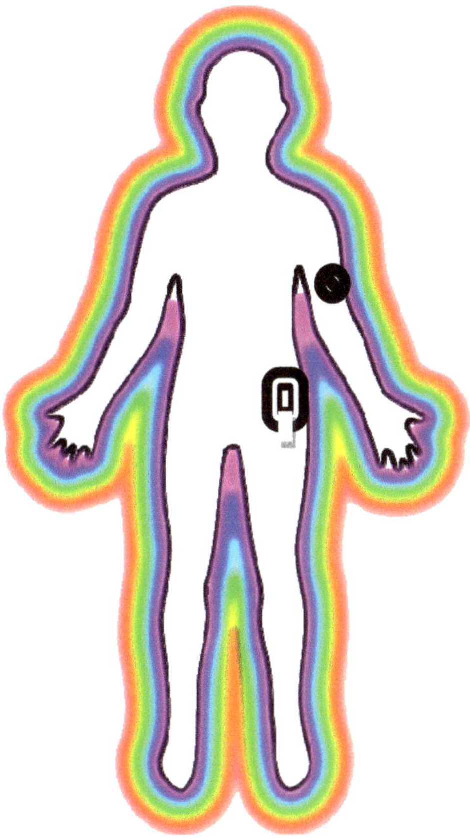

Figure 14.6 Continuous subcutaneous glucose monitoring system

14.6 Respiratory airflow monitoring sensor

The diaphragm is significantly responsible for respiratory activities in the human body. The carbon dioxide and oxygen are the two gases transported to the different parts of the body with the help of circulatory system. The abnormalities in the respiratory system are identified in two ways, they are:

- Increased respiratory rate (tachypnea)
- Decreased respiratory rate (bradypnea)

It can be seen in different situations such as anemia, fever, pneumonia, pulmonary embolism, head trauma, and metabolic abnormalities. The conventional nursing observation is inaccurate and unreliable in respiratory monitoring system [43]. Respiratory sensors must be capable in detecting respiratory airflow, measuring parameters related to blood gas monitor, and respiratory gas flow. The sensors meant for tracking respiration activities must satisfy some criteria such as patient safety, specificity, sensitivity, and reproducibility. It should also meet few specific criteria that include miniaturized, less expensive, rapid response, and minimal interference [44]. Reusability features can be promoted by eventually considering the cleanliness/disinfection of a sensor device without transmitting any hazardous infections. The airflow patterns of respiration are detected via nose and mouth breathing that completely varies at the time of inspiration and expiration (Figure 14.7). Some of the detection modalities for detecting the respiratory airflow [45] are as follows:

- The gas flow velocity is measured with the pressure measurement or fluctuation.
- If the exhaled air is warmer than the ambient temperature, it can be detected using a temperature sensor.

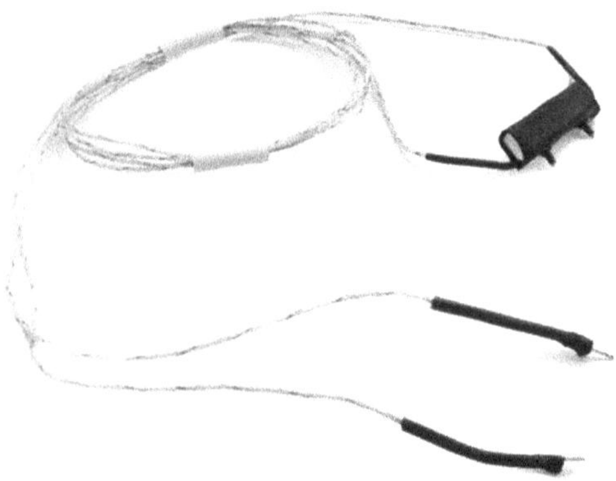

Figure 14.7 Airflow sensor

Table 14.2 Properties for respiratory flow sensing

Parameters	Signal range	Background range	Sensitivity × specificity
Temperature	5 °C	2 °C	Low
Humidity	30%	5%	Moderate
Carbon dioxide	4%	0.05%	High

- If it is more humid than the ambient temperature, it can be detected using a humidity sensor.
- If it contains carbon dioxide (CO_2), it can also be detected.

Some of the characteristics of respiratory flow sensing are given in Table 14.2.

14.6.1 Pressure and acoustic sensing devices

The pressure sensing can be performed to measure the flow by following Bernoulli's equation:

$$\Delta P = \frac{\rho v^2}{2}$$

where "ΔP" denotes the differential pressure based on the flow, ρ refers to the density, v denotes the flow velocity. It requires high dynamic range instead of linear response.

The pressure fluctuations are due to the convergence of ambient and expired air. The silicon microphones or standard electret can be exploited to manage the fluctuations.

Some of the characteristics of acoustic sensing devices are as follows:

- Silicon-based sensor devices are available in miniaturized size either as packaged units or in chip form.
- Rapid response.

Acoustic respiratory tracking devices are designed to pick up the signal using thin polymer for transmitting the acoustic signal to the remote devices. The device is completely subjected to conduct various types of clinical testing/assessment for respiratory monitoring. The market of acoustic wave sensor is forecasted to grow from USD 526.47 million in 2018 to USD 1195.89 million by the year 2026. The Compound Annual Growth Rate (CAGR) is expected to be 10.8% during the year between 2019 and 2026 (Figure 14.8).

14.6.2 Thermal flow sensors

Anemometry principle is commonly followed to measure the airflow velocity because of its robustness, simplicity, and range coverage. Temperature sensor devices such as thermocouples, thermistors, and pyroelectric materials are commonly used for respiratory flow detection [45]. The response time is entirely

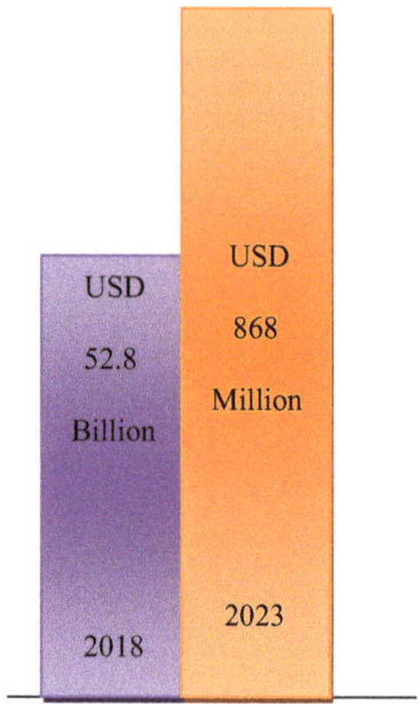

Figure 14.8 Acoustic wave sensor market

Figure 14.9 Temperature sensor

dependent on the thermal flow sensors and many of the sensors suffer due to slow response. The response time is completely adequate at a normal flow rate whereas it is inadequate if velocity is low [46]. The solution is to have the miniaturized sensor by having a thin pyroelectric material and minimizing the thermal mass. The thin films can be fabricated by standard silicon chips using surface micromachining (Figure 14.9) [47].

14.6.3 Humidity sensors

Exhaled air is completely saturated due to the presence of mucus membranes in the air [48]. A conventional method to detect the expired air is to locate a glass mirror near the object to root cause the condensation of vapor on the surface (Figure 14.10) [49,50].

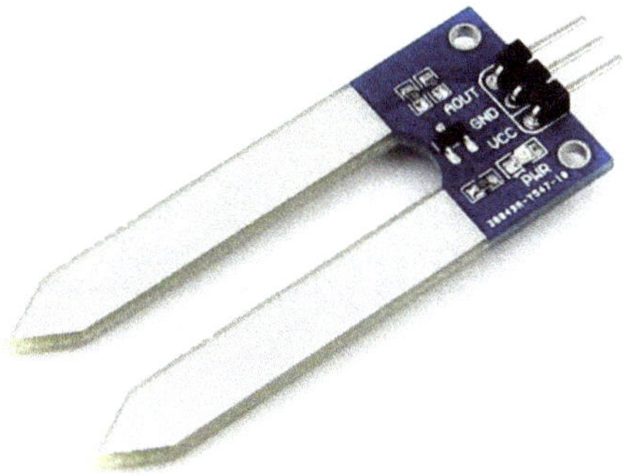

Figure 14.10 Humidity sensor

14.6.4 CO_2 sensors

The major difference in the concentration of CO_2 between ambient and expired air resulted in specificity and sensitivity. CO_2 is strongly associated with the respiration process and it is considered as an ideal variable commonly in respiratory monitoring activities. In fact, the methods to measure CO_2 are of infrared and mass spectroscopy [51,52]. Black body radiators are exploited as emitters and pyroelectric devices are often used as detectors. Diffractive gratings and interference filters are generally used for wavelength dispersion.

14.6.5 Indirect sensors

Attaching equipment surrounding, the airways is always impractical and it disturbs the patient, and they prefer to leave the airways free. Wearing mouthpiece or facemask interrupt the breathing pattern by reducing the respiratory rate. Flow-based sensors necessitate patient's cooperation and limit mobility. Hence, numerous indirect respiratory monitoring approaches are introduced. In many circumstances, the utilization of indirect sensors result in practical benefits, not basically leading to accuracy of records [53]. The devices summarized do not measure the actual respiratory flow, rather the parameters included in the respiratory drive are abdominal/chest motion, intrathoracic pressure effects, respiratory muscle activities, etc. The Table 14.3 shows the details of indirect respiratory sensor devices.

14.6.6 Torso devices

The respiratory systems such as magnetometer, respiratory inductance plethysmograph (RIP), strain gauge, and transthoracic impedance plethysmograph (TTI) are developed for measuring abdominal and/or thoracic impedance [54–57]. Respiration is caused due to the interaction among the diaphragm and respiratory muscles of abdomen and thorax. Abdominal and thoracic impedance are measured simultaneously [58,59].

Table 14.3 Indirect sensor devices of respiration

Sensor devices	Abdominal/ chest motion	Intrathoracic pressure effects	Respiratory muscle activities
Torso devices			
Magnetometer	Yes		
Respiratory inductance plethysmograph (RIP)	Yes		
Strain gauge	Yes		
Transthoracic impedance plethysmograph (TTI)	Yes	Yes	
Electrocardiographic (ECG) sensor		Yes	
Electromyographic (EMG) sensor			Yes
Photoplethysmographic (PPG) sensor		Yes	
Invasive sensor		Yes	
Pressure sensor	Yes	Yes	

- The tidal volumes can be continuously monitored quantitatively or qualitatively after suitable calibration.
- The paradoxic motion of abdomen and rib cage linked with apnea can be sensed.

14.6.7 Magnetometry

The abdominal and thoracic diameters are measured by sensing the magnetic fields that are induced by two receiver coils on the back. The respiratory movement can be correctly sensed by measuring the induced current in the receiver coil. If the distance variation is small among the separation of coils, then it is considered as a linear relation. The tidal volumes can be measured based on Konno-Mead relation by positioning the pair of coils across the abdomen. It measures the distance between coils and not on cross-sectional zone. This makes sense based on body distortion with respect to breathing pattern, motion, and posture variation, Nowadays, magnetometers are not frequently used in clinical situations.

14.6.8 Respiratory inductance plethysmograph

There are two elastic bands located around the abdomen and thorax in respiratory inductance plethysmograph (RIP), respectively. The band consists of an insulated coil placed in a zig-zag manner. The main benefit of RIP is in measuring the cross-sectional zone rather the distance. Moreover, it forms the output which is less sensitive among two compartments at the time of posture change or motion. The technique is found superior and reliable to monitor the tidal volume of the breathing patient. It is a widely used technique in clinics; unfortunately, it is difficult to handle, covers larger body area, and is relatively expensive.

14.6.9 Strain gauge

It is a device used to measure the circumference of abdomen/chest. Mercury-in-silastic is a circumference monitoring strain gauge and it measures the resistance of a mercury level in a stretchable tube located around the abdomen or thorax. Once the gauge is being stretched, automatically the resistance of mercury level increases as it turns into longer and thinner. Stretch-sensitive coils are also applied and generally strain gauges are exploited for tracking respiratory rate that too in a qualitative manner. In electromagnetic noisy environment, [31] presents the system that measures the loss of light intensity which is transmitted via a fiber loop. If the abdominal or chest belt is stretched, intensity decreases and loop inclined more bent. By considering the radius "r" of a circular loop, the intensity loss "I_1" for each curved length is denoted by

$$I_1 = \frac{C_1}{\sqrt{r}} e^{-C_2 r}$$

where C_1 and C_2 refer to the characteristics of optical fiber.

The disadvantage of strain gauge is that the complex procedure for positioning a gauge limits the clinical acceptance.

14.6.10 Transthoracic impedance plethysmograph

It is a widespread technology that employs standard electrodes on the skin surface to monitor the respiration activities. Transthoracic impedance plethysmograph (TTI) measures the impedance variation that occurred between the blood and air in the thoracic part. Impedance systems can be jointly practiced with ECG to monitor the respiratory rate. Practically, some authors found that it is possible to monitor tidal volumes using TTI [34,36,39,42,43] and others declared that it is too decisive [44,45]. Later, it is concluded that acceptable estimates are made during normal breathing, not precise ample for individual breathing. TTI is highly sensitive to the artifacts originated in the cardiac pulsation, posture change, patient motion, and especially in the interface.

14.6.11 Electrocardiographic sensor

Electrocardiographic (ECG) is a noninvasive widely used technology to measure and record the fluctuations in the cardiac potential. It is an effective diagnostic technique that has been utilized for many years to detect heart-related issues like various kinds of arrhythmias. However, not all arrhythmias are life threatening and may end in cardiac arrest if not handled properly. Once the inconsistency is detected in cardiac activity, ambulatory monitoring is required in the earlier stage itself. Some serious arrhythmias include hypertrophic cardiomyopathy, QT syndrome, Brugada syndrome, etc. and that too can be detected only on prolonged monitoring. Respiratory parameters could be extracted by determining Respiratory Sinus Arrhythmia (RSA). In particular, the heart rate increases and decreases at the time of inspiration and expiration. As per Nyquist theorem, the sampling rate is caused to be inadequate if the respiratory rate becomes higher than the half of the heart rate. Both T-wave and R-peak of ECG could be used to augment the frequency range. The advantages of ECG occur in the possibility of cardiac monitoring and simultaneous respiration.

14.6.12 Electromyographic sensors

Intercostal electromyography is measured in which the diaphragm electrical activities could be sensed. When electromyographic (EMG) is measured in the esophagus, the good signal quality is obtained and the standard skin electrodes are exploited for monitoring. These signals can easily get disturbed by any electrical activity or electrode motion due to its low signal level. The significant benefit of EMG is the early identification of fatigue in the respiratory muscle, although the techniques are still limited for several clinical applications.

14.6.13 Photoplethysmographic sensor

Photoplethysmographic (PPG) is a technique used for tracking respiratory rate and blood gas level in pulse oximetry. Reflective mode sensor is utilized which emits infrared light into the patient's skin. Infrared light is exploited as an absorption technique in tissue and blood that causes a huge penetration of the light. Such penetration reaches the photodetector depending on blood flow and volume in the skin, firmly influencing the scattering and absorption. The respiratory variations could be extracted by employing digital filtering with the reflection mode monitoring device. The reflection mode makes the possibility to fix the sensor anywhere on the human body. The benefits of PPG sensor are the possibility of measuring the heart rate, respiratory rate, oxygen saturation level, and its simplicity in use. The disadvantages of PPG are quality loss during motion and disturbance in signal transfer processing.

Wearable technologies improve convenience, comfort, quality, and security of human life by monitoring their health conditions regularly. It assists in early diagnosis of different diseases, therapy, individual care of every person's welfare, continuous monitoring of people's health, and real-time monitoring of treatment process. Moreover, wearable electronic devices receive a great attention due to its interaction with the human body in monitoring wrist pulse rate, heart rate, intraocular pressure, glucose, motion, temperature, etc. The wearable devices provide continuous, real-time data that are recorded related to health conditions periodically. Furthermore, in order to reduce the utilization of rigid batteries and to maximize the wearable and portable benefits of wearable devices, low power consumption devices were integrated. The features of wearable devices automatically improve user's compliance with the medication schedule and medical instructions. The information gathered by the wearable devices will be transmitted to the centralized node (microcontroller or mobile phone) and then it can be sent to the medical center for processing.

Wearable systems consist of wearable materials, smart sensors, actuators, communication media, power supplies, processing units, user interface, and algorithms required to extract the data to make decisions efficiently. Hence, the patients' physiological data such as blood pressure, glucose level, temperature, stress and biomolecules, and ions in the blood stream can be monitored. Moreover, the smart sensors should be equipped with flexible substrates and the electrodes which are embedded must be light-weight, ultrathin, highly flexible, low modulus, and stretchable. In general, smart sensors are fabricated using hybrid materials, hybrid structures, nanomaterial, and carbon nanofibers in order to meet the requirements. In particular, inkjet, electrohydrodynamic (EHD), and 3D printing methodologies are exploited to enrich the electrodes with high performance and high resolution.

14.7 Conclusion

An accurate diagnostic method for disease is important for a successful recovery of suffering patient. The diagnostic methodology should be sensitive and it can able to find various biomarkers that are available at low concentrations in biological fluids. Biosensors can satisfy these conditions. Biosensors are becoming the most popular discipline because of their highly sensitive, rapid, easy, and highly selectivity characters. This chapter overviewed biosensors, its techniques, and application in the field of healthcare monitoring especially for blood glucose monitor and respiratory monitoring system. Glucose biosensor is considered as the most reliable, accurate, and compact device. Monitoring is a vital part in patient management with acute respiratory failure. Unlike other monitoring organs and functions, respiratory monitoring is not easy. But respiratory biosensors can make it easy and accurate.

References

[1] Ibrahim, U., Ozsos, M., and Al-Turjman, F., "Futuristic CRISPR-based biosensing in the cloud and Internet of Things era: An overview." Multimedia Tools and Applications. 2020. doi. 10.1007/s11042-020-09010-5.

[2] Clark, L. C. Jr., and Lyons, C. "Electrode systems for continuous monitoring in cardiovascular surgery." Annals of the New York Academy of Sciences. 1962; 102: 29–45.

[3] Updike, S. J., and Hicks, G. P. "The enzyme electrode." Nature. 1967; 214: 986–988.

[4] Hiratsuka, A., Fujisawa, K., and Muguruma, H. "Amperometric biosensor based on glucose dehydrogenase and plasma-polymerized thin films." Analytical Sciences. 2008; 24: 483–486.

[5] Solanki, P. R., and Kaushik, A. "Nanostructured metal oxide-based biosensors." NPG Asia Materials. 2011; 3: 17–24.

[6] Scheller, F. W., Wollenberger, U., Warsinke, A., and Lisdat, F. "Research and development in biosensors." Current Opinion in Biotechnology. 2001; 12: 35–404.

[7] North, S. H., and Lock, E. H. "Critical aspects of biointerface design and their impact on biosensor development." Analytical and Bioanalytical Chemistry. 2010; 397: 925–933.

[8] Ravindra, N.M., and Prodan, C. "Advances in the manufacturing, types, and applications of biosensors." JOM. 2007; 59: 37–43.

[9] Queiros, R. B., Noronha, J. P., Marques, P. V. S., and Sales, M. G. F. "Emerging (bio)sensing technology for assessing and monitoring freshwater contamination: Methods and applications." Voudouris, K. (ed.), *Ecological water quality—Water treatment and reuse*. 1st ed. Rijeka, Croatia: InTech; 2012. pp. 65–94.

[10] Ansari, A. A., Kaushik, A., Solanki, P. R., and Malhotra, B. D. "Nanostructured zinc oxide platform for mycotoxin detection." Bioelectrochem. 2010; 77: 75–81.

[11] Tusa, J. K., and He, H. "Critical care analyzer with fluorescent optical chemosensors for blood analytes." Journal of Materials Chemistry. 2005; 15: 2640–2647.

[12] Osaka, T., Komaba, S., Seyama, M., and Tanabe, K. "High-sensitivity urea sensor based on the composite film of electroinactive polypyrrole with polyion complex." Sensors and Actuators B: Chemical. 1996; 36: 463–469.

[13] Jijun, T., Jie, H., Zhongchao, H., Min, P., and Yuquan, C. "A novel lactate biosensor." *Proceedings of the 27th Annual International Conference of the IEEE Engineering in Medicine and Biology Society*. Shanghai, China, January 17–18, 2006; pp. 252–254.

[14] Hang, T. C., and Guiseppi-Elie, A. "Frequency dependent and surface characterization of DNA immobilization and hybridization." Biosensors and Bioelectronics. 2004; 19: 1537–1548.

[15] Monosik, R., Stredansky, M., and Strurdik, E. "Application of electrochemical biosensor in clinical diagnosis in clinical diagnosis." Journal of Clinical Laboratory Analysis. 2012; 26: 22–34.

[16] Hofmann, U., Michaelis, S., Winckler, T., Wegener, J., and Feller, K. H. "A whole-cell biosensor as in vitro alternative to skin irritation tests." Biosensors and Bioelectronics. 2013; 39: 156–162.

[17] Newman, J. D., and Turner, A. P. *Biosensors: Principles and practice*. Vol. 27. London, UK: 1992; pp. 147–159.

[18] Habermuller, K., Mosbach, M., and Schuhmann, W. "Electron-transfer mechanisms in amperometric biosensors." Fresenius' Journal of Analytical Chemistry. 2000; 366: 560–568.

[19] Pearson, J. E., Gill, A., and Vadgama, P. "Analytical aspects of biosensors." Annals of Clinical Biochemistry. 2000; 37 (Pt 2): 119–145.

[20] Thevenot, D. R., Toth, K., Durst, R. A., and Wilson, G. S. "Electrochemical biosensors: Recommended definitions and classification." Biosensors and Bioelectronics. 2001; 16: 121–131.
[21] Price, C. P. "Point-of-care testing in diabetes mellitus." Clinical Chemistry and Laboratory Medicine. 2003; 41: 1213–1219.
[22] D'Costa, E. J., Higgins, I. J., and Turner, A. P. "Quinoprotein glucose dehydrogenase and its application in an amperometric glucose sensor." Biosensors. 1986; 2: 71–87.
[23] Clark, L. C., Jr. "Monitor and control of blood and tissue oxygen tensions." Transactions of the American Society for Artificial Internal Organs. 1956; 2: 41–48.
[24] Drury, M. I., Timoney, F. J., and Delaney, P. "Dextrostix: A rapid method of estimating blood glucose levels." Journal of the Irish Medical Association. 1965; 56: 52–53.
[25] Updike, S. J., and Hicks, G. P. "Reagentless substrate analysis with immobilized enzymes." Science. 1967; 158: 270–272.
[26] Wang, J. "Electrochemical glucose biosensors." Chemical Reviews. 2008; 108: 814–825.
[27] Liu, J., and Wang, J. "Improved design for the glucose biosensor." Food Technology and Biotechnology. 2001; 39: 55–58.
[28] Chaubey, A., and Malhotra, B. D. "Mediated biosensors." Biosensors and Bioelectronics. 2002; 17: 441–456.
[29] Hilditch, P. I., and Green, M. J. "Disposable electrochemical biosensors." Analyst. 1991; 116: 1217–1220.
[30] Matthews, D. R., Holman, R. R., Bown, E., *et al.* "Pen-sized digital 30-second blood glucose meter." Lancet. 1987; 1: 778–779.
[31] Rubio Retama, J., Lopez Cabarcos, E., Mecerreyes, D., and Lopez-Ruiz, B. "Design of an amperometric biosensor using polypyrrole-microgel composites containing glucose oxidase." Biosensors and Bioelectronics. 2004; 20: 1111–1117.
[32] Albisser, A. M., Leibel, B. S., Ewart, T. G., *et al.* "Clinical control of diabetes by the artificial pancreas." Diabetes. 1974; 23: 397–404.
[33] Rebrin, K., and Steil, G. M. "Can interstitial glucose assessment replace blood glucose measurements?" Diabetes Technology & Therapeutics. 2000; 2: 461–472.
[34] Poscia, A., Mascini, M., Moscone, D. *et al.* "A microdialysis technique for continuous subcutaneous glucose monitoring in diabetic patients (part 1)." Biosensors and Bioelectronics. 2003; 18: 891–898.
[35] Nielsen, J. K., Freckmann, G., Kapitza, C. *et al.* "Glucose monitoring by microdialysis: Performance in a multicentre study." Diabetic Medicine. 2009; 26: 714–721.
[36] Vaddiraju, S., Burgess, D. J., and Tomazos, I., *et al.* "Technologies for continuous glucose monitoring: current problems and future promises." Journal of Diabetes Science and Technology. 2010; 4(6): 1540–1562.

[37] Zhang, W., Du, Y., and Wang, M. L. "Noninvasive glucose monitoring using saliva nano-biosensor." Sensing and Bio-Sensing Research. 2015: 4; 23–29.

[38] Oliver, N. S., Toumazou, C., Cass, A. E., and Johnston, D. G. "Glucose sensors: A review of current and emerging technology." Diabetic Medicine. 2009; 26: 197–210.

[39] MacKenzie, H. A., Ashton, H. S., Spiers, S., *et al.* "Advances in photoacoustic noninvasive glucose testing." Clinical Chemistry. 1999; 45: 1587–1595.

[40] Goetz, M. J., Jr., Cote, G. L., Erckens, R., March, W., and Motamedi, M. "Application of a multivariate technique to Raman spectra for quantification of body chemicals." IEEE Transactions on Biomedical Engineering. 1995; 42: 728–731.

[41] Gabriely, I., Wozniak, R., Mevorach, M., Kaplan, J., Aharon, Y., and Shamoon, H. "Transcutaneous glucose measurement using near-infrared spectroscopy during hypoglycemia." Diabetes Care. 1999; 22: 2026–2032.

[42] Larin, K. V., Eledrisi, M. S., Motamedi, M., and Esenaliev, R. O. "Noninvasive blood glucose monitoring with optical coherence tomography: A pilot study in human subjects." Diabetes Care. 2002; 25: 2263–2267.

[43] Haborne, D. "Measuring respiratory rate." Archives of Emergency Medicine. 1992; 9: 377–378.

[44] Semmes, B. J., Tobin, M. J., Snyder, J. V., and Grenvik, A. "Subjective and objective measurement of tidal volume in critically ill patients." Chest. 1985; 87: 577–579.

[45] Sackner, M. A. "Monitoring of ventilation without a physical connection to the airway." Lung biology in health and disease, diagnostic techniques in pulmonary disease. Vol. 16 (Part 1). New York: Marcel Dekker; 1980. pp. 503–536.

[46] Cobbold, R. S. C. *Transducers for biomedical measurements: Principles and applications*. New York, Wiley; 1974. pp. 57–111.

[47] Gardner, J. W. *Microsensors: Principles and applications*. New York: Wiley; 1994.

[48] Al-Turjman, F., and Lemayian, J. "Intelligence, security, and vehicular sensor networks in internet of things (IoT)-enabled smart-cities: An overview." Computers & Electrical Engineering. 2020; 87: 106776.

[49] Larsson, C., and Staun, P. "Evaluation of a new fibre-optic monitor for respiratory rate monitoring." Journal of Clinical Monitoring and Computing. 1999; 15: 295–298.

[50] Tatara, T., and Tsuzaki, K. "An apnea monitor using a rapid-response hygrometer." Journal of Clinical Monitoring. 1997; 13: 5–9.

[51] Lenz, G., Heipertz, W., and Epple, E. "Capnometry for continuous postoperative monitoring of nonintubated, spontaneously breathing patients." Journal of Clinical Monitoring. 1991; 7: 245–248.

[52] Kavanagh, B. P., Sandler, A. S., and Turner, K. E. "Use of end-tidal PCO_2 as noninvasive measurement of arterial PCO_2 in extubated patients recovering from general anesthesia." Journal of Clinical Monitoring. 1992; 8: 226–230.

[53] Gilbert, R., Auchincloss, J. H., Brodsky, J., and Boden, W. "Changes in tidal volume, frequency, and ventilation induced by their measurement." Journal of Applied Physiology. 1972; 33: 252–254.

[54] Deebak, D., Al-Turjman, F., Aloqaily, M., and Alfandi, O. "An authentic-based privacy preservation protocol for smart e-healthcare systems in IoT." IEEE Access. 2019. doi: 10.1109/ACCESS.2019.2941575.

[55] Al-Turjman, F., and Alturjman, S. "Context-sensitive access in Industrial Internet of Things (IIoT) healthcare applications." IEEE Transactions on Industrial Informatics. 2018; 14(6): 2736–2744.

[56] Al-Turjman, F., Ulusar, U., and Nawaz, M. "Intelligence in the Internet of Medical Things era: A systematic review of current and future trends." Elsevier Computer Communications Journal. 2020; 150(15): 644–660.

[57] Cohn, M. A., Rao, A. S. V., Broudy, M., *et al.* "The respiratory inductive plethysmograph: a new non-invasive monitor of respiration." Bulletin of European Physiopathology Respiratory. 1982; 18: 643–658.

[58] Al-Turjman, F., Zahmatkesh, H., and Mostarda, L. "Quantifying uncertainty in Internet of Medical Things and big-data services using intelligence and deep learning." IEEE Access. 2019; 7(1): 115749–115759.

[59] Al-Turjman, F., and Baali, I., "Machine learning for wearable IoT-based applications: A survey." *Wiley Transactions on Emerging Telecommunications Technologies*. 2019. doi. 10.1002/ett.3635.

Chapter 15

Swarm intelligence-based medical diagnosis systems

Punitha Stephan[1], Fadi Al-Turjman[2] and Thompson Stephan[3]

The automatic disease diagnosis system plays a vital role in the detection of diseases in early stages. Computer-aided systems constructed using swarm intelligence metaheuristic approaches have proved their efficiency in diagnosing the diseases due to their strong exploration and exploitation capabilities which help them to find out the optimal solutions. These medical systems that are based on swarm intelligence techniques eliminate the demerits of manual diagnosis systems such as human errors, huge labor and high computational time. These systems have proven to decrease the death rate all over the world by helping the doctors as a second opinion process in complex decision-making situations.

15.1 Introduction

It is an artificial intelligence approach which is based on collective behaviors of systems which are self-organized and decentralized. In 1989, Beni and Wang introduced swarm intelligence which consists of a population of agents that interact with each other within themselves and with the environment. The interactions have no centralized control and it is based on intelligent behaviors like swarming, flocking, schooling, and following.

These self-organized systems have properties such as positive feedback that accepts better solutions by assigning the agents to them. The negative feedback rejects the solutions that are worst and have the same behavior and state. The amplifications followed by these systems are the random walks and randomness. These self-organized systems have a good number of social interactions between them. There are many computationally intelligent systems that are built based on different swarm intelligence and it is used for solving various optimization

[1]Department of Computer Science and Engineering, Karunya Institute of Technology & Sciences, Coimbatore, India
[2]Artificial Intelligence Department, Research Center for AI and IoT, Near East University, Nicosia, Mersin 10, Turkey
[3]Department of Computer Science & Engineering, Amity School of Engineering and Technology, Amity University Uttar Pradesh, Noida, India

problems for real time applications as applied in [1,2]. A lot of metaheuristic approaches were used for the medical diagnosis system which is demonstrated in [3–5]. Health care applications have gained attention in recent researches due to the need for early diagnosis of various diseases as demonstrated in [6–11]. Some of the swarm technologies that are used for medical diagnosis system to overcome the disadvantages of the manual diagnosis system are as follows:

1. Particle swarm optimization
2. Ant colony optimization
3. Artificial bee colony optimization
4. Bacterial foraging optimization

15.1.1 Particle swarm optimization

Particle swarm optimization (PSO) is inspired by the behavior of swarming of bees, gathering of birds, and schooling of fishes [12]. The main advantage of PSO is that it is simple with very few control parameters and it is a flexible metaheuristic search algorithm which can be well suited for medical diagnosis applications. The algorithm is framed based on the particle position and velocity. Each particle represents the solution to an optimization problem. Let $X_i(t)$ is the particle position at time interval t in problem space, the new position of the particle can be generated using (15.1) where $Y_i(t)$ is the velocity of the particle calculated using (15.2):

$$X_i(t+1) = X_i(t) + Y_i(t+1) \tag{15.1}$$

$$Y_i(t) = Y_i(t-1) + a_1 r_1(\text{local}(t) - X_i(t-1)) + a_2 r_2(\text{global}(t) - X_i(t-1)) \tag{15.2}$$

where c_1 and c_2 are the acceleration coefficients and r_1 and r_2 are random vectors.

The PSO optimization algorithmic steps are as follows.

1. Initialize initial parameters such as population, position, and velocity of the particle.
2. Calculate the particle fitness (local best).
3. Pick the best solution till the current iteration (global best).
4. Calculate the velocity using local best and global best.
5. Update the velocity and position of the particle.
6. Continue till the maximum number of iterations.

15.1.1.1 Particle swarm optimization for medical diagnosis

A cancer diagnosis system using entropy measures and an artificial neural network is proposed where initial weights were optimized by PSO [13]. A multilayer perceptron neural network is used for classification. Weights were adjusted using back propagation algorithm with PSO. The system is evaluated using a breast cancer dataset. Classification accuracy of 97.51% is achieved using PSO and 98.83% is achieved while using Levenberg–Marquardt algorithms.

A cancer diagnosis system is proposed using velocity bounded particle swarm optimization [14]. The system acts as a computer-aided diagnosis system for diagnosing liver and kidney diseases. It concentrates on the features selection problem and its efficiency is proved using different benchmark functions. The proposed system served as a computationally intelligent system for automatic diagnosis of live and kidney diseases by selecting an appropriate input dataset for the classification. The best classification accuracy gained for kidney disease is 77.14% and 90% when a probabilistic neural network and a support vector machine classifier are used, respectively. The best classification accuracy gained for liver disease is 77.14% and 82.86% when a probabilistic neural network and a support vector machine classifier are used, respectively.

15.1.2 Ant colony optimization

Ant colony optimization (ACO) is introduced in 1991 by Dorigo *et al.* and it is population-based and inspired by real ant manners [15]. It is based on ant foraging which shows how ants do shortest path discovery from nest and food source. Ants search for food around the neighboring place and they leave a substance called pheromone during searching. The other ants select the strongest pheromone concentrated way. After the food is found, the ant takes the food and returns back to the nest by calculating the quality of the food.

The pseudo-code for ACO algorithm is given as follows:

1. Parameters and pheromone trail initialization
2. Generate initial population
3. Calculate the fitness of each solution
4. For each ant, calculate the best position
5. Calculate best global ant
6. Update pheromone trail
7. Continue till terminating condition

15.1.2.1 Ant colony optimization for medical diagnosis

A medical diagnosis system developed based on the integration of ACO and cuckoo search algorithm is proposed [16]. The local search of ACO is improved using exploitative capabilities of cuckoo search. The proposed system selects feature sets which are optimal from 78 GLCM texture features of a digital mammogram. MIAS digital mammograms are used as the input dataset. A set of five features were selected as an optimal set. The accuracy was 94% for the proposed medical diagnosis system. This medical diagnosis system uses SVM as a predictive model. It is compared with PSO and ACO where the proposed diagnosis system that uses hybrid ACO and cuckoo search shows the high accuracy of 4% and 2% when compared with PSO and ACO techniques.

ABC and ACO were proposed for the diagnosis of various diseases [17]. The approach is evaluated on UCI benchmark datasets for optimal feature subset selection process. The proposed approach yielded an accuracy of 99.07% for breast cancer diagnosis under the Wisconsin breast cancer dataset. The proposed algorithm is only used for feature selection and it has not been used for selecting the optimal initial parameter selection.

15.1.3 Artificial bee colony optimization

Artificial bee colony (ABC) is a swarm-based technique which is inspired by the foraging behavior of honey bees. It is a population-based and metaheuristic search algorithm [18]. It is based on three factors as follows:

1. Food source: Each solution is represented by a food source and its value depends on the quality of the food source.
2. Employed foragers: They concentrate on exploitation and exploit the available food sources.
3. Unemployed foragers: They discover new food sources that are exploited by employed foragers.

Generate food sources positions using (15.3):

$$E^n_m = E^n_{\min} + \text{random}\,(0,1) * \left(E^n_{\max} - E^n_{\min}\right) \tag{15.3}$$

E^n_m represents food source with parameter m and $m = 1, 2, \ldots, N$ and N is food source size. $N = 1, 2, \ldots, X$ and "X" indicates the dimension of the optimization problem. Random $(0, 1)$ is the step size between 0 and 1. $E^n_{\max}$ is a maximum of nth parameter and $E^n_{\min}$ is a minimum of nth parameter.

Exploit the food source by employee bees using (15.4):

$$X^n_m = E^n_m + \text{random}[-1,1]\left(E^n_m - E^n_k\right) \tag{15.4}$$

E_k is the random food source, $k \in \{1, 2, \ldots, N\}$. "$n$" is the random integer and $m = \{1, 2, \ldots, X\}$ and "m" should not be equal to "k."

Exploit better food source by onlooker bees using (15.4) and (15.5):

$$Y_i = \frac{\text{fitness}(E_m)}{\sum_{i=1}^{N} \text{fitness}(E_i)} \tag{15.5}$$

where $\text{fitness}(E_m)$ indicates the quality of the source E_i.

Discover new food source using scout bees by (15.3).

It is a simple and robust search technique which is used for solving multi-dimensional optimization problems. Each bee represents a solution to the problem. The algorithm is as follows:

1. Set the initial value of parameters.
2. Calculate the nectar value of the food source.
3. Exploit the food sources by the employee bees.
4. Exploit better food sources by onlooker bees.
5. Eliminate unimproved solutions and discover new food sources by scout bees.
6. Save the best food source.
7. Continue till terminating conditions.

15.1.3.1 Artificial bee colony optimization for medical diagnosis

A medical diagnosis system is proposed that uses ABC and decision tree gradient boosting model [19]. It used the breast cancer dataset and Haberman's survival dataset. The proposed system selects features from these two datasets. A regression tree is used for diagnosis purposes. The classification accuracy is 74.3% using Haberman's cancer dataset and 92.8% using the breast cancer dataset.

ABC with differential evolution optimization is used for proposing an automatic system [20]. The standard optimization of ABC is improved using the differential evolution properties. The proposed system is used for optimal feature selection from UCI benchmark medical datasets. This system attained F-measure of 92.2, 96.4, and 97.6 for decision tree classifier, Naive Bayes classifier, and RBF networks classifier, respectively, for breast cancer dataset.

15.1.4 Bacterial foraging optimization

Bacterial foraging optimization (BFO) is a metaheuristic search algorithm which is swarm based [21]. It is inspired by the foraging process of bacteria where the bacteria search for highly nutritious areas for survival. Each bacteria tumble in a random direction and then swims in that direction if the fitness is more than the old one. Four phases are involved in BFO.

1. **Chemotaxis**

   Escherichia coli (*E. coli*) tumbles and swims until it reaches the reproduction stage using (15.6). The new position of nth *E. coli* is calculated using $\theta^n(k+1, m, n)$, where c is the chemotaxis step, m is the reproduction loop, and n is the elimination dispersal loop:

   $$\theta^n(k+1,\ m,n) = \ \theta^n(k,m,n) + a(i)\mathrm{r}(j) \tag{15.6}$$

   $a(i)$ indicates step size in the random direction and $r(j)$ indicates direction given by tumble.
2. **Reproduction**

   Calculate the cost function represented by (15.7):

   $$P^n_{\text{fitness}} = \sum_{n=1}^{C} p^n \tag{15.7}$$

   where c represents the number of swim steps. Less fitness bacteria die and the good fitness bacteria indulge itself in reproduction where they are divided into two.

3. **Elimination and dispersal operation**

   The dispersion operation occurs when the bacteria gets randomly positioned in the surroundings. The attraction and repelling is represented using (15.8):

$$R_{cc}(\theta, P(k,m,n)) = \sum_{i=1}^{S} R_{cc}\left(\theta, \theta^{i}(k,m,n)\right)$$
$$= \sum_{i=1}^{S}\left[-d_{\text{attr}} \exp\left(-\omega_{\text{attr}}\sum_{x=1}^{D}\left(\theta_x - \theta_x^i\right)^2\right)\right] \tag{15.8}$$
$$+\sum_{i=1}^{S}\left[h_{rep} \exp\left(-\omega_{rep}\sum_{m=1}^{D}\left(\theta_m - \theta_m^i\right)^2\right)\right]$$

   where $R_{cc}\ (\theta^{i}, \theta)$ is the cost value, S is the bacteria size, and P is the parameters optimized. d_{attr}, ω_{attr}, h_{rep}, and ω_{rep} are coefficients selected.
4. **Swarming**

Each bacterium communicates with each other by sending signals and they form groups. They maintain a safe distance between them. They have gravitation and repulsion in between them. Bacteria generate information such that each one travels toward the center and forms groups maintaining distance based on the repulsive information. The computational complexity of BFO depends on the dimension of the problem space and the application on which it is applied. The BFOA algorithm is as follows:

1. Initialization of variables
2. For each elimination dispersal loop
3. For each reproduction loop
4. For each chemotactic loop
   (i) Compute fitness of each bacterium
   (ii) Generate tumble direction
   (iii) Swim in tumble direction
   (iv) Calculate fitness
   (v) Swim in the same direction and evaluate fitness
5. Continue till chemotactic loop stops
6. Reproduction
   (i) Calculate the cost for the bacterium.
   (ii) Sort according to the cost in increasing order.
   (iii) The high cost bacterium dies and the remaining bacterium splits for reproduction.
7. Repeat till the reproduction loop's terminating condition
8. Elimination and dispersal
   (i) Eradicate and scatter each bacterium
9. Repeat till the elimination dispersal's terminating condition is satisfied.

15.1.4.1 Bacterial foraging optimization-based medical diagnosis

Medical diagnosis system based on bacterial foraging is proposed for breast cancer diagnosis [22]. The standard BFO is improved by modifying the elimination dispersal and swarming phases. Classification is done using SVM with the help of features from the Australian dataset for breast cancer which contains 14 features and 690 instances. The classification accuracy is 87.3 % with a number of features (average) as 8.2. Using feed forward, the neural network classification rate is 86.2%. The proposed medical diagnosis system is evaluated in terms of computational time and complexity.

An intelligent medical diagnosis system is proposed using chaotic-based BFO for Parkinson's disease [23]. This system uses a fuzzy-based KNN approach which is enhanced using the proposed chaotic BFO optimization. These novel algorithms are tested using UCI medical benchmark datasets. The parameters of the fuzzy-based KNN classifier are tuned using the chaotic BFO. The proposed diagnosis system achieved an accuracy of 96.97%, sensitivity of 96.87%, and specificity of 98.75% using Oxford dataset and an accuracy of 83.68%, sensitivity of 96.92%, and specificity of 33.33% using Istanbul dataset.

15.2 Discussions

The medical diagnosis systems that are summarized in this work can be used to overcome all the demerits of the manual diagnosis systems. The performance of the classifiers used in the aforementioned approaches can be increased by successfully generating optimal parameters for the successful diagnosis by increasing the prediction rate. The main advantages of such systems are (1) reduced time consumption, (2) increased prediction rate, (3) reduced labor, (4) chances of reducing human error, (5) avoiding unwanted biopsy procedures, (6) reduced treatment cost by elimination unwanted radiation therapies, and (7) acts as the second opinion for radiologists in complex decision-making cases.

These systems are usually measured using various performance metrics such as the sensitivity, specificity, accuracy, miss prediction rate, positive predictive value or precision, negative predictive value or recall, false-negative rate or miss rate, false-positive rate or fall out, false discovery rate, false omission rate, F1 score, and confusion matrix.

15.3 Conclusion

The widely used swarm-based metaheuristic approaches such as particle swarm optimization, ant colony optimization, artificial bee colony optimization, and the bacterial foraging optimization are summarized in detail. The medical diagnosis systems that are built based on the aforementioned algorithms that are demonstrated in various existing research are summarized and it has been proved to be effective in diagnosing various diseases efficiently and accurately. Hence, they can be used as the replacement of the radiologist's opinions when there is a high demand for diagnosis.

References

[1] Thompson S., and Joseph K. S. "Particle swarm optimization-based energy efficient channel assignment technique for clustered cognitive radio sensor networks." *The Computer Journal.* 2018; 61(6): 926–936.

[2] Thompson S., and Joseph K. S. "PSO assisted OLSR routing for cognitive radio vehicular sensor networks." *ACM International Conference on Informatics and Analytics.* August 2016, pp. 1–8. DOI: 10.1145/2980258.2980457

[3] Punitha S., Amuthan A., and Joseph K. S. "Enhanced monarchy butterfly optimization technique for effective breast cancer diagnosis." *Journal of Medical Systems.* 2019; 43(7): 1–14. DOI: 10.1007/s10916-019-1348-8

[4] Punitha S., Amuthan A., and Joseph K. S. "Benign and malignant breast cancer segmentation using optimized region growing technique." *Future Computing and Informatics Journal.* 2018; 3(2): 348–358. DOI: 10.1016/j.fcij.2018.10.005

[5] Punitha S., Ravi S., Anousouya Devi M., and Vaishnavi J. "Particle swarm optimized computer aided diagnosis system for classification of breast masses." *Journal of Intelligent & Fuzzy Systems.* 2017; 32(4): 2819–2828. DOI: 10.3233/JIFS-169224

[6] Al-Turjman F., Ulusar U., and Nawaz M. "Intelligence in the Internet of Medical Things era: A systematic review of current and future trends." *Elsevier Computer Communications Journal.* 2020; 150(15): 644–660. DOI: 10.1016/j.comcom.2019.12.030

[7] Al-Turjman F., Zahmatkesh H., and Mostarda L. "Quantifying uncertainty in Internet of Medical Things and big-data services using intelligence and deep learning." IEEE Access. 2019; 7(1): 115749–115759.

[8] Jin J., Sun W., Al-Turjman F., Khan M., and Yang X. "Activity pattern mining for healthcare." IEEE Access. 2020; 8: 56730–56738. DOI: 10.1109/ACCESS.2020.2981670

[9] Deebak D., Al-Turjman F., Aloqaily M., and Alfandi O. "An authentic-based privacy preservation protocol for smart e-healthcare systems in IoT." IEEE Access. 2019. doi: 10.1109/ACCESS.2019.2941575

[10] Campioni F., Choudhury S., and Al-Turjman F. "Scheduling RFID networks in the IoT and smart health era." *Journal of Ambient Intelligence and Humanized Computing.* 2019. doi: 10.1007/s12652-019-01221-5

[11] Al-Turjman F., and Alturjman S. "Context-sensitive access in Industrial Internet of Things (IIoT) healthcare applications." *IEEE Transactions on Industrial Informatics.* 2018; 14(6): 2736–2744. DOI: 10.1109/TII.2018.2808190

[12] Kennedy J., and Eberhart R. "Particle swarm optimization." *Proceedings of IEEE International Conference on Neural Networks*, IV 1942–1948. 1995. DOI: 10.1109/ICNN.1995.488968

[13] Huang M.-L., Hung Y.-H., and Chen W.-Y. "Neural network classifier with entropy based feature selection on breast cancer diagnosis." *Journal of Medical Systems*. 2009; 34(5): 865–873.

[14] Gunasundari S., Janakiraman S., and Meenambal S. "Velocity bounded Boolean particle swarm optimization for improved feature selection in liver and kidney disease diagnosis." *Expert Systems with Applications*. 2016; 56: 28–47.

[15] Dorigo M., Maniezzo V., and Colorni A. "Positive feedback as a search strategy." Technical Report, Report No. 91-016. Politecnico di Milano, Italy, 1991.

[16] Jona J., and Nagaveni N. "Ant-cuckoo colony optimization for feature selection in digital mammogram." *Pakistan Journal of Biological Sciences*. 2014; 17(2): 266–271.

[17] Shunmugapriya P., and Kanmani S. "A hybrid algorithm using ant and bee colony optimization for feature selection and classification (AC-ABC Hybrid)." *Swarm and Evolutionary Computation*. 2017; 36: 27–36.

[18] Karaboga D. "An idea based on honey bee swarm for numerical optimization." Technical Report TR06. Computer Engineering Department, Engineering Faculty, Erciyes University, 2005.

[19] Dhamodharan U., SunilGavaska P., Al-Turjman F., Sathiyaraj R., and Balusamy B. "Artificial bee colony method for identifying eavesdropper in terrestrial cellular networks." *Transactions on Emerging Telecommunications Technologies*. 2020. DOI: 10.1002/ett.3941.

[20] Zorarpacı E., and Özel S. A. "A hybrid approach of differential evolution and artificial bee colony for feature selection." *Expert Systems with Applications*. 2016; 62: 91–103.

[21] Passino K. M. "Biomimicry of bacterial foraging for distributed optimization and control." IEEE Control Systems. 2001; 22(3): 52–67.

[22] Chen Y.-P., Li Y., Wang G., *et al.* "A novel bacterial foraging optimization algorithm for feature selection." *Expert Systems with Applications*. 2017; 83: 1–17.

[23] Nagasubramanian G., Sankayya M., Al-Turjman F., and Tsaramirsis G. "Parkinson data analysis and prediction system using multi-variant stacked auto encoder." *IEEE Access*. 2020; 8(1): 127004–127013.

Chapter 16

An extraocular muscle stimulation system based on EOG and FES

Maram Arto[1], Aamnah Fannoush Alabboud[1], Fadi Al-Turjman[2], Ilker Ozsahin[1,3] and Dilber Uzun Ozsahin[1,3]

A device and system for treating misaligned eye positions sourced for any type of eye movement disease lead to checking for performance errors, by which the extraocular muscles or nerves are electrically stimulated when the eye movement of the controller's eye is detected non-invasively. The stimulation signal is sent to the electrode via a wireless system while the electrode is implanted in the stimulation selected area.

16.1 Introduction

The development of neuroscience has broadened our understanding about the mystery of the brain, and our knowledge of how the brain works through electrical signals has enabled us to communicate with this mysterious organ – it is the language through which we analyse questions related to brain functions. We are even able to pick up signals from the brain or send signals to it, and through this technology, we can develop medical devices that serve to solve health issues. The aim of our study is to use the concept of signal capture and transmission to solve eye conditions when not performing well together because the success of the role of the eyes depends on the optic nerve and the movement of the eyes, and any malfunction in these leads to vision conditions. In some cases, the movement of one eye is disrupted, causing each eye to move in a different direction. For this reason, in a study that depends on making a healthy eye is the controller and the affected eye is controlled, and recording the nerve signals of healthy eye movements and sending them to the affected eye to carry out the directional movement taken by the healthy eye. In this study, there are three main parts: (1) Detecting, which will be connected

[1]Department of Biomedical Engineering, Near East University, Nicosia, North Cyprus
[2]Department of Artificial Intelligence Engineering, Research Center for AI and IoT, Near East University, Nicosia, Turkey
[3]DESAM Institute, Near East University, Nicosia, North Cyprus

with the healthy eye. The aim of this part is to detect the electrical activity of the healthy eye; this process helps to determine the correct eye movement. This part allows the study to choose a perfect start of the stimulation process. The success of the subsequent steps depends on the success of this step. (2) Working as a signal processor for the signal we detect in the first part. This step is very important because we need to reprocess the signal before we use it in the third step. The signal needs amplification and duplication while performing this step. (3) Working to send orders to the affected muscle or muscles; this part will innervate the extraocular muscle instead of its nerve. However, in this step, the muscle will receive the order from the movement of the healthy eye, not from the brain at all.

16.2 Literature review

Fundamentally, to relate the literature reviews together by defining the correlation between experimental procedures leads to establishing basic knowledge to identify execution methodology of an evoked eye movement by stimulation. Stimulation Processes are applied to the selected area to effectively elicit eye movement. Studies by Upadhyaya, Meng and Das (2016) and Stryker and Schiller (1975) investigated the effect of stimulation in the superior colliculus on eye movement. In the lateral rectus muscle, the feasibility of electrically stimulation was investigated by Velez *et al.* (2009). In another experiment, studies of abnormalities in paramedian pontine reticular formation are associated with saccade disconjugacy in strabismus (Walton *et al.*, 2013). The Frontal Eye Fields has shown by Blum *et al.* (1982) which induced by Electrical Stimulation. The final area is the selective saccular nerve, which was investigated by Goto *et al.* (2004).

16.2.1 Subjects and surgical procedures

The total number of in vivo samples used in the studies is 38, including 27 monkeys, 14 cats and 1 rabbit. Figure 16.1 shows the number of samples for each reference. All experiments were conducted in conformity with Animal care Institute in various research centres. The subjects underwent surgical preparation for an implantable device or denervate the eye muscle to conducting experiments, implantation containing leads, electrodes, stimulators, or other parts required for the experiment, denervation methods vary from each experiment to provide a sample case worthy of testing the stimulation to apply it on the experimental condition [1–9]. Denervation of the extraocular muscle by rearing the subject using an optical prism-viewing paradigm was conducted, and during the rearing, the subject wore the prism goggles for the first days or month after birth rustling strabismus or misalignment eye movement [1,3]. Another way to denervate is chemically by injection botulinum toxin A (Botox) into the extraocular muscle, where the neuromuscular junction starts degradation after several weeks [2,5]. Denervation also occurs when surgical immobilisation leads to paralysed muscles [8].

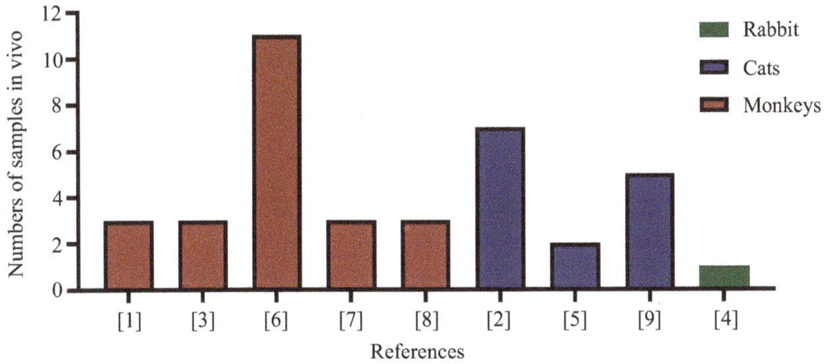

Figure 16.1 Number of in vivo samples in each reference

16.2.2 Eye movement measurements

Measurement of eye movement identifies the stimulating value or quantifies the degree of eye deflection. Eye movement recorded using the magnetic search coil method that was surgically implanted was calibrated before the experiment procedure, performed in each eye independently under monocular viewing by means of requiring the subject to fixate series of targets [1,3]. Eye deflection generated by the stimulation was measured via a force–displacement transduce powered by power supply [2,5]. Another method for recording eye position is electrooculography (EOG) [6,8,9].

16.2.3 Stimulation procedures and experimental tools

The stimulation concept is applying stimulation current to elicit neuromuscular junction in the target area to supplement or improve nerves or muscle function. Table 16.1 exhibits the investigation of evoked eye movement in a different positions and stimulation types except [10]* studied the stimulation in the lacrimal gland.

16.2.4 Stimulation parameters

16.2.4.1 Frequency

As can be seen in Figure 16.2, the frequency range is applied in various areas, from 300 to 400 Hz range of the stimulation elicited in the superior colliculus [1,8]. Wide-range frequency used in the paramedian pontine reticular formation was between 100 and 400 Hz [3]. Among the range of foregoing, lateral rectus muscle was greater than 175 Hz frequency [2,5], and in levator palpebrae superioris area 250 Hz is adjusted [4] while unilateral saccular (SAC) nerve used 300 Hz [9].

16.2.4.2 Time

The time indicates to the pulse duration of stimulation applied to a specific area; as shown in Figure 16.3, the time unit is the millisecond, in the same area with

Table 16.1 Types of stimulation related to the elicited position in each reference

Reference	Position of stimulation	Type of treatment
[1]	Superior colliculus	Electrical microstimulation methods
[2]	Lateral rectus muscle	Functional electrical stimulation
[3]	Paramedian pontine reticular formation	Microstimulation
[4]	Levator palpebrae superioris	Direct electrical stimulation
[5]	Lateral rectus	Functional electrical stimulation
[6]	Frontal eye fields (FEF)	Electrical stimulation
[8]	Superior colliculus	Electrical stimulation
[9]	Unilateral saccular (SAC) nerve	Electrical stimulation
[11]	The ocular recti	Electrical stimulation
[12]	Lateral rectus muscle	
[13]	Cranial nerve	Electrical stimulation
[14]	Ciliary muscles	Electrical stimulation
[15]	One or more extraocular muscles and/or to one or more cranial nerves that innervate the extraocular muscles	Electrical stimulation and/or drug stimulation
[10]*	The lacrimal gland	Passive stimulation

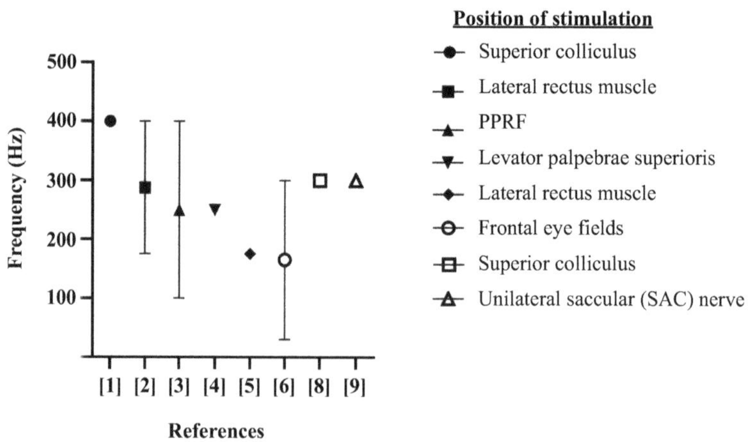

Figure 16.2 The frequency range of stimulation applied in various areas

antithesis intervals, long-duration 500 ms in the superior colliculus area in contrast 0.5 ms short duration [1,8], in the lateral rectus muscle, long-duration 10 s and 0.5 to 2.0 ms is a short duration [2,5]. The other areas elicited 100 ms in the paramedian pontine reticular formation [3], 0.2–0.5 ms in the frontal eye fields (FEF)

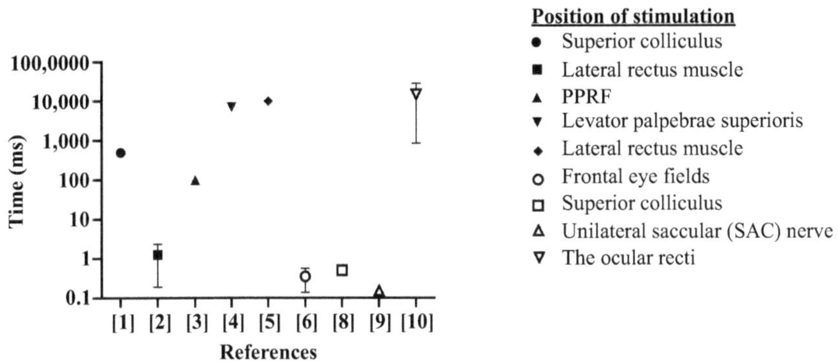

Figure 16.3 The pulse duration of stimulation applied in various areas

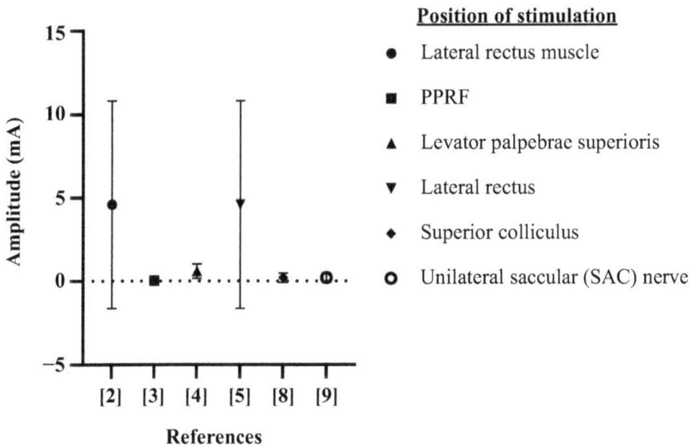

Figure 16.4 The amplitude value of stimulation applied in various areas

[6] and 150 μs unilateral saccular (SAC) nerve [9], while range 5–25 s in the ocular recti [11].

16.2.4.3 Amplitude

The pulse amplitude is shown in Figure 16.4, where the value in the lateral rectus muscle is 0.2–9 mA [2,5]. While in the other areas, it was tiny values 30–50 μA in the paramedian pontine reticular formation [3], 0.3–0.9 mA in levator palpebrae superioris [4], 1–400 μA in superior colliculus area [8] and 10–400 μA in unilateral saccular (SAC) nerve [9].

16.2.5 Experimental procedures

To conduct the experiment of evoking eye movement, devices used to generate the pulse stimulation parameters were adjusted to suitable values to evoke eye position or used the deliver pulses to selective area via electrode implanted or noninvasive delivery. In some experiments, they provide signal processing; Table 16.2 lists the devices used in the literature review.

16.2.6 Comparison between related patent and our study

Table 16.3 shows the composition of each device using electrical stimulation to solve eye movement condition relevant in the literature. Patent numbers [11,13,15] comprise stimulator and stimulation electrode to treat ocular misalignment such as strabismus or improving cranial nerve function, where [13] non-invasively applying stimulation [11] and [13] implanted electrodes. Patent numbers [12] and [14] include additional components from the previous eye movement detector and analyser unit. Meanwhile, the targeted application in [12] has been designed for lateral rectus muscle in paralysed or non-innervated condition. Moreover, the authors in [14] aim at improving the ocular focusing at near points in case of the presbyopia condition. Our proposal comprises the previous components in addition to the feedback movement analysis unit to provide an accurate stimulation process by study the movement and location of the eyes after transmitting the stimulation pulse to the electrode implanted stimulation area.

Table 16.2 Devices and tools used in experimental procedures

Reference	Pulse generator	Electrodes	Signal processing
[1]	BITS# visual stimulus generator and Psychtoolbox 3	Epoxy-coated tungsten electrodes	Antialiasing filters
[2]	LabVIEW 7.0	Sleeve-type bipolar electrode	
[3]		Glass-coated tungsten microelectrodes	Antialiasing, 6-pole Bessel filter (200 Hz)
[4]	Implantable pulse generator (IPG)	LPS electrode	
[5]	LabVIEW 7.0	Sleeve electrode	
[6]		Platinum–iridium ball electrodes	
[7]		Parylene-insulated tungsten microelectrode	
[8]	Grass PSIU-6 photoelectric constant current stimulus isolation unit	Platinum–iridium microelectrodes	
[9]	PC nerve stimulation	Silver electrodes	

Table 16.3 Comparison between related patent and our study

Reference	Eye movement detector	Analyser unit	Stimulator	Deliver stimulation electrode	Feedback movement analyser unit	Target application
[11]			√	√		Ocular misalignments
[12]	√	√	√	√		Paralysed or non-inervated lateral rectus muscle
[13]			√	√		Improving cranial nerve function to improve muscle function and thereby overcome visual/percep tual dysfunction of a user
[14]	√	√	√	√		IMPROVING OCULAR FOCUSING AT NEAR WSON POINTS **presbyopia**
[15]			√	√		Treating a patient with Strabismus
Our system	√	√	√	√	√	Treating misalignments eye movement

16.3 Methodology

16.3.1 Background of the study

The present study is directed to a device and system for effectively treating misaligned eye movement by evoking eye movement via stimulation treatment to fix eye position and prevent complex eye conditions such as diplopia and amblyopia. Where the eye movement is controlled by three antagonistic pairs of muscles and innervated via three of the cranial nerves.

The misalignment eye movement is the deflection in one or two of the eyes from the looking direction causing double vision or lazy eye or even configurable deformation. A person having misalignment eye movement is suffering from vision conditions such as diplopia, amblyopia and another side effect such as headache and dizziness.

The misalignment eye movement believed to occur as a result of weakens or strengthens in eye muscles or damaged in cranial nerves that supply the extraocular muscles. Eye misalignment positions may present as a symptom of other diseases.

The treatments of misaligned eye movement may in some cases seem ineffective, and eyeglasses and surgery are useless when the misalignment occurs because of nervous disease. To overcome this drawback, the treatment by electrical stimulation evokes the movement to fix eye positions.

16.3.2 Summary of the study

The present study explores a device and system for effectively treating misaligned eye movement that satisfies the need for aligning the eyes position together resulting from any deactivate movement reasons. A device and system are provided which align with the eyes using functional electrical stimulation on the selected area; when they fail to move together in the same direction while the eyes perform together, the system avoids applied stimulation.

In a preferred embodiment of the study, the eye movement is recorded non-invasively by electrooculogram, electrodes are placed around the eyes and one electrode for delivering the stimulation pulse is implanted in extraocular muscle or the selective area. The analyser unit outside the body controls the system and sends the pulse stimulator via wireless.

When the patient's eyes move, the EOG recording the direction then sends it to the analyser unit to distinguish between the two eye positions, and if the two eyes are in the same location, no stimulation process occurs. But if the eyes are in a different location, the stimulation process occurs.

It is an object of the present study to provide an improved system and apparatus for treating misalignment eye movement by functional electrical stimulation and electrooculogram for recording eye movement.

It is a further object of the present study to provide a new apparatus for analysis and stimulate the movement of the eyes in a loop system to adjust the parameter of stimulation or transmit the stimulation pulse.

It is a further object of the present study to minimise the number of surgeries needed for stimulation procedures by recording the position of the eyes non-invasively and the stimulator transmit the impulse wireless to implanted electrode in stimulation selected area.

16.3.3 Detailed description of the device and system

Referring to the framework presented in Figure 16.5, the system is comprising of three main devices, which are electrooculogram, a new designing device for electrical stimulation and electrode implanted in stimulation selected area. The number (1)

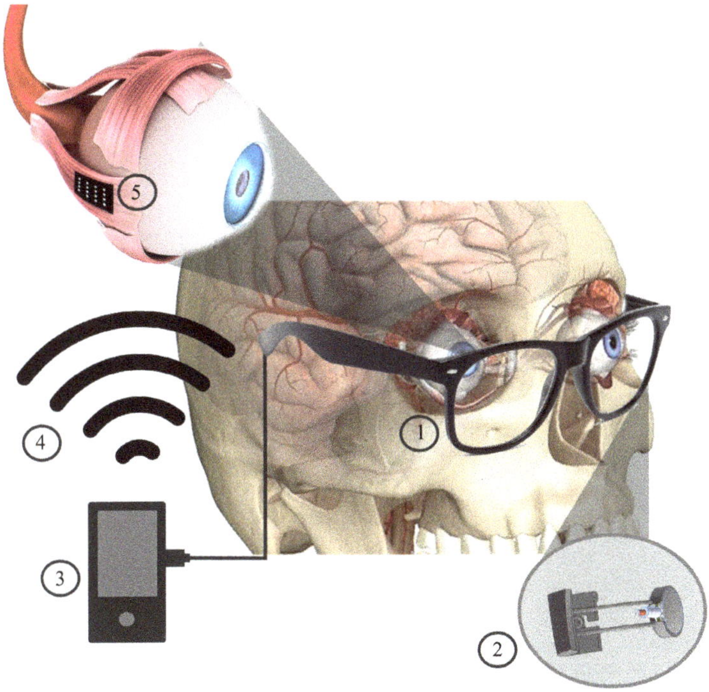

Figure 16.5 The diagram of the system

in Figure 16.5 is eyeglass or a holder frame for EOG electrodes where electrodes are presented in number (2); the purpose of EOG is to provide recording data for eye movement non-invasively where each eye is recorded individually to identify the degree of deflection. Electrooculogram works an initial step of the system and in the feedback analyser unit which studies conducting stimulation process by recording the eye positions after applying stimulation pulse. The number (3) in Figure 16.5 is a new designing device for electrical stimulation accompanying with analyser unit and feedback system. The device receives data recorded from EOG to analyse eye movements and their locations then move the results to a stimulator unit, transmitting stimulation pulse via wireless where electrical stimulation signal referred by number (4) in Figure 16.5. When one eye stimulated the recording of the eye movement, it triggers the movement of both eyes together and the feedback system stores the associated parameters of the stimulated in the memory. The electrode shown as number (5) in Figure 16.5 receives the stimulation pulse from stimulator unit to evoke eye movement. The electrode is implanted in selected extraocular muscles area component from biocompatible materials and contains a receptor for the stimulation signal.

Referring to the flowchart of the system process shown in Figure 16.6, the process starts when the patient wearing eyeglasses holds the electrodes of the eye movement detector. The patient may suffer from a defect in eye movement caused

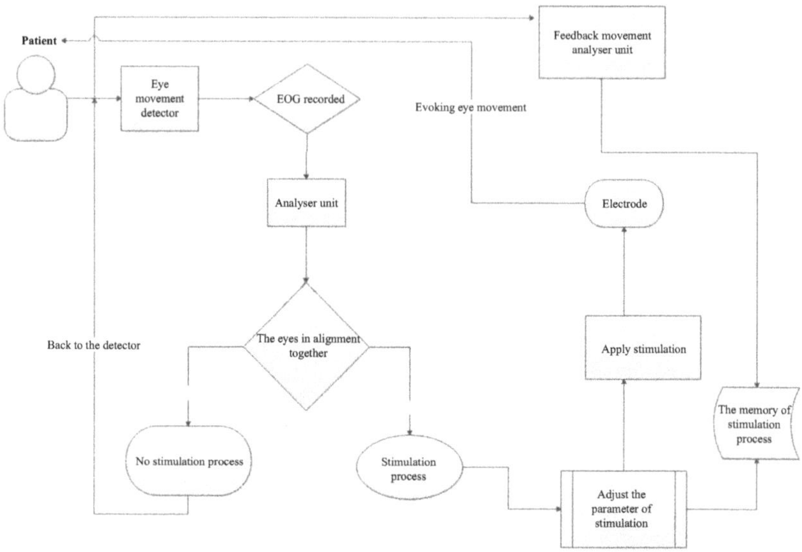

Figure 16.6 The flowchart of system process

by muscle or nerve conditions. The eye movement detector in this paradigm is electrooculogram which records the direction of each eye separately to discriminate between the eye affected and the controller's eye, and then the data of EOG recorded goes to process in the analyser unit and the system analysis with the position of the eyes to identify if the eyes align together or not; in case of the eye alignments, the system ends to the back of the eye movement detector without applying stimulation pulse. In contrast, the eyes have misalignments movements conducting the stimulation process occurs. To decide the parameters of the stimulation pulse, the system adjusted according to results of EOG recording analysed in a subprocess. After adjustment of parameters stimulation signal transmitted to the electrode and evoking eye movement, the system provides feedback movement analysis unit which studies of locations of the eyes after applying stimulation then sends the parameters that perform useful in the alignment of eyes to the memory and loops of the system start again.

16.4 Conclusions and future work

This study had an initial step to designing a device and system for extraocular muscle stimulation based on EOG and FES. The present study proposed treatment using electrical stimulation applied in selected extraocular muscles, which studied this proposal in relevant work in the literature to derive the execution methodology of an evoked eye movement by stimulation with various positions and parameters. From the foregoing knowledge presented, the present study achieved design for treating misalignment eye movement, in which details of the device and system are explained with the background of the study. Future work of this study is to examine the working performance of the proposed treatment using a new designing device and system based on EOG and FES. The examination may contain simulation using the software in the computer or in experimental in vivo and hardware to test the procedure of stimulation.

References

[1] Upadhyaya S, Meng H, and Das VE. Electrical stimulation of superior colliculus affects strabismus angle in monkey models for strabismus. J Neurophysiol. 2017; 117(3): 1281–1292. DOI:10.1152/jn.00437.2016.

[2] Velez FG, Isobe J, Zealear D, *et al.* Toward an implantable functional electrical stimulation device to correct strabismus. J AAPOS. 2009; 13(3): 229–35.e1. DOI:10.1016/j.jaapos.2008.08.013.

[3] Walton MM, Ono S, and Mustari MJ. Stimulation of pontine reticular formation in monkeys with strabismus. Invest Ophthalmol Vis Sci. 2013; 54 (10): 7125–7136. DOI:10.1167/iovs.13-12924.

[4] Scott AB, Miller JM, and Danh KK. Electrical stimulation of nerves to the levator and oculorotary muscles. 2000 Van Ness Ave, San Francisco CA, 94109.

[5] Isobe J, Velez F, Lee H, Patnode S, Judy J, and Rosenbaum A. In vivo functional electrical stimulation of feline lateral rectus: relationship between stimulation parameters and eye rotation in denervated muscle. [Online] Available from researchgate.net/publication/237757104.

[6] Blum B, Kulikowski JJ, Carden D, and Harwood D. Eye movements induced by electrical stimulation of the frontal eye fields of marmosets and squirrel monkeys. Brain Behav. Evol. 1982; 21: 34–41.

[7] Gamlin PD and Miller JM. Extraocular muscle motor units characterized by spike-triggered averaging in alert monkey. Journal of Neuroscience Methods 2012; 204: 159–167.

[8] Stryker MP and Schiller PH. Eye and head movements evoked by electrical stimulation of monkey superior colliculus. Exp. Brain Res. 1975; 23: 103–112.

[9] Goto F, Meng H, Bai R, *et al.* Eye movements evoked by selective saccular nerve stimulation in cats. Auris Nasus Larynx. 2004; 31(3): 220–225. DOI: 10.1016/j.anl.2004.03.002.

[10] Ackermann DM, Loudin JD, Kuzma J, Palanker D, and Wetenkamp SF. United States Patent (2017) Patent No. US 9821159 B2.

[11] Harry G. Friedman and Plymouth, Minn. United States Patent (1981). Patent No. 4271841.
[12] Stephen N. Lipsky. United States Patent (1996) Patent No. 5496355.
[13] Mary R. Fisher. United States Patent (1994) Patent No. 5360438.
[14] Ben Israel. United States Patent (1998) Patent No. 5782894.
[15] Allison M. Foster. United States Patent Application Publication (2009) (0) Pub. No.: US 2009/0005756A1.

Chapter 17
Smart system for the blind

Yousaif Esaam Ismaeel[1], *Mohammed Bin Merdhah*[1], *Abdullah Omar Alani*[1], *Fadi Al-Turjman*[2,3], *Ilker Ozsahin*[1,4] *and Dilber Uzun Ozsahin*[1,4]

Blindness is a major problem that a person may experience in life, and blindness may be due to eye disease, or blindness occurs genetically. A blind person's life can be difficult due to the inability to move only with the help of people or the use of simple sticks. After studying and researching this subject, we have presented a new idea that might help change the life of the blind person even in a simple way. The smart system consists of two parts. The first part is use of the smart glove, where the blind person wears it with his hand; the second part includes use of smart shoes, where the blind person wears it with his feet. By using this system, the blind person can gain better mobility and independence. The glove contains an ultrasonic sensor for distance measurement and light-emitting diode (LED) with an Light dependent resistor (LDR) sensor. The smart shoes will contain an ultrasonic sensor, water sensor and LDR sensor with LED and Global Positioning System (GPS). All these parts work by the Arduino to facilitate the blind person's movement and mobility easily and without the help of others.

17.1 Introduction

17.1.1 Internet of Things

The Internet of Things (IOT) is a platform that provides an enormous access to the network communication the world has emerged into an evolution that affected and enhanced the way human beings live [1–3]. Throughout decades and centuries, these information technologies have continuously interchanged positively to ultimately benefit the world with endless solutions and improvement of technologies, which has also led to numerous employment opportunities. Furthermore, IOT has affected the human life in many ways, from the way a driver navigates a car to a

[1]Department of Biomedical Engineering, Near East University, Nicosia / TRNC, Mersin-10, Turkey
[2]Department of Artificial Intelligence, Near East University, Nicosia / TRNC, Mersin-10, Turkey
[3]Research Center for AI and IoT, Near East University, Nicosia / TRNC, Mersin-10, Turkey
[4]DESAM Institute, Near East University, Nicosia / TRNC, Mersin-10, Turkey

person who easily can purchase and shop online within seconds, and the smartwatches have a lot of services that life and health activities can be monitored and displayed; on the other hand, modern houses have remarkably collecting smart appliances such as smart room temperature adjusting along with air conditioners, smart washers and smart TVs. In addition, the IOT is embedded in any item that consists of the important facts which are electronic, software and sensors. Where IOT acts as data storage that data can be stored, managed, remoulded and shared in the benefits of the consumer and increasing data efficiency. The data can be collected from the sensors, and the work can be done afterward. A perfect example that can simplify the work of IOT is with the air conditioner, which contains an embedded sensor with health and temperature information that can be sent to the IOT platform, where feedback services alert the consumer in case of replacement or emergency. IOT can assist the way of lifestyle and how a human being can connect with technology where the era of IOT is continuously developing the world will hugely benefit from that improvement [3–6].

17.1.2 Definition of blindness

Blindness is defined as having either permanent or partial vision loss or having difficulties in seeing. The World Health Organization has estimated that 39 million human beings are blind, and the organization published that 1.3 billion people around the world have some form of visual impairment. The main leading cause of blindness or even visual loss is either refractive error or cataracts. Diabetic eye disease can damage the eyes and lead to serious vision loss or even blindness. In the United States, people with diabetes are 25 times most likely to be blind more than people who do not have diabetes. In the United States, people aged 40 years or older are estimated to be at 119 million people, and of those, 3.4 million people are blind or have vision impairment. From age 65 and older, one in three Americans have vision loss symptoms. In contrast, with this large amount of people that nearly comes to 3.5 million, it is estimated to cost the United States government more than $4 billion. Blindness has affected people aged 50 years or older, and 90% of people with blindness or affected by visual loss symptoms are living in the low industrial base and low human development index countries. In contrast, approximately 80% of people with vision impairment are considered to be cured due to early detection and effective interference with the eye disease and excessive treatment. For instance, the refractory errors are healed by frequently wearing glasses for the patient in daily life. In addition, applying cataract surgery can heal poor vision. What is sufficient enough for enhancing the irreversible vision impairment is vision rehabilitation. A great plan intended from the World Health Organization will develop and enhance eye care services, including tools such as an assessment tool, diabetic eye disease tool and developing rehabilitation services and systems. It will help in changing the global concept of vision and achieving goals that includes improving the life of blind people or vision impairment people. Blindness is the lack of vision or complete loss of vision. As it is known, the process of vision consists of the stages of light coming into the eyes, passing through the anterior segment, falling into the retina and transferring it to the brain with the optic nerve or optic nerve. On the other hand, visual impairment, which means having a poor vision, can be assessed through several treatments such as wearing glasses, applying cataract surgery, or even contact lenses. Blindness can differ from one person to another causing several types of blindness among people. First, we can identify night blindness, which is the complex process of vision of a decay of illumination, and it can be gained either genetically or attained, although a larger part of people who suffer night blindness crisis are most likely to see or to function adequately in a normal lighting state. However, it is not a case of sightlessness. Second, colour blindness is where a person fails to differentiate or sense the various types of colours, especially green and red. This type of blindness is mostly gained genetically from parents and affects males more than females in 8% for the men and 1% for the female. Third, snow blindness is the process of losing sight due to heavy amounts of ultraviolet (UV) light emitted directly to the eyes. This type of blindness can occur due to the swelling corneal surface cells. Although it is a complex type of blindness, the blind person is able to be cured and perceive movements. The origin of blindness can also differ from one person to another depending on each individual's genes or age. Also, it can differ between developing countries to high developed index countries. Furthermore, people aged 40 years and older have different main causes, starting with glaucoma, which is one of the most dominant causes of blindness in the United States, and it is the most common cause in African Americans. In addition, nearly 3 million people can be affected by glaucoma, but it will get diagnosed after period of time where it is difficult to be cured. Glaucoma is known as a series of damage to the eye's optics caused by a high pressure from the fluid around the eye that eventually will cause vision loss. Though glaucoma at first has no symptoms, a slow process of losing sight will occur, starting from losing side vision. Once it still untreated, the patient will continuously lose the side and the corners of the eyes where it is most likely to be a tunnel vision for the patient. Moreover, the patient will lose central vision that will lead to full vision loss, which is a dead end with no possibility to be cured and cannot be restored. Diabetic retinopathy, on the other hand, is one of the most common causes of blindness, where it can affect people around the world who are diagnosed with diabetes. The retina is composed of thin tissues, and it is an important component of the eye, where it is located near the optic fibre. It functions as a receiver for the light when the lens start focussing, and it will be converted into nerve signals sent to the brain for visual recognition. Diabetic retinopathy can affect the eyes in several ways, for instance, blurry vision or vision loss. It can be cured once there is an early stage of treatment where it can prevent vision loss. On the other hand, diabetic retinopathy can cause full blindness if it is not treated.

17.2 Related work

The evolution of technology has emerged in many fields of science, and it helped create a path to a new device or instrument to provide a healthy and more sophisticated community around the world. Thus it makes the invention of instrument as wide as it takes, resulting in a wide range of devices with similarity among the projects. For instance, in [7] the project aims to help blind people via smart shoes making the blind person to walk easily without any kind of help, and it focused on dealing with batteries in any kind of instrument that needs a power supply, where it used a hardware that does not consume power as high as usual. Furthermore, the main goals for this project are its portability, simplicity of control and cost effectiveness. This system is designed be portable where it can be lifted with no hardness and made with an Android application, which is easy to monitor and supplied by an eco-battery. On the other hand, the project referenced in [8] worked on the same ideas that occurred on the previous project, where it must be power supplied by batteries along with Arduino that connected to an ultrasonic sensor and buzzer. The unique part of this project is including a wireless charging part, which it does ease the way a blind person can recharge his smart shoe, and this wireless charges has included a charging pad that acts as a transmitter and each shoe has its own charge by just placing it on the charging pad making the full battery available within half an hour to an hour. Moreover, in the project from [9], we identify that it acts as simple as two previous projects, where it contains the same Arduino along with breadboards, jumping wires, ultrasonic, battery and nylon cable ties. The only exception it made is that it was applied by a smart phone application called "1Sheild" where it gives the project its ability to communicate with the mobile device, which can replace external components. Unlike the aforementioned projects, our smart system contributed in both limbs of the body, starting with the hands with the smart gloves in the system, where it consists of an Arduino, ultrasonic sensor, LDR sensor, vibration sensor, jumper wires and battery. On the other hand of the system, smart shoes consist of the same electric components as mentioned in the smart gloves, and in addition it has a GPS tracker that can locate the blind person area by typing the longitude and latitude on Google Maps, and directly it will direct the sponsors to the area mentioned.

17.2.1 Comparisons

Project	Arduino	Ultrasonic sensor	Water sensor	GPS	LDR sensor	Vibration motor	Extra (Smart) instrument
Smart System for the blind	Mega Uno	YES	YES	YES	YES	YES	YES (smart gloves)
[a]	Nano	YES	NO	NO	YES	NO	NO
[b]	Uno	YES	YES	YES	NO	YES	NO
[c]	Uno	YES	NO	NO	NO	NO	NO

[a]Smart shoes for visually impaired/blind people.

[b]Design of Arduino-based shoe for blind with wireless charging.

[c]Smart shoes using Arduino Uno and 1Shield.

17.2.2 Results

17.2.2.1 Performance evaluation of Case Study 1

The quality and the great benefit of the ultrasound sensor have emerged in this project via multiple sensors that have been installed in both smart gloves and smart shoes, which the detection accuracy of the sensor must be aimed high and efficient enough to detect the objects nearby. We wanted the sensor to vibrate once an object is in front of the sensor and with that goal we had to measure various distance values to ensure that the consumer acknowledge the absolute distance value for the ultrasound to be vibrated and alert the blind person. We started detecting the accuracy of the sensor starting from 250 cm away from the instruments and placed an object in front, which results to us that the distance is exaggerated and we had to minimise it 50 cm less than before. The result that is the sensor accuracy has marginally changed to 10%, which led us to minimise the distance 50 cm less than the main distance that is 250. So on the 150 cm, the sensor finally started to slightly detect the object better than the last value by 25%. We also made the object a bit closer to the sensor to 100 cm and the sensor is now able to identify the object by 50%. Furthermore, the object had placed 50 cm near to the sensor, and the percentage of accuracy has improved to 75%. Lastly, we had to make it to highest point by placing the object to an accurate distance value which is 35 cm that resulted to 100% accuracy.

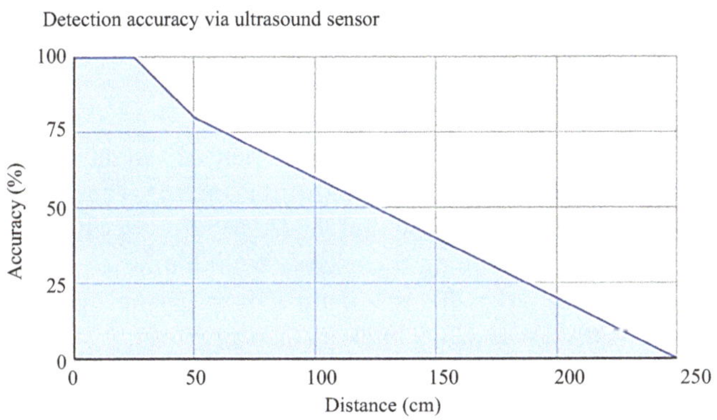

17.2.2.2 Performance evaluation of Case Study 2

In the second case, we started to examine the light intensity at various times of the day with 24 h experiment; the aim of LED is the blind person can be identified by the pedestrians and the surrounding, especially in the dark areas. We begin the test of the LDR sensor at 4 a.m. to check if it will emit light in dawn, which is proportional to the time between 4 a.m. and 5 a.m., which had to be the sunrise period that is making the LDR sensor to partially emit light by 50% of LDR sensitivity.

More of the experiment, as we mentioned earlier, that the sensor is specially designed to function only in the dark; thus it meant it has 0% of LDR sensitivity in the day. Furthermore, in the time of sunset between 6 p.m. and 7 p.m., the sensor responded with 50% as it improved to 100% in the night. With the second experiment, we examined the water sensor that can detect any fluid on the floor, which will directly alert the blind person for caution; this sensor can function as an important addition to the project to avoid any slide for the person's body. In addition, it can detect any liquid on the floor with more than 1.5 cm and the sensor will directly vibrate.

Periods of the day	**Light emitted**	**LDR sensor**
Dawn	YES	50%–10%
Day	NO	0%
Dusk	YES	50%–10%
Night	YES	100%

Water sensor vibration	**Water surface height**
OFF	<1.5 cm
ON	>1.5 cm

17.3 Smart system for the blind

17.3.1 Overview

Throughout the years, the blind person has experienced upgrades and inventions that will assist them in their daily life due to their vision loss. They had to learn how to read the dots, which is the replacement of the letters; also they had to use a simple stick to walk around and preventing themselves from hitting an object in front of them or nearby. A helpful move that will make a blind person's life much easier is the usage of instruments designed for limbs such as smart shoes which will assist in the lower extremities and the gloves which will assist in the upper extremities. Both instruments will help the blind person to simply start wearing smart shoes and comfortable gloves, and the instruments will provide a series of features such as preventing the blind person from hitting an object, an alarm for the presence of water on the floor, and global positioning system (GPS), which will help the blind person's sponsors to detect their locations. In addition, in some cases, the blind person can easily detect the instrument's place via an alarm to direct the blind person. A helpful smart instrument will help prevent injuries and help with daily life.

As mentioned earlier, the smart system will consist of two parts: smart gloves, which will include an ultrasonic sensor connected to the Arduino through which the blind person will be able to sense the obstacles around him (such as the wall or the door, etc.); and smart shoes, as easy to wear as normal shoes, which also contain an ultrasonic sensor to enable the blind person to know the obstacles in front of him and around him while walking. The smart shoes also contain a water sensor that will alert the blind person if there is any water around him—for example, a water pond. The last content of smart shoes will be the GPS device, which is the most important part of this project, by using this technology, it will be easy to locate the blind person, where it will be easy for anyone (his family or friends) to reach the blind person if any emergency happened by using Google Maps in the phone or computer by entering the code of the GPS device installed on the smart shoes.

17.3.2 Methodology of the project

17.3.2.1 Arduino Uno

Arduino is an open source platform that is used to build electronic projects. Arduino consists of a programmable circuit board (called a microcontroller), as well as a programmable part of an integrated development environment (IDE) that runs on the computer and is used to write and load code from the computer to the Arduino board, Arduino Uno has 14 digital inputs and outputs and 6 analogue inputs, and we can use 6 of them as pulse width modulation (PWM) outputs. It can be said that any Arduino painting should include the central microcontroller for the motherboard, digital pins in and out, analogue pins, power connector and socket, which is connected with the computer and the motherboard is programmed as shown in Figure 17.1. We will use two Arduino in our project. The first Arduino will control the first circuit (ultrasonic sensor) and will install on the glove, and the second Arduino will control the second circuit (GPS, ultrasonic sensor and water sensor) that will be installed on the shoes.

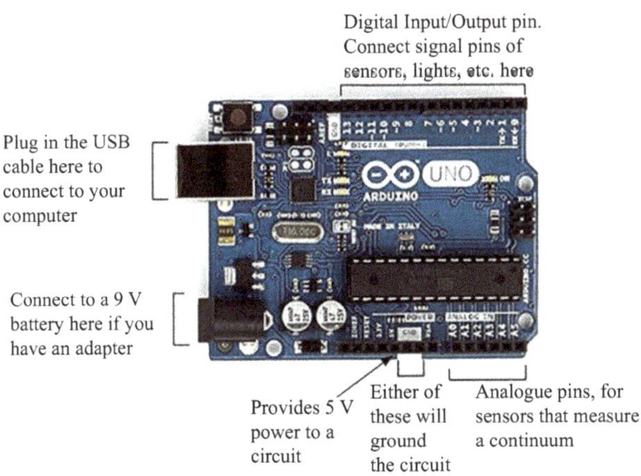

Figure 17.1 Arduino Uno

17.3.2.2 Arduino Nano

The Arduino Nano is small, complete, flexible and compatible, and it has more or less the same functionality of the Arduino Uno but smaller. It can work at 5 V, but also it can work with 7–12 V. Arduino Nano contains 14 digital pins, 8 analogue pins, 6 power pins and 2 reset pins shown in Figure 17.2. And for each digital and analogue pins, there are multiple functions, but the main function of the pins is to work as inputs or outputs. The pins work as input when they are connected with sensors; otherwise they are used as output. Despite its small size, the Arduino Nano can perform all functions that can be performed by the Arduino Uno and is characterised by the addition of two additional analogue inputs and the possibility to connect directly on the breadboard, and its price is cheaper compared to the Arduino Ono, making it ideal for fans of electronic projects and robotics manufacturers.

17.3.2.3 Global positioning system

GPS is a satellite navigation system that provides location and time information in all weather conditions anywhere on the ground. The system operates with the use of a GPS receiver, where the device attracts the signals of four satellites at a minimum and can calculate the distance separating the device from the satellites, which leads to the geographical location of the device. The operation accurately ranges from 3 to 50 m. In our project, we will use GY-GPS6MV2 type of GPS as shown in Figure 17.3, where we will connect it and program it with Arduino where it will facilitate tracking the blind person and find its location in case of an emergency by entering the code of this type of GPS to Google Maps where the site will identify the location easily. This is included in the instrument, which plays an important role for the blind person to find his own instrument in case of loss. It can be tracked by the blind person's sponsors to detect location.

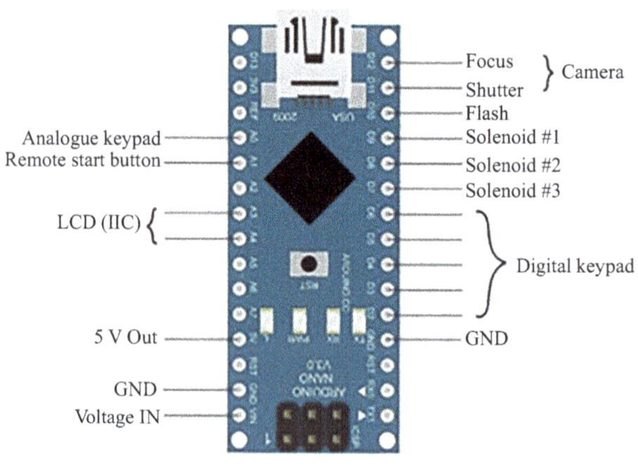

Figure 17.2 Arduino Nano

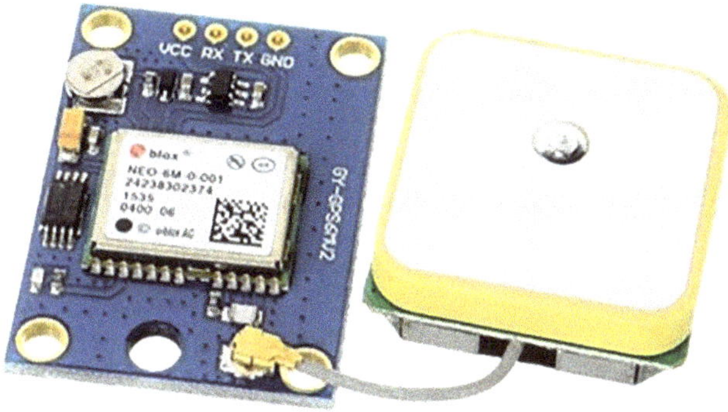

Figure 17.3 GY-GPS6MV2

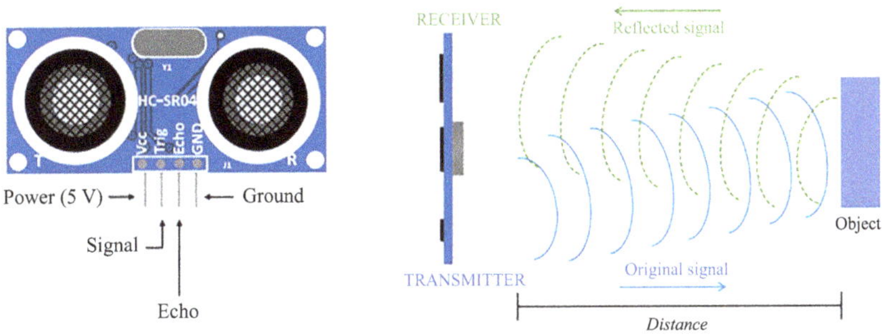

Figure 17.4 Ultrasonic sensor

17.3.2.4 Ultrasonic sensor

An ultrasonic sensor will be used, which is an essential part of the instruments where it measures the distance via ultrasound waves. The sensor contains two speakers: one to send the ultrasonic wave and the other one to receive the signal back from the object that the ultrasonic sensor will emit those waves to the object and will return back to the sensor. The time will be measured for emitting and receiving the wave. We will use an ultrasonic sensor in our project to alert the blind person about an obstacle in his way by generating a sound or vibration (Figure 17.4).

17.3.2.5 Vibration motor

It is a motor available in different sizes and types that is alerted by vibrations; also we will connect it and the water sensor with the Arduino. This motor will vibrate when the water sensor senses (Figure 17.5).

Figure 17.5 Vibration motor

Figure 17.6 LED and LDR sensors

17.3.2.6 LDR sensor and LED

The light-emitting diodes that appear as small light will be programmed with Arduino to light up in the dark only by using LDR in the circuit. The goal is to alert people around the blind person and will be installed on the smart shoes (Figure 17.6).

17.3.2.7 Buzzers

The buzzer consists of two pins, and the outside case is shown in Figure 17.7, one for power and the other one for ground. We will connect the buzzer and the ultrasonic sensor with Arduino, where the sensor will alert by the buzzer, so when carry is applied to the buzzer the ceramic disc will vibrate. The goal of using the buzzers is to enable the blind person to hear the sensor alerts in this smart system.

17.3.2.8 Water sensor

This sensor designed to detect the water by sound waves which will be reflected from the water or other liquid surface (Figure 17.8), and water sensors send an information signal to the controller to which they are connected when detection water. We will use the water sensor in the smart shoes that will detect and alert the presence of water.

Figure 17.7 Buzzer

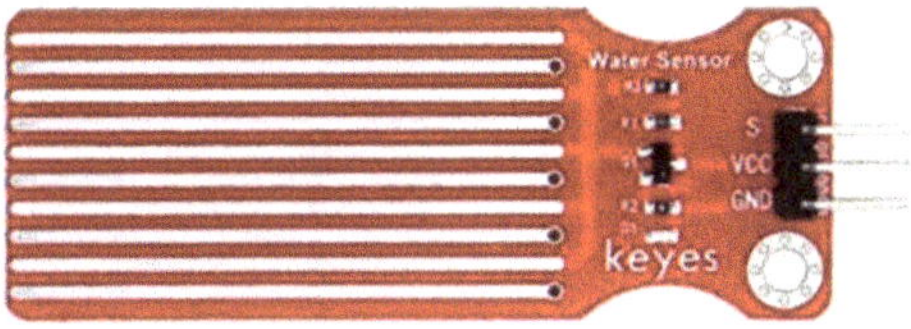

Figure 17.8 Water sensor

Figure 17.9 Jumper wires

17.3.2.9 Jumper wires

Jumpers are connecting wires used to transfer current. They are available in different sizes and colours as shown in Figure 17.9. We will use it for making our connections between Arduino and items on the breadboard.

17.3.2.10 Breadboard

The breadboard, which is an essential part in the electric circuit, is solderless and is the part where the jumper wires will be connected from it to the Arduino such as from board to power or ground due to its shape and the distribution of bus strips in two sides of the board (Figure 17.10).

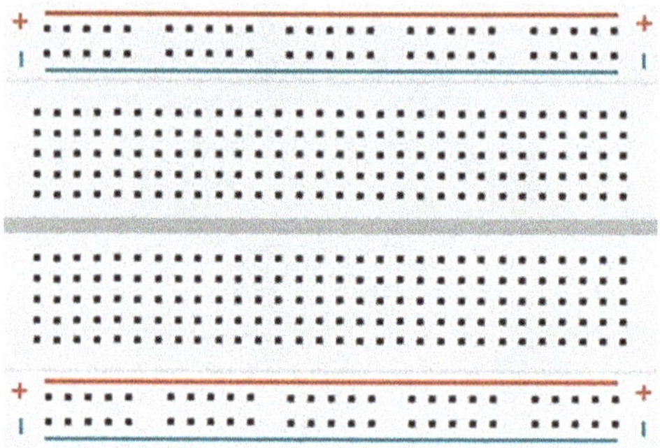

Figure 17.10 Breadboard

Figure 17.11 Battery 9 V

17.3.2.11 Battery

Battery power is available in different sizes, types and voltages. We will use it to supply the Arduino with electrical power. We will use two batteries in our system to power up the Arduino for the smart shoes and the second Arduino for the smart gloves (Figure 17.11).

17.4 The working principle of the smart system materials

17.4.1 LDR sensor and LED circuit

The circuit consists of the LDR, LED, two resisters, Arduino, five jumper wires and the breadboard. After installing the LED and the LDR on the breadboard, we connect the first wire from the +5 V in the Arduino to the positive side of the breadboard, and the second wire from the ground pins (GND) in the Arduino to the negative side of the breadboard. By this, we were able to get two full rows on the breadboard connected with +5 V and GND from Arduino, and after that we connect the third wire from the positive side of LDR to the positive side of breadboard, the fourth wire connect from negative side of LDR to the A5 in the Arduino, the fifth wire from the positive side of LED to the pin 7 in the Arduino and the negative side of LED connect with the negative side of breadboard by the resistance (Figure 17.12). After completing these connections, we connect the Arduino to the computer via USB cable and write the code via Arduino IDE program and then send the code to Arduino.

17.4.2 Ultrasonic sensor and buzzer

The electronic circuit consists of Arduino Mega and ultrasonic sensor and buzzer, and we will make two similar circuits. The first circuit will be associated with the smart gloves and the second circuit will be associated with the smart shoes. And thus, the blind person can feel the objects in front of him while walking or moving. The ultrasonic sensor will sense any object away from the person at a distance of 20–60 cm as the buzzer will issue an alarm that the person can hear. The connections of the first ultrasonic and buzzer circuit will be shown in Figure 17.13: one wire from the voltage common collector (VCC) of the ultrasonic sensor to the 5 V of the Arduino, another wire from the GND of the ultrasonic sensor to the GND of

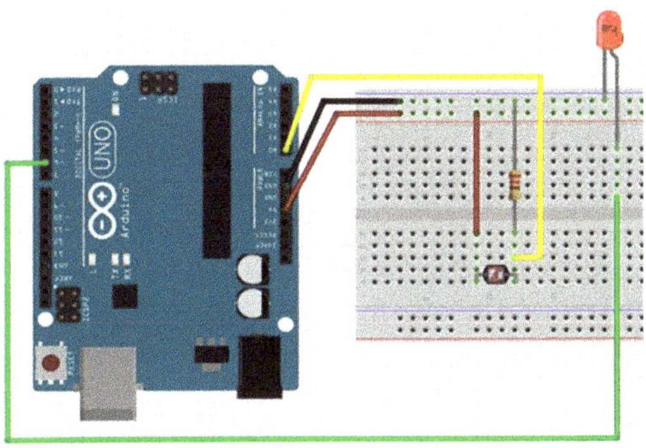

Figure 17.12 LDR sensor and LED circuit

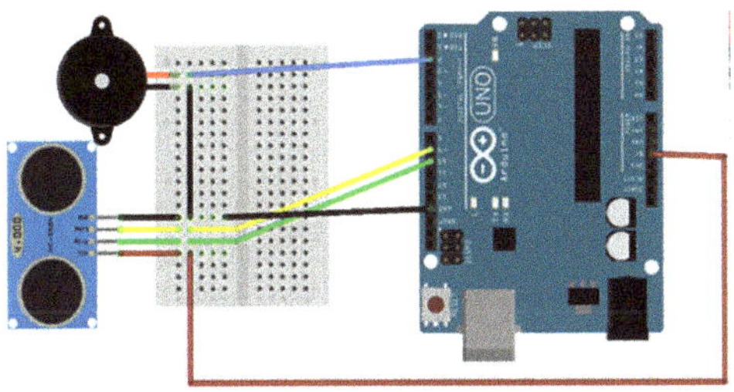

Figure 17.13 Ultrasonic sensor and buzzer circuit

the Arduino, another wire from the trig of the ultrasonic sensor to the pin number 7 of the Arduino, one wire from the echo of the ultrasonic sensor to the pin number 8 on the Arduino, and one wire from the negative side of the buzzer to the GND of the Arduino and another wire from the positive side of the buzzer to the pin number 6 on the Arduino Mega. The connections of second ultrasonic and buzzer circuit are shown in Figure 17.13: one wire from the VCC of the ultrasonic sensor to the 5 V of the Arduino, another wire from the GND of the ultrasonic sensor to the GND of the Arduino, another wire from the trig of the ultrasonic sensor to the pin number 7 of the Arduino, one wire from the echo of the ultrasonic sensor to the pin number 8 on the Arduino, and one wire from the negative side of the buzzer to the GND of the Arduino and another wire from the positive side of the buzzer to the pin number 9 on the Arduino Uno. After completing these connections, we connect the Arduino to the computer via USB cable and write the code via Arduino IDE program and then send the code to Arduino.

17.4.3 Water sensor and vibration motor circuit

The electronic circuit consists of Arduino Mega, water sensor, and vibration motor. We will use the water sensor and vibration motor in the smart shoes that will detect and alert the presence of water for the blind person. The connections of water sensor and vibration motor circuit will be shown in Figure 17.14. One wire connects from the positive side of the water sensor to the 5 V of the Arduino, another wire from the negative side of the water sensor to the GND of the Arduino, another wire from the S of the water sensor to the pin number A0 of the Arduino, one wire from the negative side of the vibration motor to the GND of the Arduino, and another wire from the positive side of the vibration motor to the pin number 5 on the Arduino mega. After completing these connections, we connect the Arduino to the computer via USB cable and write the code via Arduino IDE program and then send the code to Arduino.

17.4.4 GPS circuit

The mechanism of operation of the system is through the operation of a GPS receiver, where the device attracts the signals of four satellites at a minimum, and can calculate the distance separating the device from the satellites, which leads to the geographical location of the device. The operation accurately ranges from 3 to 50 m. The connections of GPS with Arduino will be as shown in Figure 17.15: one

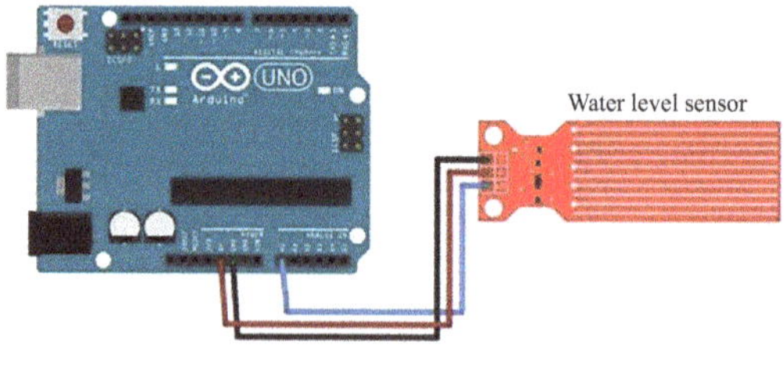

Figure 17.14 Water sensor circuit

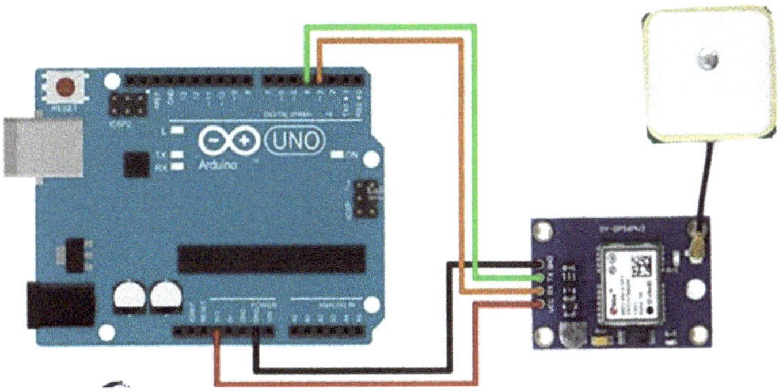

Figure 17.15 GPS sensor circuit

wire from the VCC of the GPS sensor to the 5 V of the Arduino and another wire from the GND of the GPS sensor to the GND of the Arduino and another wire from the RX of the GPS sensor to the pin number 9 of the Arduino, and one wire from the TX of the GPS sensor to the pin number 8 on the Arduino. After completing these connections, we connect the Arduino to the computer via USB cable and write the code via Arduino IDE program and then send the code to Arduino. After that, we open the serial monitor, which is located within the Arduino program, where we will see the appearance of repeated numbers on the serial monitor. These numbers represent the latitude and longitude that determine the location. We enter these numbers that represent latitude and longitude to Google Maps via mobile or computer. We can show the real location of the blind person.

17.5 The working principle of the smart system

17.5.1 Smart gloves circuit

The circuit of the smart gloves consists of Arduino Uno, LDR sensor with LED and the ultrasonic sensor with buzzer. The connections of smart gloves circuit will be as shown in Figure 17.16. We connect the first wire from the +5 V in the Arduino to the positive side of the breadboard and the second wire from the GND in the Arduino to the negative side of the breadboard. In this way, we were able to get two full rows on the breadboard connected with +5 V and GND from Arduino, and after that we connect the third wire from the positive side of LDR to the positive side of breadboard, the fourth wire connect from negative side of LDR to the A5 in the Arduino, the fifth wire from the positive side of LED to the pin 7 in the Arduino and the negative side of LED connect with the negative side of breadboard by the resistance.

One wire from the VCC of the ultrasonic sensor to 5 V of the Arduino and another wire from the GND of the ultrasonic sensor to the GND of the Arduino and another wire from the trig of the ultrasonic sensor to the pin number 9 of the Arduino, and one wire from the echo of the ultrasonic sensor to the pin number 8 on

Figure 17.16 Smart gloves circuit

Figure 17.17 Smart shoes circuit

the Arduino, and one wire from the negative side of the buzzer to the GND of the Arduino, and another wire from the positive side of the buzzer to the pin number 6 on the Arduino.

17.5.2 Smart shoes circuit

The electronic circuit of the smart shoes consists of Arduino Mega, GPS, water sensor with vibration motor, LDR sensor with LED and the ultrasonic sensor with buzzer. The connections of smart shoes circuit are shown in Figure 17.17. One wire from the VCC of the GPS sensor to 5 V of the Arduino, another wire from the GND of the GPS sensor to the GND of the Arduino, another wire from the RX of the GPS sensor to the pin number 9 of the Arduino, one wire from the TX of the GPS sensor to the pin number 8 on the Arduino, one wire from the positive side of the water sensor to the 5 V of the Arduino, another wire from the negative side of the water sensor to the GND of the Arduino, another wire from the S of the water sensor to the pin number A0 of the Arduino, one wire from the negative side of the vibration motor to the GND of the Arduino, another wire from the positive side of the vibration motor to the pin number 5 on the Arduino mega, one wire from the VCC of the ultrasonic sensor to the 5 V of the Arduino, and another wire from the GND of the ultrasonic sensor to the GND of the Arduino. The connections continue with another wire from the trig of the ultrasonic sensor to the pin number 9 of the Arduino, one wire from the echo of the ultrasonic sensor to the pin number 8 on the Arduino, one wire from the negative side of the buzzer to the GND of the Arduino, and another wire from the positive side of the buzzer to the pin number 6 on the Arduino Mega. We connect one wire from the +5 V in the Arduino to the positive side of the breadboard and one wire from the GND in the Arduino to the negative side of the breadboard, and in this way we were able to get two full rows on the breadboard connected with +5 V and GND from Arduino. After that we connect one wire from the positive side of LDR to the positive side of breadboard, one wire connect from negative side of LDR to the A5 in the Arduino, one wire from the positive side of LED to the pin 7 in the Arduino, and the negative side of LED connect with the negative side of breadboard by the resistance.

17.6 The smart system design

17.6.1 Smart gloves design

The design of the smart glove is simple. It is a normal glove that has an electronic circuit on its surface that will help the blind person move easily. The glove will be able to alert the blind person to anything that exists when moving his hand in any direction. The electronic circuit contains an ultrasonic sensor connected with the buzzer both of them connected with the Arduino Uno and also contains an LDR sensor and LED that light up in the dark only to alert people around the blind person. In Figure 17.18, this design is simple and practical and will be very helpful for the blind person.

17.6.2 Smart shoes design

The smart shoe is the main part of the smart system, where its design will be very distinctive and useful to the blind person as it will help him move easily without using any kind of blind sticks. In the front of the shoe, there is an ultrasonic sensor connected with the buzzer. In the right side of the shoe, there is a water sensor connected to the vibration motor, and in the left side of the shoe, there is an LDR sensor, LED, and GPS. All these electronic parts are connected to the Arduino Mega shown in Figure 17.19.

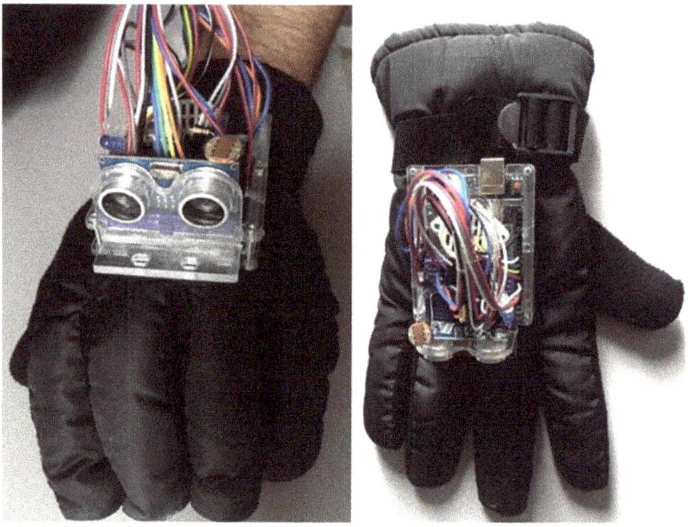

Figure 17.18 Smart gloves design

Figure 17.19 Smart shoes design

17.7 Code of the smart system

17.7.1 Smart gloves code

```
byte led = 6;
int triggerPin = 7; //triggering on pin 7
int echoPin = 8; //echo on pin 8
int buzzer = 9;
void setup() {
pinMode(led, OUTPUT);
pinMode(buzzer, OUTPUT);
pinMode(triggerPin, OUTPUT); //defining pins
pinMode(echoPin, INPUT);
}
void loop() {
int light = analogRead(A1);
if(light<20)
digitalWrite(led,HIGH);
else
digitalWrite(led,LOW);
delay(500);
int duration, distance;
//Adding duration and distance
digitalWrite(triggerPin, HIGH);
//triggering the wave(like blinking an LED)
delay(100);
digitalWrite(triggerPin, LOW);
duration = pulseIn(echoPin, HIGH);
//a special function for listening and waiting for the wave
distance = (duration / 2) / 29.1;
//transforming the number to cm (if you want inches, you have to change the
29.1 with a suitable number)
if (distance < 25 )
{
digitalWrite(buzzer, HIGH);
delay(65);
digitalWrite(buzzer, LOW);
}
else {
digitalWrite(buzzer, LOW);
}}
```

17.7.2 Smart shoes code

```
const int waterSens = A0;//define water sensor to pin A0
const int speaker = 5;//define speaker to pin 5
```

```
int SensorValue;//create sensor data variable
byte led = 6;
int triggerPin = 7; //triggering on pin 7
int echoPin = 8; //echo on pin 8
int buzzer = 9;
void setup() {
pinMode(speaker, OUTPUT);//set speaker as an output
pinMode(waterSens, INPUT); //set water sensor as an input
pinMode(led, OUTPUT);
pinMode(buzzer, OUTPUT);
pinMode(triggerPin, OUTPUT); //defining pins
pinMode(echoPin, INPUT);
}
void loop() {
int sensorValue = analogRead(waterSens);//read the water sensor value
if(sensorValue >= 50) {
tone(speaker, 800, 800);
delay(200);
tone(speaker, 600, 800);
delay(200);
}//if the sensor senses water then play an alarm
int light = analogRead(A1);
if(light<20)
digitalWrite(led,HIGH);
else
digitalWrite(led,LOW);
delay(500);
int duration, distance; //Adding duration and distance
digitalWrite(triggerPin, HIGH); //triggering the wave(like blinking an LED)
delay(90);
digitalWrite(triggerPin, LOW);
duration = pulseIn(echoPin, HIGH); //a special function for listening and
waiting for the wave
distance = (duration / 2) / 29.1;
//transforming the number to cm (if you want inches, you have to change the
29.1 with a suitable number)
if (distance < 35 )
{
digitalWrite(buzzer, HIGH);
delay(350);
digitalWrite(buzzer, LOW);
}}
```

17.8 Recognition

The process of producing an end product that is unique in the field of science in which it can be a new idea of brand new instrument or an improvement on a specific device is a helpful move that can aim to have a better life situation for patients or even stopping the spread of specific disease. In general, it aims for public awareness. This new smart system introduced to help with blindness in particular has specific advantages for improving daily life for people with blindness, beginning with auto detection, where the device can process and detect daily obstacles automatically, for example, walking through a crowded room or street where it can be inconvenient to walk through using a stick. In normal situations, a person with blindness would need another person's help to assist them in inconvenient places such as navigating city streets or a shopping mall. Instead, a buzzer will directly alert the blind person of the best route for that specific situation. A smart system consists of two instruments, where it contains a series of features. In contrast is a simple, uncomplicated system where anyone can simply wear the two instrument system. The smart gloves are simple and convenient for both elderly and young people to wear. That the instrument will be placed on top of the glove and only raising the hand is enough to alert the blind person and preventing them from hitting nearest object such as a wall, door, car, etc. In addition, a specific LED sensor is used to alert the person with blindness to their surroundings, especially in the dark, where it is a useful feature. The second instrument is the smart shoes, which add features such as an ultrasonic sensor, water sensor, buzzer and GPS and are placed sufficiently and comfortably for the blind person shoes, where it is easy and simple to wear. By wearing normal shoes and only placing the instrument on top and around the shoes, where it contains series of features, is one way to maximise comfort and ease of use. The water sensor is useful for a blind person who lives in cool areas or in rainy countries where it rains frequently throughout the day and night. It can assist them in walking the street knowing that the smart shoes will alert them to the presences of water by vibrating the shoes, allowing them to directly sense the presence of water earlier. The continuous evolution of the technology can help with daily life. As a result, saving the hundreds or even thousands of lives in an annual perspective, also hostilities and hurtful organisms such as bacteria or viruses. Computers and mobile phones are now essential items in everyday life, and this project aims to provide a steady and convenient life and aims for making the blind person more independent. The smart shoes are helped by the evolution of GPS technology, which is considered to be one of our major advantages in the project. It contributes to giving the blind person and his relatives location technology. The idea is to provide an easy way to help the person with blindness be located if needed.

17.9 Future goals

The progress of this project can be improved in various, starting with the smart gloves, which can become exposed to water. A special fabric with water resistance can be used rather than standard fabrics. One potential development in the project is to use water-resistant fabrics so the user does not have to avoid rainy areas or any situation that water may fall on the gloves and the instrument. One water-resistant fabric is polyurethane laminate (PUL). PUL is used with a water-resistant coating, and the fabric is coated with a plastic backing. It features essential qualities such as the fabric is completely water proof. Also, it is durable and breathable and can be easily and sufficiently placed next to the skin due its flexibility; in addition, it is comfortable. On the other hand, the most important part of the project placed on the smart shoes is the GPS, which could be improved by adding several features such as helping the user navigate their own way and moving from one place to another with the use of GPS. The shoes could then be used to navigate which side of the road or which direction to move to with the help of a buzzer. The buzzer vibrates the right shoe if the navigation is pointed to the right, for example, and this can occur simultaneously for both shoes for either direction. Moreover, the blind person can become more independent with the continuous evolution of the project's technologies.

17.10 Conclusion

The basic idea of our project is to support blind people in general and to support young people with blindness in particular, where the person will be able to walk comfortably with the use of this smart system. The smart system consists of gloves and shoes, which contain a set of sensors that will alert the user about the existence of any obstacle and also contains a GPS system, which will facilitate access to the person at time and in any emergency. This system will also help the person move and walk easily without using a walking stick.

References

[1] F. Al-Turjman, "Intelligence and Security in Big 5G-oriented IoNT: An Overview", Elsevier Future Generation Computer Systems, vol. 102, no. 1, pp. 357–368, 2020.

[2] F. Al-Turjman, U. Ulusar, and M. Nawaz, "Intelligence in the Internet of Medical Things Era: A Systematic Review of Current and Future Trends", Elsevier Computer Communications Journal, vol. 150, no. 15, pp. 644–660, 2020.

[3] F. Al-Turjman, H. Zahmatkesh, and L. Mostarda, "Quantifying Uncertainty in Internet of Medical Things and Big-Data Services Using Intelligence and Deep Learning", IEEE Access, vol. 7, no. 1, pp. 115749–115759, 2019.

[4] J. Jin, W. Sun, F. Al-Turjman, M. Khan, and X. Yang, "Activity Pattern Mining for Healthcare", IEEE Access, vol. 8, no. 1, pp. 56730–56738, 2020.

[5] A. Barua, Z. Zhang, F. Al-Turjman, and X. Yang, "Cognitive Intelligence for Monitoring Fractured Post-Surgery Ankle Activity Using Channel Information", IEEE Access, vol. 8, pp. 112113–112129, 2020.

[6] F. Al-Turjman, and K. Killic, "LaGOON: A Simple Energy-aware Routing Protocol for Wireless Nano Sensor Networks", IET Wireless Sensor Systems, vol. 9, no. 3, pp. 110–118, 2019.
[7] B. Craft. Arduino projects for dummies. John Wiley & Sons, 2013.
[8] Arduino, Store Arduino. "Arduino." Arduino LLC. http://web.csulb.edu/~hill/ee400d/Technical%20Training%20Series/02%20Intro%20to%20Arduino.pdf
[9] E. R. Melgar and C. C. Diez. Arduino and kinect projects: Design, build, blow their minds. Apress, 2012.

Index

www.ingramcontent.com/pod-product-compliance
Lightning Source LLC
LaVergne TN
LVHW051050210926
848486LV00001B/55
9781839530562